I0051655

Microbial Fuel Cells
Materials and Applications

Edited by

**Inamuddin[1,2,3], Mohammad Faraz Ahmer[4],
and Abdullah M. Asiri[1,2]**

[1]Centre of Excellence for Advanced Materials Research, King Abdulaziz University,
Jeddah 21589, Saudi Arabia

[2]Chemistry Department, Faculty of Science, King Abdulaziz University,
Jeddah 21589, Saudi Arabia

[3]Department of Applied Chemistry, Faculty of Engineering and Technology,
Aligarh Muslim University, Aligarh-202 002, India

[4]Department of Electrical Engineering, Mewat College of Engineering and Technology,
Mewat-122103, India

Copyright © 2019 by the authors

Published by **Materials Research Forum LLC**
Millersville, PA 17551, USA

All rights reserved. No part of the contents of this book may be reproduced or transmitted in any form or by any means without the written permission of the publisher.

Published as part of the book series
Materials Research Foundations
Volume 46 (2019)
ISSN 2471-8890 (Print)
ISSN 2471-8904 (Online)

Print ISBN 978-1-64490-010-9
ePDF ISBN 978-1-64490-011-6

This book contains information obtained from authentic and highly regarded sources. Reasonable efforts have been made to publish reliable data and information, but the author and publisher cannot assume responsibility for the validity of all materials or the consequences of their use. The authors and publishers have attempted to trace the copyright holders of all material reproduced in this publication and apologize to copyright holders if permission to publish in this form has not been obtained. If any copyright material has not been acknowledged please write and let us know so we may rectify this in any future reprints.

Distributed worldwide by

Materials Research Forum LLC
105 Springdale Lane
Millersville, PA 17551
USA
http://www.mrforum.com

Manufactured in the United States of America
10 9 8 7 6 5 4 3 2 1

Table of Contents

Preface

Nowadays, a renewable and clean source of energy is the need for our industrialized world. The industrial developments have increased the socioeconomic status of the people. On the other hand, it is leading to a global energy crisis, excessive environment pollution and depletion of fossil fuels. An alternative source of energy may somehow encounter these human threatening problems. Microbial fuel cells (MFCs) are one of the new renewable sources of energy that are based on the direct conversion of organic or inorganic matters to electricity by utilizing dynamic microorganism as a biocatalyst. Supported advancements and consistent improvement endeavours have recognized the practicality of microbial fuel cells for power generation in various specialized applications that require only low power, for example, ultracapacitors, small 3-wagon toy train, portable electronic gadgets, meteorological buoys, remote sensors, digital wristwatches, smartphones, hardware in space and robots. Wastewater treatment beyond electricity generation is also possible by using microbial fuel cells. Therefore, the values added applications of microbial fuel cells has drawn the wise attention of research and development specialist of various disciplines including, engineers, biotechnologists biologists, environmentalist, material scientists and mechanical engineers. The research in the area of microbial fuel cells has been in progress towards the development of practically viable technologies. Thus, microbial fuel cells based devices have an incredible future but still more research and development studies are needed for commercializing them at a large scale.

Microbial Fuel Cells: Materials and Applications explores the various aspects of microbial fuel cells, including fuel cells electrochemistry, characterization techniques and operating conditions. The progress of microbial fuel cells is also discussed in brief. The use of different types of materials for the construction of anode and cathode are also reported. Wastewater treatment, desalination and biofuel production by using microbial fuel cells are also described in details.

We are appreciative to all the contributing authors and their co-authors for their nice chapters. We may like to thank all publishers and authors who had given permission to use their figures, tables, and schemes.

Inamuddin[1,2,3], Mohammad Faraz Ahmer[4] and Abdullah M. Asiri[1,2]
[1]Centre of Excellence for Advanced Materials Research, King Abdulaziz University, Jeddah 21589, Saudi Arabia
[2]Chemistry Department, Faculty of Science, King Abdulaziz University, Jeddah 21589, Saudi Arabia
[3]Department of Applied Chemistry, Faculty of Engineering and Technology, Aligarh Muslim University, Aligarh-202 002, India

[4]Department of Electrical Engineering, Mewat College of Engineering and Technology, Mewat-122103, India

Microbial Fuel Cells: Materials and Applications Materials Research Forum LLC
Materials Research Foundations **46** (2019) 1-20 doi: http://dx.doi.org/10.21741/9781644900116-1

Chapter 1

Microalgae–Microbial Fuel Cell

Sabeela Beevi Ummalyma[1,*], Dinabandhu Sahoo[1], Ashok Pandey[2], Kooloth Valapil Prajeesh[3]

[1]Institute of Bioresources and Sustainable Development (IBSD), A National Institute under Department of Biotechnology Govt.of India, Takyelpat, Imphal-795001 Manipur, India

[2]CSIR-Indian Institute for Toxicology Research, 226-001, Lucknow, India

[3]CSIR-National Institute for Interdisciplinary Science and Technology (CSIR-NIIST), Trivandrum-695 019, India

* sabeela25@gmail.com

Abstract

Pollution of the environment associated with increased population along with energy consumption and the projected reduction of fossil fuels highlights the necessities for sustainable, cost-effective eco-friendly bio-energy sources. The latest research on microalgae revealed that algal biomass has promising technologies for biofuel production, high-value product development, carbon sequestration and wastewater treatment. However, the latest application of microalgal biomass is its use as microbial fuel cells (MFCs). Microalgae-based microbial fuel cells (mMFCs) are used as a device that can convert energy from sunlight into electrical energy through biological pathways. This chapter is aimed to highlight the advantages of microalgae for power generation in MFCs, factors influencing electricity production from algae, and future perspectives of mMFCs.

Keywords

Microalgae, Microbial Fuel Cells, Bioenergy, Algal Biomass

Contents

1.   Introduction

The increase in alarming signals of global warming associated with the energy crisis throughout the world lead researchers to focus on alternative renewable energy sources. There are different types of resources which have been investigated for bio-energy applications ranging from corn, lignocellulosic biomass, agricultural and other industrial wastes [1]. Microalgae are known as photosynthetic green microscopic plants and resources for third generation biofuels. The advantageous features of algal biomass include high growth rate, availability throughout the year, cultivation on non-agricultural land and are non-competitive with food. The photosynthesis process is initiated with the photon induction in an algal cell similar to higher plants and carbon fixation into different storage compounds such as lipids, proteins and carbohydrates. Many types of research have proven that algal biomass can be used for biofuel production. However, the process technology has not been commercialized in a full-scale operation due to certain challenges associated with energy intensive and costly process. Algal biomass harvesting from liquid suspension itself is 30% of the cost of the whole process [2]. Consequently, a possible way of algal biomass utilization is where drying and harvesting are not involved in the process. Hence the coupling of algal biomass in microbial fuel cells represents a possible, cost effective, alternative environmental friendly technology for sustainable biorefineries. Microalgae can be mass cultivated in seawater, wastewater and rivers as valuable biomass enriched with carbohydrates, lipids and protein. They can be exploited for a wide range of applications of bio-energy with different routes for bio-oils, biodiesel, bioethanol, biomethane, hydrogen and even electricity as well [3-5]. The microalgal biomass production has been initiated with suitable algal strain selection and cultivation, dewatering of biomass from suspended water via harvesting methods followed by subsequent thickening, drying, post-processing for oils and other product extractions [6]. For mass cultivation of algae, biomass is conducted either in phototrophic or mixotrophic cultivation systems. Algal growths under phototrophic conditions are limited by the light sources. Under high light intensities, the growth of cells is limited; whereas penetration of light can be affected at high cell density, penetration of light is limited to few millimetres of depth from the surface. However, photobioreactor with a large surface area per unit volume and appropriate light intensities are recommended [7]. Microorganisms' cells have simple cellular construction for fast growth; usually grow in motile unicellular states which help for an easy harvest [8]. Algae-based photosynthetic microbial fuel cell (PMMFC) is an attractive technology, algae are utilized to provide organic substrates in the anodic compartment, production of oxygen for the cathodic compartment, carbon dioxide capture, biofuel production and wastewater treatment [1, 9-16]. This chapter summarizes the microalgal microbial fuel cells along with their advantages and limitations. Future perspectives for the commercial applications are also discussed.

2. Microbial fuel cells

The microbial fuel cell (MFC) functions on the catalytic activity of certain microorganisms which utilize organic compounds as a substrate for generating electrons at anode [17]. The electrons present in the anode chamber travelled through an external circuit for electricity generation. These electrons are reduced in the cathode chamber and finally, redox reaction is completed. Proton exchange membranes (PEMs) are used for separating anode and cathode chambers [18]. The ordinary MFCs operate by using a variety of substrates such as alcohols [19], glucose [20], acetates [21] and organic compounds [22]. These fuel cells are gaining more attraction recently owing to their possibility as a renewable source of energy coupled with wastewater treatment [23, 24]. MFCs technology based on the microbiological process, where some bacteria oxidize organic compounds under anaerobic condition and power is generated as a result of electron transport via an electrical circuit. In addition, as a result of the metabolism of substrates, secondary byproduct like hydrogen is produced which can act as fuel and oxidized to produce electrons [25].

The main limitation of MFC is that it works on connection with wastewater treatment which reduces system performance and hence decreasing the output of power. Even though, for more power the substrate feeding rate needs to be increased which is not a cost effective process [26, 27]. The main problems of MFCs for practical use are the cost associated with high installation/operation along with the use of costly membranes and the Pt-implemented cathode. Research is undergoing to lower the fuel cell cost with the refinement of the architecture of MFCs [28] for applying single chamber design to proton exchange membrane. However, microbial fuels cells can be replaced or linked with microalgae-based fuel technology to reduce the cost and resultant algal biomass can be utilized for biofuels and other high-value products formation.

3.   Microalgal microbial fuel cell (mMFC) technology

Recent investigations on microalgae in relation to MFCs have been extensively reported by many researchers throughout the world [9-12, 28]. The mMFCs utilize photosynthetic microorganisms potential for generating electricity from solar energy via metabolic reactions of photosynthetic microorganisms along with bacteria [13]. In addition to that mMFCs also have the potential for CO_2 sequestration from the atmosphere and the removal of nitrogen contaminants from wastewaters [1, 13]. Exploitation of energy from sunlight by microalgae in a microbial fuel cell for generation of electricity is a significant achievement for sustainable production of power. Microalgae can be used in either cathode chamber for supplying oxygen or as substrate at anode for the multiplication of bacteria. It has been noticed that most of the mMFCs generated more power in the dark. Oxygen produced in the light during photosynthesis reaction reduces the power production [29]. Many types of research are undergoing for improving power production in the light phase as well and exploitation of better microorganisms for improving the efficiency with photosynthetic microalgae [29-31]. MFCs are mainly used for wastewater treatment systems.

3.1   Role of microalgae at the anode

The utilization of microalgae as a substrate at the anode is primarily linked with wastewater treatment. The electrogenic bacteria present in the chamber utilize algal biomass, but fresh algal biomass does not produce enough electrons for high power density [32-34]. Hence algal powder and live cells have been exploited for this purpose. Deoiled algal biomass, activated sludge and pretreated biomass have also been utilized, but these resulted in low power output [35]. The algae produce proton and electrons in the anodic chamber during light phase and algae perform photosynthesis with the help of light and CO_2. During this process products such as algal biomass, organic substrates and oxygen are liberated. Microorganisms present in the anodic chamber utilize/ degrade algal waste and excreted solute for producing electrons and protons [36]. Some researchers have used algae as a source of electron donor. It has been reported that marine microalgae produced maximum electricity [37]. Microalgae *Tetraselmis gracilis*, *Mougeotia*, *Scenedesmus*, *Chlorella sp.*, blue-green algae, and mixed cultures have been exploited by several researchers for electricity generation. The electron extraction from algae and electricity production are a novel idea. Oxygen liberation during the photosynthesis process negatively affects the power output which is one of the limitations in this kind of fuel cells. In order to overcome this issue, some reports showed that purging of nitrogen gas, as well as use of concentrated salt solution, can minimize the problem [38]. For this condition, a salt-tolerant alga is needed to balance the normal functioning of the cell without affecting its physiology.

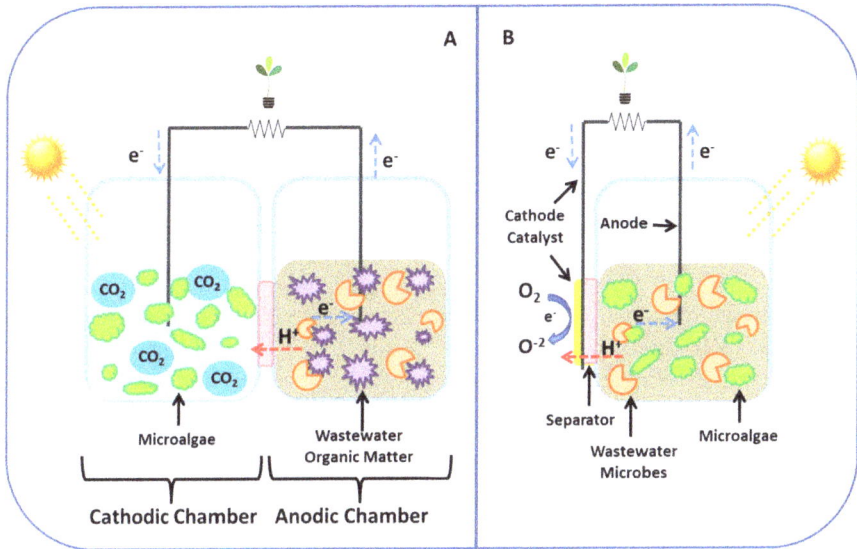

Figure1: Representing the role of algae at the cathode and anodic chambers (A), microalgae used as substrates for microorganisms in the anode chamber of fuel cells. (Adapted from Baicha et al. [15]

3.2   Role of algae at the cathode

Interest in using photosynthetic microorganisms as biocathode of MFC has increased due to their potential for oxygen production and capturing of generated CO_2. Photosynthetically produced oxygen can be used at cathode whereas carbon dioxide consumed as a carbon source for the photosynthetic process. Sometimes cathodic chamber of MFCs chemicals is used which is not a viable choice as it increases the cost as well as environmental pollution. Hence, the air cathode has been preferred where oxygen is utilized as a terminal electron acceptor, but it needs costly catalyst such as platinum for redox reactions and energy-intensive process to maintain the dissolved oxygen at its optimum level [39, 40]. The biocathode approach offers microorganisms to act as a biocatalyst for enabling electrochemical reduction at the cathode [41, 42]. Integration of microalgal photosynthesis in MFCs is a viable way to reduce or maintain consumption of energy [43, 44]. The released oxygen in the cathode chamber via the photosynthetic process helps in running MFCs and utilization of produced biomass for various applications. The utilization of algae at the cathodes allows expensive catalysts normally used for reduction of oxygen to be replaced with biomaterials [38, 40]. Figure 1 illustrates the role of algae on both cathode and anode for electricity generations.

4. Factors affecting bioelectricity generation from microalgae

There are several factors influencing the MFCs and power output. Some of the important parameters affecting the functioning of fuel cells are discussed below:

4.1 pH

pH is the main physiological factor affecting the proper functioning of MFCs. Low concentrations of ions and neutral pH in the anode chamber are required for proper metabolic activity of microorganisms. In this regard, the cathode chamber pH is almost similar to that of the anode chamber [45]. Any fluctuation in proton, electron or oxygen concentrations leads to a pH gradient and decreases the power output by perturbing the physiology of the cell. The travelling of electrons to anode leads to elevated proton concentration in the anodic chamber. Zhang et al. [46, 47] reported that a higher electric current can also result in decreased pH at the anode. The chemical and physical atmosphere in the MFCs harshly affects the kinetics and thermodynamics of oxygen reduction. In the case of oxygen reduction in the cathode chamber leads to a shifting of alkaline pH. The physical and chemical environments of MFCs adversely affect the kinetics and thermodynamics of the electrocatalytic reduction of oxygen. Reduction of oxygen at the cathode resulted in alkaline pH [97]. Alkaline pH influences cathode and anode reactions of the single chambered membrane. Even though high pH (8-10) inhibits the activity of microorganism at the anode, which can be favoured for cathodic reaction thus improving overall efficiencies. Hence, the current density at the cathode can be improved by using low-cost catalysts [48]. In order to maintain the pH in the original level, buffers like phosphate, carbonates, carbon dioxide, borax and zwitterions are used [40, 49-50]. For large scale operation, a carbonate buffer is the best. It has been proven that carbonate is more beneficial than phosphate for regulating pH because of inorganic carbon readily available in wastewaters and more diffusion coefficient in water which enable its transports via biofilm. However, the benefits of the zwitter ionic buffer are its less toxicity, non-interference with biochemical activity and slightly high pKa (6-8) than the normal pH buffer [51-53].

4.2   Temperature

For the proper functioning of MFCs, high operation temperatures are beneficial in terms of the mass transfer, reaction kinetics, low resistance, high current density and better columbic efficiency. The high-temperature increase the rate of biochemical reaction, which enhances the growth of microorganisms due to more substrates assimilation and helps in high ionic conductivity which reduces the resistance [54]. Many researchers have conducted experiments to assess the effect of a wide range of temperature for the performance of MFCs [55, 57]. It has been reported that 35°C, but can be varied with the type of organisms is optimum.   The conductivity of membrane enhanced with the increased in temperature [56]. Zhang et al. [58] evaluated the influence of day/light temperature ranges and showed that more power density was obtained at 18/30°C.

4.3   Light and photosynthesis

Photosynthetic microalgae are productive when compared with terrestrial plants due to their efficiency in photosynthesis. Sometimes photosynthetic efficiency of algae is reduced due to reduced kinetic coupling between light harvesting system for captured light, electron transfer processes and down-stream photochemical reactions [58, 59]. It was reported that theoretical maximum microalgal productivity was within 170-190 g $DWm^{-2} d^{-1}$ [60]. The real efficiencies in photobioreactors and ponds obtained with present technology with available strains ranged from 20-35 g $DWm^{-2} d^{-1}$ [61, 62]. The efficiency of photosynthesis can be enhanced by a different approach such as modulation of growth condition, tuning of light harvesting systems and reducing the size of antenna per reaction complex [50, 63-66]. When high light intensity condition, electron transfer saturation can be lowered by an optical cross section of antennae complexes which further reduces the shading effect of cells and further enhances the diffusion of photosynthetically active radiation (PAR) to larger depth [40]. Research on metabolic engineering approach for deregulation of light harvesting protein translation to study the effect of photosynthesis in algae is progressing [67, 68]. It has been reported that microalgae lacking peripheral light-harvesting complex II, which helps in increasing the photosynthetic rate, however in certain cases algae can achieve photosynthetic efficiency via reduced size of peripheral antennae under an autotrophic growth mode [64, 69, 70]. Therefore researchers have fabricated photosynthetic apparatus for increasing the process efficiency [71]. It has been reported that from microbial solar cells, average light/electricity production was estimated to be 0.03% of power conversion efficiency (PCE) [72]. Reports of syntrophic binary cultivation of *Chlamydomonas reinhardtii* along with *Geobacter sulfurreducens* which is an iron-reducing bacteria in microbial solar cells anode chamber produced 0.1 % of PCE [73], but organic acids produced as a result of fermentation of photosynthetically produced intracellular starch were found to be a bottleneck for light/electricity conversion. Other studies on algae digesting consortium of *Lactobacillus* and *Geobacter* used for bioelectricity from phototrophic grown *Chlamydomonas* cells showed 0.47% of power conversion efficiency [74]. Usually, algae-based MFCs, alteration of algal biomass into organic acids is directly proportional to PCE. The photosynthetic fuel cell from light to electricity is promising and results showed that conversion efficiency of 3% was similar to organic solar cells [75-77]. Mixed culture predominating *Chlorella* showed the efficiency of PAMFC converting PAR light energy to electricity as 0.1% per day [30]. Another parameter affecting algae-assisted cathodes is light intensity. Wu et al. [78] evaluated influence of various intensities of light on photo-microbial fuel cells with cathode containing *Desmodesmus sp* and results showed that resistance of cathode and anode were strongly influenced by variation in light intensity and hence the voltage. Gonzalez del Campo et al. [79] achieved maximum power output in continuous flow mode compared to that of sequencing batch mode.

4.4   Nature of substrate and its load.

MFCs are able to use a broad range of substrates from pure to complex along with wastewaters. The substrates type and nature has been significantly affecting the system performance. The fast and simple digestible substrates such as glucose, molasses and volatile fatty acids have been used for electricity generation in MFCs [79-82]. Different wastewaters also tested as a substrate in MFCs. The major constraints of choosing a substrate are its nature and biodegradability. Some of the wastewaters contain large quantities of organic carbon that cannot easily be degraded into simple molecules meanwhile; simple molecules are easily utilized by a microorganism and produce a large number of reducing equivalents which help in better system performance. Therefore, the selection of the substrates for MFCs depends on their type and nature that are critical for the proper performance of the system [82]. In addition to these substrates, loading rates and retention time significantly influence the power and Coloumbic efficiency of microbial fuels cells performance which especially depends on substrates used as a fuel and also affecting the bacterial growth and morphology of biofilm [83]. Power density depends on the substrates loading rates. Reports showed that increase in power density from $1884 Mw/m^3$ to 2981 Mw/m^3 obtained from substrates loading was increased from 1.9-3.8 g/Ld [84].

4.5   Membrane material

The type of membrane materials used for cathode and anode chambers affects the transfer rates of chemical species of the reaction in fuel cells which will affect the fuel cells [85]. Cations of high conductivity and low internal resistance need to be transferred from anode to cathode chamber and their transfers happen as a result of potential gradients generated inside the cells [86-88, 59]. Good quality membrane material has the capacity to overcome pH splitting. The resistance permeability for oxygen is required to reduce the cost of MFC [89-91]. For this, different membrane filtration systems such as salt bridge, bipolar membrane, an ultrafiltration membrane (UFM), anion exchange membrane (AEM), cation exchange membrane (CEM), glass fibre and ceramic membrane, etc. have been used [92-95]. For MFCs ion exchange membranes, cationic and anionic membranes have been preferred. The cation exchange membranes (CEMs) have fixed negatively charged groups ($-COO-$, SO_3 -, PO_3 $^{2-}$) in the backbone of the membrane for selective permeability of cations [96, 97]. Membranes used as cation exchanger are Nafion, bipolar membranes, polystyrene, and glass wool and microfiltration membranes. The most commonly chosen one for MFC is Nafion which helps in providing more specific conductivity of proton, enhanced Columbic efficiency of fuel cells and helps in microorganisms in active state [98,99] but its thickness and hydration level affect the system performance. The anion exchange membrane (AEMs) containing positive charges such as PR3 +, $-SR+$ and $-NH3 +$ attached to the membrane transfers anions [99]. Another type of exchanger is a bipolar membrane containing both AEM and CEM which helps to conduct proton and hydroxide ion. Separator like bipolar membrane (BPM) contains both CEM and AEM, which conduct both proton and hydroxide ion. The single unit MFC is not economical and hence for various studies, stacks of MFCs have been used to make the technology possible on a large scale [97].

Future perspective and conclusion

For the production of electricity, biomass is the best choice and algae are a readily available biomass for power generation. Microalgal based microbial fuel cells (mMFCs) have proven successful for the treatment of effluents and simultaneous electricity generation. However, mMFCs have some barriers and limitations for their commercialization. These types of fuel cells are influenced by seasonal variation. Mohan et al. have proven low fuel cell activity in summer compared to winter seasons [100]. Better designs for microalgal bacterial fuel cells are still not available. Symbiotic microorganisms can be beneficial for the proper functioning of fuels cells, but extensive researches are needed on this aspect. Biomass production coupled with electricity generation is attractive but lots of research and experimental proof are required for overcoming the obstacles for economically efficient and viable mMFCs. Sustainability of the technology strongly depends on the performance of cathodic and anodic reactions. Moreover, future research work on MFCs can be focused on technologies for developing low-cost energy input in mMFCs and exploitation of wet algal biomass for boils and biochar production. Hence mMFCs represent a novel, promising, environmental friendly cost-effective technology for sustainable energy production.

Acknowledgement

Sabeela Beevi Ummalyma is grateful to the Institute of Bioresources and Sustainable Development, A National Institute under Department of biotechnology Govt. of India for providing financial support and help for this work.

References

[1] M.J. Salar-Garcia, I. Gajda, V.M. Ortiz-Martinez, J. Greenman, M.M. Hanczyc, A.P. delos Rios, I.A. Leropoulos, Microalgae as substrate in a low cost terracotta-based microbial fuel cells, Bioresour. Technol. 209 (2010) 380-385. https://doi.org/10.1016/j.biortech.2016.02.083

[2] S.B. Ummalyma, E. Gnansounou, R.K. Sukumaran, R. Sindhu, A. Pandey, D. Sahoo, Bioflocculation: An alternative strategy for harvesting of microalgae-An Overview, Bioresour. Technol. 242 (2017) 227-235. https://doi.org/10.1016/j.biortech.2017.02.097

[3] S.B. Ummalyma, R.K. Sukumaran, Cultivation of the fresh water microalga Chlorococcum sp. RAP13 in sea water for producing oil suitable for biodiesel, J. Appl. Phycol. 27 (2015) 141–147. https://doi.org/10.1007/s10811-014-0340-4

[4] S.B Ummalyma, R.K Sukumaran, Cultivation of microalgae in dairy effluent for oil production and removal of organic pollution load, Bioresour. Technol. 165 (2014) 295-301. https://doi.org/10.1016/j.biortech.2014.03.028

[5] A. Bahadar, M.B. Khan, Progress in energy from microalgae: A review, Renew. Sust. Energ. Rev. 27 (2013) 128 118, https://doi.org/10.1016/j.rser.2013.06.029

[6] C.Y Chen, K.L. Yeh, R. Aisyah, D.J. Lee, J.S. Chang, Cultivation, photobioreactor design and harvesting of microalgae for biodiesel production: a critical review, Bioresour. Technol. 102 (2011) 71-81. https://doi.org/10.1016/j.biortech.2010.06.159

[7]   J.H. Lin, D.J, Lee, J. S. Chang, Lin, Lutein in specific marigold flowers and microalgae, J. Taiwan Inst. Chem. Eng. 49 (2015) 90–94. https://doi.org/10.1016/j.jtice.2014.11.031

[8]   K.Y. Show, D.J. Lee, J.S.C. Show, Algal biomass dehydration, Bioresour. Technol. 135 (2013) 720–729. https://doi.org/10.1016/j.biortech.2012.08.021

[9]   Z. Yang, H. Pie, Q. Hou, L. Jiang, L. Zhang, C. Nie, Algal biofilm assisted microbial fuel cells to enhance domestic waste water treatment: nutrients, organics removal and bioenergy production, Chem. Eng. J. 332 (2018) 277-285. https://doi.org/10.1016/j.cej.2017.09.096

[10] A. Khandelwal, A. Vijay, A. Dixit, M. Chhabra, Microbial fuel cell powered by lipid extracted algae: a promising system for algal lipid and power generation, Bioresour. Technol. 247 (2018) 520-527. https://doi.org/10.1016/j.biortech.2017.09.119

[11] L. He, P. Du, Y. Chen, H. Lu, X. Cheng, B. Cheng, Z. Wang, Advances in microbial fuel cells for waste water treatment, Renew. Sust. Energ. Rev. 71 (2017) 388-403. https://doi.org/10.1016/j.rser.2016.12.069

[12] Y. Dong, Y. Qu, C. Li, X. Han, J.J. Ambuchi, J. Liu, Y. Yu, Y. Feng, Simultaneous algae polluted water treatment and electricity generation using a biocathode coupled eletrocoagulation cell (bio-ECC), J. Hazard. Mater. 340 (2017) 104-112. https://doi.org/10.1016/j.jhazmat.2017.06.055

[13] A.S. Commault, O. Laczka, N. Siboni, B.Tamburic, J.R. Crosswe, J.R Seymour, P.J Ralph, Electricity and biomass production in a bacteria- Chlorella based microbial fuel cell treating waste water, J. Power. Sources 356 (2017) 299-309. https://doi.org/10.1016/j.jpowsour.2017.03.097

[14] X.A. Walter, J. Greenman, B. Taylor, I.A. Ieropoulos, Microbial fuel cells continuously fuelled by untreated fresh algal biomass, Algal Res. 11 (2015)103–107. https://doi.org/10.1016/j.algal.2015.06.003

[15] Z. Baicha, M.J Salar-García, V.M. Ortiz-Martínez, F.J. Hernández-Fernández, A.P. de los Ríos, N. Labjar, E. Lotfi, M. Elmahi, A critical review on microalgae as an alternative source for bioenergy production: a promising low cost substrate for microbial fuel cells, Fuel Process. Technol. 154 (2016)104–116. https://doi.org/10.1016/j.fuproc.2016.08.017

[16] X. Hu, B. Liu, J. Zhou, R. Jin, S. Qiao, G. Liu, CO2 fixation, lipid production, and power generation by a novel air-lift-type microbial carbon capture cell system,

Environ. Sci. Technol. 49 (2015) 10710–10717.
https://doi.org/10.1021/acs.est.5b02211

[17] M. Zhou, T. Jin, Z. Wu, M. Chi, T. Gu, Microbial fuel cells for bioenergy and bioproducts, Bioenergy Bioprod. Part Ser. Green Energy Technol. 88 (2011) 131–171.

[18] Z. Ghassemi, G. Slaughter, Biological fuel cells and membranes, Membranes 7 (2017) 3. https://doi.org/10.3390/membranes7010003

[19] J. Li, L.G. Liu,R.D. Zhang, Y. Luo , C.P. Zhang, M.C. Li, Electricity generation by two types of microbial fuel cells using nitrobenzene as the anodic or cathodic reactant, Bioresour. Technol. 101 (2010) 4013–4020.
https://doi.org/10.1016/j.biortech.2009.12.135

[20] V.F. Passos, V. Fabiano, S. Aquino Neto, A.R. Andrade, V. Reginatto, Energy generation in a microbial fuel cell using anaerobic sludge from a wastewater treatment plant, Sci. Agric. 73 (2016) 424–428. https://doi.org/10.1590/0103-9016-2015-0194

[21] J.R. Kim, S.H. Jung, J.M. Regan, B.E. Logan, Electricity generation and microbial community analysis of alcohol powered microbial fuel cells, Bioresour. Technol. 98 (2007) 2568–2577. https://doi.org/10.1016/j.biortech.2006.09.036

[22] H. Luo, G. Liu, R. Zhang, S. Jin, Phenol degradation in microbial fuel cells. Chem. Eng. J. 147 (2009) 259–264. https://doi.org/10.1016/j.cej.2008.07.011

[23] J. Li, L.G. Liu, R.D. Zhang, Y. Luo, C.P. Zhang, M.C. Li, Power generation from glucose and nitrobenzene degradation using the microbial fuel cell, Environ. Sci. 31 (2010) 2811–2817.

[24] V.G. Gude, Wastewater treatment in microbial fuel cells – an overview, J. Clean. Prod. 122 (2016) 287–307. https://doi.org/10.1016/j.jclepro.2016.02.022

[25] J. Khera, A. Chandra, Microbial fuel cells: recent trends, Proc. Natl. Acad. Sci. India. Sect A: Phys. Sci. 82 (2012) 31–41. https://doi.org/10.1007/s40010-012-0003-2

[26] D.F. Juang, P.C. Yang, H.Y. Chou, L.J. Chiu, Effects of microbial species, organic loading and substrate degradation rate on the power generation capability of microbial fuel cells, Biotechnol. Lett. 33 (2011) 2147–2160.
https://doi.org/10.1007/s10529-011-0690-9

[27] S. Mahesh, D. Tadesse, A. Melkamu, Evaluation of photosynthetic microbial fuel cell for bioelectricity production, Indian J. Energy 2 (2013) 116–120.

[28]   Y. Chisti, Biodiesel from microalgae beats bioethanol. Trends Biotechnol. 26 (2008) 126-131. https://doi.org/10.1016/j.tibtech.2007.12.002

[29]   E. Bazdara, R. Roshandela, S. Yaghmaeib, M.M. Mardanpour, The effect of different light intensities and light/dark regimes on the performance of photosynthetic microalgae microbial fuel cell, Bioresour. Technol. 261 (2018) 350–360. https://doi.org/10.1016/j.biortech.2018.04.026

[30]   D.P.B.T.B. Strik, H. Terlouw, H.V.M. Hamelers, C.J.N. Buisman, Renewable sustainable biocatalyzed electricity production in a photosynthetic algal microbial fuel cell (PAMFC), Appl. Microbiol. Biotechnol. 81 (2008) 659–668. https://doi.org/10.1007/s00253-008-1679-8

[31]   D.F. Juang, C.H. Lee, S.C. Hsuc, Comparison of electrogenic capabilities of microbial fuel cell with different light power on algae grown cathode, Bioresour. Technol. 123 (2012) 23–29. https://doi.org/10.1016/j.biortech.2012.07.041

[32]   G.P.M.K. Ciniciato, F.L. Ng, S.M. Phang, M. Jaafar, A.C. Fisher, K. Yunus, Investigating the association between photosynthetic efficiency and generation of biophotoelectricity in autotrophic microbial fuel cells, Sci. Rep. 6 (2016) 31193. https://doi.org/10.1038/srep31193

[33]   S. Kondaveeti, K.S. Choi, R. Kakarla, B. Min, Microalgae Scenedesmus obliquus as renewable biomass feedstock for electricity generation inmicrobial fuel cells (MFCs), Front. Environ. Sci. Eng. 8 (2014) 784–791. https://doi.org/10.1007/s11783-013-0590-4

[34]   A.M. Lakaniemi, O.H. Tuovinen, J.A. Puhakka, Anaerobic conversion of microalgal biomass to sustainable energy carriers-a review, Bioresour. Technol. 135 (2013) 222–231. https://doi.org/10.1016/j.biortech.2012.08.096

[35]   C.C. Fu, T.C. Hung, W.T. Wu, T.C. Wen, C.H. Su, Current and voltage responses in instant photosynthetic microbial cells with Spirulina platensis, Biochem. Eng. J. 52 (2010) 175–180. https://doi.org/10.1016/j.bej.2010.08.004

[36]   N. Rashid, Y.F. Cui, M. Saif Ur Rehman, J.I. Han, Enhanced electricity generation by using algae biomass and activated sludge in microbial fuel cell, Sci. Total Environ. 456 (2013) 91–104. https://doi.org/10.1016/j.scitotenv.2013.03.067

[37]   S. Mateo, A. Gonzalez del Campo, P. Ca-izares, J. Lobato, M.A. Rodrigo, F.J. Fernandez, Bioelectricity generation in a self-sustainable microbial solar cell, Bioresour. Technol. 159 (2014) 451–454. https://doi.org/10.1016/j.biortech.2014.03.059

[38] G.R. Ramanathan, S. Birthous, D. Abirami, Efficacy of marine microalgae as exoelectroge in microbial fuel cell system for bio-electricity generation, J. Fish Mar. Sci. 3 (2011) 79–87.\

[39] C. Xu, K. Poon, M.M. Choi, R. Wang, Using live algae at the anode of a microbial fuel cell to generate electricity, Environ. Sci. Pollut. Res. Int. 22 (2015) 15621–15635 . https://doi.org/10.1007/s11356-015-4744-8

[40] V.G. Gude,B. Kokabian, V. Gadhamshetty, Beneficial bioelectrochemical systems for energy, water, and biomass production. J. Microb. Biochem. Technol. 6 (2013) 2-14.

[41] M. Shukla, S. Kumar, Algal growth in photosynthetic algal microbial fuel cell and its subsequent utilization for biofuels, Renew. Sust. Energ. Rev. 82 (2018) 402–414. https://doi.org/10.1016/j.rser.2017.09.067

[42] L. Huang, J.M. Regan, X. Quan, Electron transfer mechanisms, new applications, andperformance of biocathode microbial fuel cells, Bioresour. Technol. 102 (2011) 316–323. https://doi.org/10.1016/j.biortech.2010.06.096

[43] V. Sharma, P.P. Kundu, Biocatalysts in microbial fuel cells, Enzym. Microb. Technol. 47 (2010) 179–188. https://doi.org/10.1016/j.enzmictec.2010.07.001

[44] L. Xiao, E.B. Young, J.A. Berges, Z. He, Integrated photo-bioelectochemical system for contaminants removal and bioenergy production, Environ. Sci. Technol. 46 (2012) 11459–114566. https://doi.org/10.1021/es303144n

[45] Y. Zhang, J.S. Noori, I. Angelidaki, Simultaneous organic carbon, nutrients removal and energy production in a photomicobial fuel cell (PFC), Energy Environ. Sci. 4 (2011) 4340–4346. https://doi.org/10.1039/c1ee02089g

[46] S. Puig, M. Serra, M. Coma, Cabré, M.D. Balaguer, J. Colprim, Effect of pH on nutrient dynamics and electricity production using microbial fuel cells, Bioresour. Technol. 101 (2010) 9594–9599. https://doi.org/10.1016/j.biortech.2010.07.082

[47] E.R. Zhang, L. Liu, Y.Y. Cui, Effect of PH on the performance of the anode in microbial fuel cells, Adv. Mat. Res. 608–609 (2013) 884–888.

[48] R.A. Rozendal, H.V.M. Hamelers, C.J.N. Buisman, Effects of membrane cation transporton pH and microbial fuel cell performance, Environ. Sci. Technol. 40 (2006) 5206–5211. https://doi.org/10.1021/es060387r

[49] C.L Torres, H.S. Lee, B.E. Rittmann, Carbonate species as OH carriers for decreasing the pH gradient between cathode and anode in biological fuel cells, Environ. Sci. Technol. 42 (2006) 8773–8777. https://doi.org/10.1021/es8019353

[50]   V.B. Oliveira, M. Simões, L.F. Melo, A.M.F.R. Pinto, Overview on the developments of microbial fuel cells, Biochem. Eng. J. 73 (2013) 53–64. https://doi.org/10.1016/j.bej.2013.01.012

[51]   L. Qiang, L.J. Yuan, Q. Ding, Influence of buffer solutions on the performance of microbial fuel cell electricity generation, Environ. Sci. 32 (2011) 1524–1528.

[52]   Y. Fan, H. Hu, H. Liu, Sustainable power generation in microbial fuel cells using bicarbonate buffer and proton transfer mechanisms, Environ. Sci. Technol. 41 (2007) 8154–8158. https://doi.org/10.1021/es071739c

[53]   J.Y. Nam, H.W. Kim, K.H. Lim, H.S. Shin, Effects of organic rates on the continuous electricity generation from fermented wastewater using a single-chamber microbial fuel cell, Bioresour. Technol. 101 (2010) 533–537. https://doi.org/10.1016/j.biortech.2009.03.062

[54]   A.K. Marcus, C.I. Torres, B.E. Rittmann, Analysis of a microbial electrochemical cell using the proton condition in biofilm (PCBIOFILM) model, Bioresour. Technol. 102 (2011) 253-262. https://doi.org/10.1016/j.biortech.2010.03.100

[55]   R. Karthikeyan, A. Selvam, K.Y. Cheng, J.W. Wong, Influence of ionic conductivity in bioelectricity production from saline domestic sewage sludge in microbial fuel cells, Bioresour. Technol. 200 (2016) 845–852. https://doi.org/10.1016/j.biortech.2015.10.101

[56]   Y.L. Tang, Y.T. He, P.F. Yu, H. Sun, J.X. Fu, Effect of temperature on electricity generation of single-chamber microbial fuel cells with proton exchange membrane, Adv. Mat. Res. 393–395 (2012) 1169–1172.

[57]   M. Pérez-Page, V. Pérez-Herranz, Effect of the operation and humidification temperatures on the performance of a PEM fuel cell stack on dead-end mode, Int. J. Electrochem. Sci. 6 (2011) 492–505.

[58]   Y. Zhang, J. Sun, Y. Hu, Z. Wang, S. Li, Effects of periodically alternating temperatures on performance of single-chamber microbial fuel cells, Int. J. Hydrog. Energy 39 (2014) 8048–8054. https://doi.org/10.1016/j.ijhydene.2014.03.110

[59]   D.R. Ort, X. Zhu, A. Melis, Optimizing antenna size to maximize photosynthetic efficiency, Plant Physiol. 155 (2011) 79–85. https://doi.org/10.1104/pp.110.165886

[60]   A. Melis, Solar energy conversion efficiencies in photosynthesis: minimizing the chlorophyll antennae to maximize efficiency, Plant Sci. 177 (2009) 272–280. https://doi.org/10.1016/j.plantsci.2009.06.005

[61]   K.M. Weyer, D.R. Bush, A. Darzins, B.D. Willson, Theoretical maximum algal oil production, Bioenergy Res. 3 (2010) 204–213. https://doi.org/10.1007/s12155-009-9046-x

[62]   L. Rodolfi, G.C. Zittelli, N. Bassi, G. Padovani, N. Biondi, G. Bonini, Microalgae for oil: strain selection, induction of lipid synthesis and outdoor mass cultivation in a low-cost photobioreactor, Biotechnol. Bioeng. 102 (2009) 100–112. https://doi.org/10.1002/bit.22033

[63]   A.M. Illman, A.H. Scragg, S.W. Shales, Increase in Chlorella strains calorific values when grown in low nitrogen medium, Enzym. Microb. Technol. 27 (2000) 631–635. https://doi.org/10.1016/S0141-0229(00)00266-0

[64]   J.E. WPolle, J.R. Benemann, A. Tanaka, A. Melis, Photosynthetic apparatus organization and function in the wild type and a chlorophyll b-less mutant of Chlamydomonas reinhardtii, dependence on carbon source, Planta 211 (2000) 335–344. https://doi.org/10.1007/s004250000279

[65]   J.E.W. Polle, S. Kanakagiri, E. Jin, T. Masuda, A. Melis, Truncated chlorophyll antenna size of the photosystems — a practical method to improve microalgal productivity and hydrogen production in mass culture, Int. J. Hydrog. Energ. 27 (2002) 1257–1264. https://doi.org/10.1016/S0360-3199(02)00116-7

[66]   S. Cazzaniga, L. Dall Osto, J. Szaub, L. Scibilia, M. Ballottari, S. Purton, Domestication of the green alga Chlorella sorokiniana: reduction of antenna size improves light-use efficiency in a photobioreactor, Biotechnol. Biofuels. 7 (2014) 157. https://doi.org/10.1186/s13068-014-0157-z

[67]   Z. Perrine, S. Negi, R.T. Sayre, Optimization of photosynthetic light energy utilization by microalgae, Algal Res. 1 (2012) 134–142. https://doi.org/10.1016/j.algal.2012.07.002

[68]   P.G. Stephenson, C.M. Moore, M.J. Terry, M.V. Zubkov, T.S. Bibby, Improving photosynthesis for algal biofuels: toward a green revolution, Trends Biotechnol. 29 (2011) 615–623. https://doi.org/10.1016/j.tibtech.2011.06.005

[69]   C. Formighieri, F. Franck, R. Bassi. Regulation of the pigment optical density of an algal cell: filling the gap between photosynthetic productivity in the laboratory

and in mass culture, J Biotechnol. 162 (2012) 115–123.
https://doi.org/10.1016/j.jbiotec.2012.02.021

[70] L. Girolomoni, P. Ferrante, S. Berteotti, G. Giuliano, R. Bassi, M. Ballottari, The
      function of LHCBM4/6/8 antenna proteins in Chlamydomonas reinhardtii, J. Exp.
      Bot. 68 (2017) 627–641.

[71] T. de Mooij, M. Janssen, O. Cerezo-Chinarro, J.H. Mussgnug, O. Kruse, M.
      Ballottari, Antenna size reduction as a strategy to increase biomass productivity: a
      great potential not yet realized, J. Appl. Phycol. 27 (2015)1063–1077.
      https://doi.org/10.1007/s10811-014-0427-y

[72] A. Magnuson, S. Styring, Molecular chemistry for solar fuels: from natural to
      artificial hotosynthesis, Aust. J. Chem. 65 (2012) 564–572.
      https://doi.org/10.1071/CH12114

[73] K. Watanabe, K. Nishio, Electric power from rice paddy fields. In: Nathwani J, Ng
      A, (Eds), Paths to sustainable energy. Rijeka: In Tech, 2010, pp. 563–80.
      https://doi.org/10.5772/12929

[74] K. Nishio, K. Hashimoto, K. Watanabe, Digestion of algal biomass for electricity
      generation in microbial fuel cells, Biosci. Biotechnol. Biochem. 77 (2013) 670–
      672. https://doi.org/10.1271/bbb.120833

[75] K. Nishio, K. Hashimoto, K. Watanabe, Light/electricity conversion by defined
      cocultures of Clamydomonas and Geobacter, J. Biosci. Bioeng. 115 (2013) 412–
      417. https://doi.org/10.1016/j.jbiosc.2012.10.015

[76] T. Yagishita, S. Sawayama, K. Tsukahara, T. Ogi, Effects of intensity of incident
      light and concentrations of Synechococcus sp. and 2-hydroxy-1,4-naphthoquinone
      on the current output of photosynthetic electrochemical cell, Sol. Energy. 61
      (1997) 347–353. https://doi.org/10.1016/S0038-092X(97)00069-8

[77] H. Hoppe, N.S. Sariciftci, Organic solar cells: an overview, J. Mater. Res. 19
      (2004) 1924–1945. https://doi.org/10.1557/JMR.2004.0252

[78] Y.C. Wu, Z.J. Wang, Y. Zheng, Y. Xiao, Z.H. Yang, F. Zhao, Light intensity
      affects the performance of photo microbial fuel cells with Desmodesmus sp. A8 as
      cathodic microorganism, Appl. Energy. 116 (2014) 86–90.
      https://doi.org/10.1016/j.apenergy.2013.11.066

[79] A. González del Campo, J.F. Perez, P. Ca-izares, M.A. Rodrigo, F.J. Fernández, J.
      Lobato, Study of a photosynthetic MFC for energy recovery from synthetic

industrial fruit juice wastewater, Int. J. Hydrog. Energy. 39 (2014) 21828–21836. https://doi.org/10.1016/j.ijhydene.2014.07.055

[80]   B. Min, B.E. Logan, Continuous electricity generation from domestic waste- water and organic substrates in a flat plate microbial fuel cell, Environ. Sci. Technol. 38 (2004) 5809–5814. https://doi.org/10.1021/es0491026

[81]'   B. Logan, S. Cheng, V. Watson, G. Estadt, Graphite fiber brush anodes for increased power production in air-cathode microbial fuel cells, Environ. Sci. Technol. 41 (2007) 3341–3346. https://doi.org/10.1021/es062644y

[82]   Y. Sharma, B.K. Li, The variation of power generation with organic substrates in single-chamber microbial fuel cells (SCMFCs), Bioresour. Technol. 101 (2010) 1844-1850. https://doi.org/10.1016/j.biortech.2009.10.040

[83]   S.V. Mohan, R. Sarvanan, S.V. Raghuvulu, G.M. Krishna, P.N. Sarma, Bioelectricity production from wastewater treatment in dual chambered microbial fuel cell (MFC) using selectively enriched mixed microflora: effect of catholyte, Bioresour. Technol. 99 (2008) 596–603. https://doi.org/10.1016/j.biortech.2006.12.026

[84]   D. Pant, G. Van Bogaert, L. Diels, K. Vanbroekhoven, A review of the substrates used in microbial fuel cells (MFCs) for sustainable energy production, Bioresour. Technol. 101 (2010) 1533–1543. https://doi.org/10.1016/j.biortech.2009.10.017

[85]   J. Y. Nam, H.W. Kim, K.H. Lim, H.S. Shin, B.E. Logan, Variation of power generation at different buffer types and conductivities in single chamber microbial fuel cells, Biosens. Bioelectron. 25 (2010) 1155–1159. https://doi.org/10.1016/j.bios.2009.10.005

[86]   Z. Ghassemi, G. Slaughter, Biological fuel cells and membranes, Membranes. 7 (2017) 3. https://doi.org/10.3390/membranes7010003

[87]   C. Sund, S. McMasters, S. Crittenden, L. Harrell, J. Sumner, Effect of electron mediators on current generation and fermentation in microbial fuel cell, Appl. Microbiol. Biotechnol. 76 (2007) 561–568, https://doi.org/10.1007/s00253-007-1038-1

[88]   H. Ashoka, R. Shalini, P. Bhat, Comparative studies on electrodes for the construction of microbial fuel cells, Int. J. Adv. Biotechnol. Res. 3 (2012) 785–789.

[89] F. Harnisch, U. Schroder, Selectivity versus mobility: Separation of anode and cathode in microbial bioelectrochemical systems, Chem. Sus. Chem. 2 (2009) 921–926. https://doi.org/10.1002/cssc.200900111

[90] P.R. Motos, A. Heijne, R. Weijden, M. Shaakes, C.J.N. Buisman, H.J.A. Tom, High rate copper and energy recovery in microbial fuel cells, Front. Microbiol. 6 (2015) 527.

[91] M. Ghasemi, S. Shahgaldi, M. Ismail, Z. Yaakob ,W.R.W. Daud, New generation of carbon nanocomposite proton exchange membranes in microbial fuel cell systems, Chem. Eng. J. 184 (2012) 82–89. https://doi.org/10.1016/j.cej.2012.01.001

[92] A. Shahi, B.N. Rai, R.S. Singh, A comparative study of a biofuel cell with two different proton exchange membrane for the production of electricity from wastewater, Resour-Effic. Technol. 3 (2017) 78–81. https://doi.org/10.1016/j.reffit.2017.01.006

[93] A.T. Heijne, F. Liu, R.V Weijden, J. Weijma, C.J. Buisman, H.V Hamelers, Copper recovery combined with electricity production in a microbial fuel cell, Environ. Sci. Technol. 44 (2010) 4376–4381. https://doi.org/10.1021/es100526g

[94] K. Rabaey, G. Lissens, S.D. Siciliano, W. Verstraete, A microbial fuel cells capable of converting glucose to electricity at high rate and efficiency, Biotechnol. Lett. 25 (2003) 1531–1535. https://doi.org/10.1023/A:1025484009367

[95] J. Yan, J. Zhu, B.L. Chaloux, M.A. Hickner, Anion exchange membranes by bromination of tetramethylbiphenol-based poly(sulfone)s, Polym. Chem. 8 (2017) 2442–2449. https://doi.org/10.1039/C7PY00026J

[96] R.A. Rozendal, H.V.M. Hamelers, K. Rabaey, J. Keller, C.J.N. Buisman, Towards practical implementation of bioelectrochemical wastewater treatment, Trends Biotechnol. 26 (2008) 450–459. https://doi.org/10.1016/j.tibtech.2008.04.008

[97] M. Rahimnejad, G.D. Najafpour, A. Ghoreyshi, F. Talebnia, G. Premie, G.H. Bakeri, Thionine increases electricity generation from microbial fuel cells using, Saccharomyces cerevisiae and exoelectrogenic mixed culture, J. Microbiol. 50 (2012) 575–580. https://doi.org/10.1007/s12275-012-2135-0

[98] G.H. Flores, H.M.P. Varaldo, O.S. Feria, T.R. Castanon, E. Rios-Leal, J.G. Mayer, Batch operation of a microbial fuel cell equipped with alternative proton exchange membrane, Int. J. Hydrog. Energ. 40 (2016) 17323–17331. https://doi.org/10.1016/j.ijhydene.2015.06.057

[99] S. Peighambardoust, S. Rowshanzamir, M. Amjadi, Review of proron exchange membranes for fuel cell applications, Int. J. Hydrog. Energ. 35 (2010) 9349–9384. https://doi.org/10.1016/j.ijhydene.2010.05.017

[100] S.V. Mohan, G. Velvizhi, J.A. Modestra, S. Srikanth, Microbial fuel cell: Critial factors regulating biocatalysed electrochemical process and recent advancements, Renew. Sust. Energ. Rev. 40 (2014) 779–797. https://doi.org/10.1016/j.rser.2014.07.109

Microbial Fuel Cells: Materials and Applications
Materials Research Foundations **46** (2019) 21-52

Materials Research Forum LLC
doi: http://dx.doi.org/10.21741/9781644900116-2

Chapter 2

The Progress of Microalgae Biofuel Cells

Rajesh K. Srivastava

Department of Biotechnology, GIT, Gitam Institute of Technology and Management (GITAM) (Deemed to be University), Visakhapatnam, A. P. India

rajeshksrivastava73@yahoo.co.in

Abstract

The microalgal cell provides an alternative source for bioenergy generation with application in microbial fuel cells (MFCs) construction. Availability in an environmentally friendly nature, it has exhibited the capability to capture CO_2 gas with an accumulation of rich oil contents for biofuel. Biodiesel, bioethanol, methane or hydrogen production capabilities are found in microalgae. Microalgae-MFC has contributed in electricity generation by using the electrons released at the anode electrode. Microalgae are also useful in simultaneous electricity generation and wastewater treatment. In this chapter, our focus is on the advantages, limitations and future prospects of microalgae species, highlighting enhanced biofuel production.

Keyword

Microbial Fuel Cells (MFC), Biofuels, Microalgae, CO_2 Emissions, Wastewater Treatment, Electricity Generation

Contents

1.   Introduction

Increased use of arable land is reported for the cultivation of agricultural crops or plant biomass which is utilized for the production of first and second-generation biofuels but current researches are now focusing on the utilization of non-arable land for the cultivation of algal biomass for obtaining of third generation biofuels. During the last few years, advanced research activities have been carried out for the cultivation of algal biomasses and their use in biofuel production [1]. Applications of microalgae have been reported for conversion of CO_2 gas into potential biomasses and their ability to produce oxygen gas as strategic importance. Significant research has carried out to exploit the ability of microalgae for microbial fuel cells (MFC) development via integration with favourable facts of phototrophic organisms as *in-situ* generators of oxygen by facilitating the reaction in the cathode chamber of MFC. Microalgae have the capability of effective removal of phosphorous and nitrogen from wastewater as an important application of the MFCs [2, 3]. The potential for algal biomass production in conjunction with wastewater treatment and power generation with fully biotic MFCs has been discussed in many recent kinds of literature with anaerobic biofilm and anodic half-cell for generating current. Phototrophic biofilm on the cathode has been provided with the oxygen for the oxygen reduction reaction (ORR) and also form microalgal biomass [4]. Mass outdoor cultivation of microalgae for biofuels and co-products has shown many challenges due to low lipid productivity, contamination, inefficient CO_2 supply, and difficulties in harvesting. In this regards, stage cultivation process has been developed to address these challenges while culturing microalgae in a fermentor heterotrophically or photobioreactor

Microbial Fuel Cells: Materials and Applications   Materials Research Forum LLC
Materials Research Foundations **46** (2019) 21-52   doi: http://dx.doi.org/10.21741/9781644900116-2

mixotrophically [5]. Lipid degradation processes have been found in microalgae for their survival and cells grown under fluctuating environmental conditions. And microalgae have the capability for permanent remodelling or turnover of membrane lipids as well as rapid mobilization of storage lipids [6]. Microalgal species has the capability at first-stage for rapidly obtaining their high cell densities via inoculating a phototrophic open-pond culture and the second-stage is the nourishment of high levels of carbonate, pH, and salinity. Microalgae species have the capability to resist in phototrophic conditions and shown capability to utilize the organic carbon. A model of triacylglyceride (TAG) has been developed for rapid, non-destructive lipid quantization using liquid-state NMR. A two-stage cultivation system and a high pH-mediated auto-flocculation method have been utilized for haloalkaline-tolerant and multitrophic green microalgae (taken from soda lakes) strain ALP2 in a 1 L fermentor and 40 L open-tank which has achieved a final biomass concentration of 0.978 g DCL^{-1} with lipid content of 39.78% DC and auto-flocculation harvesting efficiency of 64.1% in unoptimized conditions [7]. There are some suitable conditions reported for microalgal biofuel (i.e. diesel) production and these are optimal bio-environment for microalgae cultivation, process design of algal biodiesel production, physicochemical properties of lipids extracted from microalgae, properties of the produced biodiesel fuel, and the transesterification process. Designs of full-scale and lab-scale photobioreactors (PBRs) have been illustrated for the cultivation of microalgae [8]. For biodiesel production, sufficient nutrients are added for enhanced microalgae biomass production. Biodiesel has become renewable and environmentally friendly energy due to its production from microalgae which have high growth rate and ability to synthesize a large quantity of lipids within their cell. Microalgae cultivation can be done in wastewater or organic compost with sufficient nutrients to maintain growth. After microalgal cell cultivation, harvesting and drying, the microalgae oil extractions are done in downstream processes during the process of microalgae-derived biodiesel production [9]. Microalgae and cyanobacteria have been utilized for production of bioethanol. The industrial bioethanol production has been found to depend on the capability of reducing production costs for the third generation of biofuels production. Production of bioethanol has carried out via three routes from microorganisms. The traditional methods are involving hydrolysis and fermentation of biomass with bacteria or yeast, the dark fermentation route and the use of engineered cyanobacteria or photofermentation [10]. A modification of CoA-dependent 1-butanol production pathway into a cyanobacterium, *Synechococcus elongatus* PCC 7942 has been done for 1-butanol production from CO_2. For this purpose, the activity of each enzyme in the pathway was done by chromosomal integration and expression of the genes. *Treponema denticola* trans-enoyl-CoA reductase (*Ter*) has utilized the NADH (as reducing power) and used for the reduction of crotonyl-CoA to butyryl-CoA by *Clostridium acetobutylicum butyryl*-CoA dehydrogenase with by-passing the need of *Clostridial* ferredoxins. Addition of polyhistidine-tag has increased the overall activity of Ter to achieve higher 1-butanol production [11]. The generation of biodiesel, fuel gas or bio-syngas, bio-oil, methane, hydrogen and different alcohols (ethanol, propanol, isopropanol and n-butanol) from microalgae biomass has demonstrated the applicability of microalgae biomass in MFCs construction. Microalgae have accumulated up to 70% of lipid content within their cells (depending on species efficiently) with high photosynthetic efficiency and mass cultivation capacity by reducing the carbon dioxide emission and impact of global warming to the atmosphere [12]. Algae culture and its optimal growth with the development of cost-effective technologies are also essential for efficient biomass harvesting, lipid extraction and biofuels production [13]. In the present chapter, the author aimed to discuss recent research developments on microalgae growth conditions with respect to biofuel production as well as effective microbial fuel cell (MFC) construction for utilization in wastewater treatment and pollution control processes.

2.   Microalgae species

Microalgae are microscopic algal organisms of single-cell (unicellular size range 1 μm to few hundred μm) of algal species with surviving capability individually or in chains or clusters. They are grown in suspended forms (i.e. free-float in a water body) or attached forms (via adhering to a submerged surface). They have the capability to produce approximately half of the atmospheric oxygen on earth by consuming vast amounts of the CO_2 gas as a component of greenhouse gas. Around 35,000 species have been identified and described as they are known to assimilate the different pollutants in natural water systems [14]. Microalgal cells growth has the capability to survive in fluctuating environmental conditions with permanent remodelling or turnover of membrane lipids in rapid mobilization of storage lipids. Lipid catabolism is associated with lipolysis (i.e. releases fatty acids and head groups by lipases at membranes or lipid droplets) and degradation of fatty acids to acetyl-CoA at peroxisome through the β-oxidation pathway in green microalgae) [15].

Necessary enzymes and regulatory proteins are found to involve in lipolysis and peroxisomal β-oxidation with highlighting of gaps in understanding of lipid degradation pathways of microalgae. Glyoxylate cycle and gluconeogenesis have been analyzed with metabolic use of acetyl-CoA products via understanding various cellular processes such as vesicle trafficking, cell cycle and autophagy of lipid turnover [6]. The microalgae have exhibited many applications in environmental biotechnology as utilized for bioremediation of wastewater and also to monitor the environmental toxicants. Microalgal biomass production occurs during wastewater treatment in the biofuel manufacturing processes. Microalgal lipid potential has forced research towards finding effective ways to manipulate biochemical pathways in lipid biosynthesis with more attention toward cost-effective algal cultivation and harvesting systems [16].

Microalgal has been used as promising feedstock for biodiesel and other liquid fuels and it has shown fast growth rate with high lipid yields as well as ability to grow in a broad range of environments. Many microalgae have been achieved maximal lipid yields only under stress conditions via hindering their growth but compositions were not found ideal for biofuel applications [17]. Metabolic engineering of algal fatty acid biosynthesis has shown the promises for the creation of certain microalgal strains with enhanced capability of economically producing fungible and sustainable biofuels. The algal fatty acid biosynthetic pathways have shown more homology to bacterial and plant systems and understand of the algal fatty acid biosynthetic pathway has gleaned from basic studies of plant or bacterial systems. Successful engineering of lipid metabolism in algae can express necessary change with the characterization of the algal fatty acid synthase (FAS) including protein-protein interactions and regulation [18]. Microalgae are capable of producing greater than 50,000 kg/acre/year of biomass and their biomass is naturally capable of accumulating energy-dense oils which can be converted into transportation fuels. It has economic parity with fossil fuels with several challenges such as identifying crop protection strategies, improving harvesting and oil extraction processes, and increasing biomass productivity as well as oil content. These challenges can be solved by genetic, molecular and synthetic biology techniques to enable the capability of microalgal biofuels to economically competitive with fossil fuels [19].

Microalgae biofuel production technologies have improved fuel security with reduction of CO emissions. Microalgal has helped in the development of photosynthetically derived fuels as renewable and potentially carbon-neutral with scalable alternative reserve. Microalgae can be produced on non-arable land as well as saline and wastewater streams with the utilization of salt or organic wastes present in it. Further, it can be used to produce a range of products such as biofuels, protein-rich animal feeds, chemical feedstocks (e.g. bioplastic precursors) and higher-value products. Selection, breeding and engineering approaches can be utilized for improved microalgal biomass production with biofuel conversion efficiencies [20].

Chlorella Vulgaris, Nannochloropsis and *Spirulina* are some reported microalgal species, utilized for algal electrical outputs at the stationary phases of algae development. And indirect correlation has been reported between absorption levels and electrical output with the investigation as a proportional increase from the stationary phase [21]. Live green microalgae *Chlorella pyrenoidosa* has been utilized at the anode of a microbial fuel cell (MFC) to act as an electron donor. High cell density has developed at controlled oxygen content and light intensity at the anode with the capability to generate electricity without any externally added substrates. Two models of algal microbial fuel cells (MFCs) have been shown with the utilization of graphite/carbon electrodes without any mediator. Model 1 algal MFC has been constructed with live microalgae grown at the anode and potassium ferricyanide at the cathode. Model 2 algal MFC has live microalgae in both the anode and cathode under different growth conditions. The maximum power densities per unit anode volume were found relatively higher (6030 mW/m^2/L) in model 1 algal MFC as compared to previously reported bacteria-driven MFC with a maximum power density of 30.15 mW/m^2 containing graphite/carbon electrode [22]. A much smaller power density (2.5 mW/m^2) has been reported in model 2 algal MFC. Increasing the algal cell permeability by 4-nitroaniline can increase the open circuit voltage with mitochondrial action where as proton leak promoting agents (i.e. resveratrol and 2, 4-dinitrophenol) has increased the electric current production in algal MFC. Introducing *Chlorella vulgaris* to the cathode chamber is used to generate oxygen in situ and this algal species is used to construct the modified microbial fuel cell (MFC) with a tubular photobioreactor (PHB) configuration as a cathode compartment [12]. And carbon paper-coated Pt is used as a cathode electrode and it has increased voltage output at a higher extent than carbon felt used as an electrode. The maximum power density of 24.4 mW/m^2 has been obtained from the MFC with algae biocathode via utilization of carbon paper-coated Pt as the cathode electrode under intermittent illumination. Maximum power density has been shown 2.8 times higher than the abiotic cathode. Continuous illumination has shortened the algal lifetime [23]. Microalgal species has the ability to grow in a freshwater and saturated saline and it can efficiently use CO_2 for global carbon fixation (more than 40%) and marine microalgae can develop the capability for more productivity due to very rapidly biomass with doubling time (6 h to one day) depending on the microalgal strain. Microalgae have more capacity to produce energy-rich oils in a natural way in total dry biomass via accumulation. Some *Botryococcus* spp. has been identified to have up to 50% of their dry mass stored as long-chain hydrocarbons. Microalgae are diatoms, green algae, golden brown, prymnesiophytes, eustigmatophytes and cyanobacteria [24, 25].

3.   Strategies for microalgal strain improvement

Groups of microalgal strains have the potential for different types of fuel production and millions of species with microalgal diversity have been identified with different production of biomass capabilities via genetic information. Various strategies are available which can be used to improve the production of biomass strains. Advances in cultivation techniques coupled with genetic manipulation of crucial metabolic networks have further promoted microalgae as an attractive platform for the production of numerous high-value compounds. And metabolic engineering has found to do the necessary change in order to achieve full processing capabilities. These have started with the development of a number of transgenic algal strains via boasting recombinant protein expression, engineered photosynthesis, and enhanced metabolism encourage as the possibility of new designed microalgae strain [26]. Recent advances in the genetic transformation as well as in genome-editing technologies have been applied for manipulation of genomes for lipid biosynthetic pathways (with the identification of gene functions and their regulation) of microalgae in order to expand their use in biotechnology such as enhanced lipid production and modifications in fatty acid composition. Microalgae are reported as a source of valuable nutritional ingredients, such as long-chain polyunsaturated fatty acids (LC-PUFA) and carotenoids, as well as precursors for biodiesel production. Focus on the biochemistry and enzymology for triacylglycerol formation in microalgal cells, can be helpful to identify some novel genes functions and cellular features for lipid metabolism in microalgae with the difference in some aspects from higher plants [15]. Some of the strategies for improvement of microalgae strains are aimed to discuss below for their cultivation of biomass and also more storage of lipid contents for biofuels production.

3.1   Genetic change in microalgae species

Engineered algal species have the ability to enhance biomass production capabilities with high storage of lipid contents in their cells and need sufficient knowledge of algal biology for conducting targeted optimized processes. Algal biotechnology for genetically modified (GM) strain needs a firm foundation of fundamental research for algal genomes information. Much biological variability is accumulated from the ancient origins of algal phyla and their early divergence from plants and animals. Much specific knowledge of algal gene regulation is required before skilful, efficient and routine genetic manipulation. The recently expanded algal gene library is available for advancement in algal genomes which is systematically mapped, curated, annotated and understood via performing a more time- consuming task than the actual sequencing. Generations of knockout mutants of all *Chlamydomonas genes* and the transcriptomic (FANTOM) approaches have been pioneered at RIKEN in Japan. FANTOM is an international research consortium established by Hayashizaki and his colleagues in 2000. It can develop the ability to quick with certain assigned biological functions to specific genes and curate algal genomes [27, 28]. Functional microRNAs in *Chlamydomonas* species have been applied for understanding the genomic biology of algae and it has expected to the involvement of molecular pattern receptors, signal transduction mechanisms and complex transcription factor-mediated feedback control of nuclear genes with many of the protein motifs. Further painstaking molecular analysis can help in final validation of proposed biochemical and information pathways for algal biosynthesis. Significant changes to cell status, such as nutrient limitation (sulphate, nitrogen, iron, copper) can lead with up-regulation of a few receptors or import proteins and coordinated changes of thousands of genes [29].

3.2 Engineering of lipid pathways

Lipid metabolism in microalgae is found to be substantially different from higher plants as well as between microalgal genera. The neutral and polar lipids, enzymes and metabolic pathways involved in their biosynthesis and catabolisms of lipids have been found with the current focus upon gene identification to enable proper metabolic engineering. *C. reinhardtii* has been the most extensively studied model for microalgal lipid metabolism and use phosphatidylcholine as a substrate in TAG synthesis or to accumulate TAG unless under stress conditions or in starch accumulation (sta) mutants [30,31]. Knowledge of specific enzymatic processes and the genes needed to do further advancement before effective metabolic engineering strategies to enhance the metabolic flux for lipid biosynthesis. Acetyl-CoA carboxylase (ACCase) transformed diatom *Cyclotella cryptica* with additional copies of the *ACC*ase gene within the TCA cycle has been found to increase the flux of carbon towards lipid biosynthesis without an increase in enzyme activity in lipid accumulation. Engineering of *C. reinhardtii* has enhanced DGAT mRNA levels without increasing intracellular TAG accumulation [32]. *C. reinhardtii* PDAT strain is involved in TAG biosynthesis with RNAi-induced PDAT knock-down mutants. RNAi, new CRISPR/Cas and TALEN technologies have offered the potential to dissect these pathways with optimization of individual catalytic steps through genetic editing and amino acid level. Incorporation of antisense and RNAi into the diatom *Thalassiosira pseudonana* has targeted a newly identified gene Thaps3_264297 for multifunctional lipase (phospholipase–acyltransferase) [33]. Knock-down mutants have increased accumulation of TAG droplets and total lipid production without negatively affecting cell division and biomass growth. Malate dehydrogenase (mdh gene), pyruvate formate-lyase (pfl gene) or the fatty acid synthase complex (FAS) has driven carbon towards fatty acid synthesis [34].

3.3 Random and insertional mutagenesis

Use of irradiation or chemical mutagenesis has resulted in a significant alteration in the behaviour of a gene, typically by partial or total deletion and resultant mutant can lead to base pair changes to a range of disturbances including altered amino acid sequence, small deletions, truncations, frameshifts and splicing defects. The functional knockouts can give rise to complex phenotypes. Biological mechanisms in *Chlamydomonas* and other algae can help in specific phenotypic screening, needed to identify relevant genes such as the multigene family of LHC genes for encoding light-harvesting chlorophyll-a/b (LHC) proteins of photosystem (PS)II [28, 35].

3.4   Genetic engineering in microalgae strains

Transformation of the nuclear genome of *C. reinhardtii* is carried out by random insertion through non-homologous end joining or by using linear DNA with the insertion of multiple copies in one locus. And resultant transformants are confirmed by phenotypic and genetic screening which minimized undesirable non-target effects of the random insertion of a transgene with disrupted genes or regulatory elements. It is used to study genes of an unknown function using high-throughput insertional mutagenesis. Targeted gene integration through homologous recombination (HR) using single-stranded transforming DNA is *C. reinhardtii* strain. High rates of homologous recombination have been reported for other green algal species *Nannochloropsis* with more promise for reverse genetics and targeted gene knockouts [36, 37]. Some of the microalgae with their genome size have been discussed with respective biofuel/ or triglyceraldyde (TAG) as shown in Table 1.

Table 1 Different microalgae reported with triglyceraldyde storage for biofuel source

Microalgal strain	Cell Structure	Bio-products	References
Chlamydomonas reinhardtii	Genome with 121Mbp	Biodiesel from triacylglycerol (TG) 46–65% of dry weight in starch accumulation (sta) mutants. Cellulosic ethanol produced from terrestrial plants	[38, 39]
Oedogonium cardiacum	Chloroplast genome	Biodiesel, Algal oil and methanol used for transesterification (92) by lipase-mediated synthesis of fatty acid monoalkyl esters (FAME, biodiesel-)	[40, 41]
Coccomyxa subellipsoidea	A genome with 49 Mbp	Saturated (C16:0 and C18:0) and monounsaturated fatty acid (C18:1) high-quality biofuels-making. Lipid productivity (from OCNL treatment) 232.37 mg/L/ day and 1.25-fold more than TBNS and as much as 5.06-fold more than OCND strategy.	[42, 43]
Volvox carteri UTEX2908	Genome with 138 Mbp	Hydrogen production (30.8 nmole H_2 mg $Chl^{-1}s^{-1}$)	[44, 45]

Red alga *Cyanidio-schyzon merolae* 10D	Genome 17 Mbp	TAG accumulation for biodiesel production at low pH (1-5)	[46, 47]
Porphyridium purpureum	Genome 20 Mbp	Accumulation of starch and triacylglycerol for production of biofuel and high added-value oil.	[48, 49]
Galdieria sulphuraria	Genome 14 Mbp	Use of acidophile-based wastewater treatment systems is done by and growing the biomass of this algae and it has been utilized as total fatty acid methyl esters (5%) with HTL biocrude yield (19%) by weight	[50, 51]
Emiliania huxleyi CCMP1516 With *Thermosipho globiformans* and *Methanocaldococc us jannaschii*	Genome 168	This microalgal served as raw materials and contained long-chain (31 and 33 carbon atoms) alkenes and very long-chain (37 to 39 carbon atoms) alkenones, in addition to phospholipids and glycolipids, generated a high yield of n-alkanes of various lengths (n-tridecane to n-pentatriacontane	[52, 53]
Attheya sp. CCMP212	Genome	Total Polyunsaturated fatty acids (PUFAs) concentration in CO_2 aerated cultures of *Attheya. longicornis* is 48.63% to 49.26%,	[54, 55]
Amphora sp. CTM 20023	Genome	FAME yield is 16 to 18% and 9 to 26% of the dry weight, in *Amphora subtropical* CTM 20013 and *Amphora sp.* CTM 20023 respectively	[56, 57]
Guillardia theta CCMP2712	Genome 87mbp	This microalga stored microalgal triglyceride (TAG) and utilized in biofuel precursor molecule production	[57]
Chroomonas mesostigmatica CCMP1168	702.9 Kbp	This algal oil or fat contained lauric acid as a constituent fatty acid	[58]

4. Application of microalgae species

4.1 In microbial fuel cell (MFC) construction

Recently research interest has developed in bioelectrochemical systems with emergence as fascinating technologies via utilization of electroactive microorganisms with the capability of consuming a variety of organic compounds and release of electrons directly on the anode electrode. And practical applications of microbial fuel cells (MFCs) have been studied with the perspective of generating electricity via removing organics pollutants from the electrolyte or from wastewater via its treatment. The performance of different designs of MFCs has been analyzed for the capability of the electric current generation and power outputs with the judging of wastewater treatment efficiency as well as removal of chemical oxygen demand (COD) with Columbic efficiency (CE) and some reported results are shown in Table 2 [59]. For a few years, electroactive microorganisms have been utilized for different applications in the development of bioelectrochemical systems (BES). In this regards, microbial fuel cell (MFC) technology has been created with widespread attention and good research extensively on electrogenic nature of certain bacteria or microalgal species (with alone or combination), used simultaneously to treat different wastewaters and produce electric power. Examining of various types of organic compounds has been done in various types of wastewaters treatment via utilized as substrates source for the feeding of bacteria in MFCs system [60, 61].

Platinum group metal-free (PGM-free) oxygen reduction reaction (ORR) catalysts from the Fe-N-C family have been recently synthesized using a sacrificial support method (SSM) technique. The effects of each synthetic parameter have been studied with the help of surface chemistry and the electrocatalytic performance in neutral media. Rotating ring disk electrode (RRDE) experiment has been utilized to increase half-wave potential and limiting current after the pyrolysis steps. Additional improvement has been found after etching and performing the second pyrolysis. A similar trend had been seen in MFCs and power output has increased from 167 ± 2 to 214 ± 5 $\mu W\ cm^{-2}$ [61]. The production of electricity in MFC has been reported with simultaneous biomass regeneration in the cathodic half-cell with dependent on the nutrient value of the anodic feedstock. The growth of algal biomass in the cathode chamber can be measured monitored and assessed against the MFC power production (charge transfer) during the process. MFC generation of electricity can be enhanced through activated cation and crossover for the formation of biomass via harvesting and reusing of microalgae as an energy source in a closed loop system. Nutrient reclamation and assimilation into new biomass of microalgae can increase the energy efficiency [8, 9].

Table 2 Application of different microalgal species with other microbial strains in MFCs construction with the capability to electricity generation

Microalgal species a lone/ with others	MFCs	Maximum Power density (mW/m^2)/ Current generation (mA)	References
Chlorella pyrenoidosa	Anode of MFCs an electron donor with two chamber	30.15- 104.06 in Model 1 algal MFCs; 2.16 in Model 2 algal MFCs -	[67, 68]
Chlorella vulgaris with *Actinobacteria* and *Deltaproteobacteria*	The cathode as an electron acceptor	23.17 - 327.67	[68]
Saccharomyces cerevisiae with methylene blue and ferricyanide as electron mediators	An anode as an electron donor with improved performance	146.71 ± 7.7	[69]
Saccharomyces cerevisae and *Clostridium acetobutylicum*	Two chambered MFC using Artificial Waste water	10.89 mA and 10.45 mA respectively	[70]
Chlamydomonas reinhardtii CC-125	MFCs with sulfur-deprived conditions by polymer-coated electrocatalytic electrodes	9 mA with a maximum hydrogen production rate of 4.1 ml h^{-1}	[71]
Scenedesmus acutus microalga as MFCs for wastewater treatment	PBI-based photosynthetic MFCs with algae-assisted biocathode	Higher volumetric power density value is 400 mWm^{-3} after more than 100 operating days	[72]
Algae biomass *Scenedesmus obliquus*	Two chamber microbial fuel cells (MFCs) with pretreated algal biomass was 102 mW·m^{-2} with lactate and acetate	951 mW·m^{-3}/ 276 mA·m^{-2} for electricity generation	[4]

Microbial biophotovoltaic (BPV) cells have been exploited by the ability of cyanobacteria and microalgae to convert light energy into electrical current using water as the source of electrons. In this regard the model cyanobacterium *Synechocystis* sp. PCC 6803 has been used extensively in BPV devices. Bioelectrochemical systems have advantages over conventional microbial fuel cells as these require the input of organic carbon for microbial growth. Innovative approaches can address scale-up issues associated with the fabrication of the inorganic (electrodes) and biological (microbe) parts of the biophotovoltaic devices [62]. It needed a simple commercial inkjet printer to fabricate a thin-film paper-based biophotovoltaic cell consisting of a layer of cyanobacterial cells on top of a carbon nanotube-conducting surface to generate a sustained electrical current both in the dark (as a 'solar bio-battery') and in response to light (as a 'bio-solar-panel') with potential applications in low-power devices [63]. Microbial electrochemical technology has been utilized for diverse applications such as wastewater treatment, biofuel production, water desalination, remote power sources and biosensors. It has shown many advantages based on the self-sustaining nature of the microorganisms with donation or acceptance of electrons from an electrode and the range of fuels can be used. Systems can be scaled up through careful consideration of electrode spacing and packing per unit volume of the reactor. MFCs can be used to produce electricity along with the use of microorganisms on the anodes or cathodes, or both electrodes for a variety of different purposes as shown in Figure 1 [64].

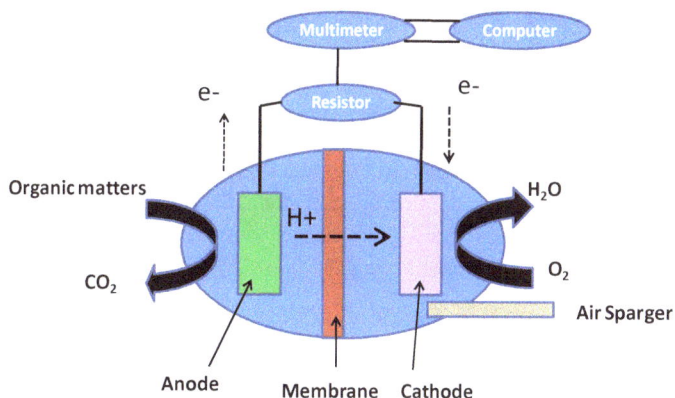

Figure 1 Schematic diagram of two-chamber microbial fuel cell (MFCs) [65]

Cyanobacteria have been found to produce electricity. Microorganisms have the capability to transfer electrons derived from the metabolism of organic matters to anode, and glucose is the most used substrate in MFC. Plants have also been used in MFC for generating electric power [66].

In an MFC, bacteria are used as catalysts to oxidize organic matters in order to generate electrical current. MFC has been found effective in the removal of chemical oxygen demand (COD) from wastewater. Removal of nitrogen and phosphorus has been reported with only bacteria containing MFC [73]. Improvement in treatment efficiency of wastewater can be achieved by a combined process consisting of MFC and microalgae cultivation. Wastewater treated with a single-chamber MFC (SMFC) has shown electricity generation (maximum power density of $268.5\,mW/m^2$) with 67% COD removal as well as partial removal of nitrogen (50%) and (34%) of total phosphorus. The treated wastewater has been further used to cultivate microalgae to remove the residual phosphorus and nitrogen (97% TP and 99% NH_4^+-N). The combined process of SMFC with cultivated microalgae has been effective and a promising candidate for treating wastewater with complete removal of TP and NH_4^+-N [74].

For power generation, various parameters have been considered in MFC technologies which include maximum power density, coulombic efficiencies and sometimes chemical oxygen demand removal rate [75]. These parameters have helped in the construction of effective MFCs devices. Application of microbes for bioremediation while at the same time generating electricity via MFC technology have a highly advantageous position for applications in various sectors of industrial, municipal and agricultural waste management. Bioelectricity production involves the generation of electricity by anaerobic digestion of organic substrates by microbes [76]. MFC is a device that converts chemical energy released as a result of oxidation of complex organic carbon sources which are utilized as substrates by micro-organisms to produce electrical energy thereby proving to be an efficient means of sustainable energy production. The electrons released due to the microbial metabolism are captured to maintain a constant power density, without an effective carbon emission in the ecosystem [77].

4.2 Bioethanol production

Microalgae (*Chlorococcum infusionum, Chlamydomonas reinhardtii Porphyridium cruentum Undaria pinnatifida* (brown seaweed) *Sargassum spp.* (brown seaweed) *Kappaphycus alvarezii* (red seaweed) *Gelidium amansii* (red seaweed) and *Ulva lactuca* (green seaweed)) have been reported as most potentially significant resources of sustainable biofuels for future of renewable energy due to high accumulation of starch/cellulose. Many developments have been made in the recent years for commercialization of algal bioethanol with some challenges of the techno-economic constraints which need to be overcome for successfully large-scale bioethanol production from microalgal species of both marine and freshwater algae [78]. Algal species do not produce any food for people. The carbohydrate content in the algae cell is available as starches and sugars to be utilized for fermentation to produce bioethanol [1].

The low hemicellulose levels and absence of lignin in microalgae (*Spirogyra*) species results in increased hydrolysis efficiency of biomass and fermentation yields as well as reduction of the cost of the bioethanol production. Algae species have the ability to take up CO_2 from the atmosphere and power the plants for use of appropriate technology options to get the good algae bioethanols to yield with GHG reductions relative to fossil and other biobased fuels [79]. Algal biomass is a crucial energy resource, utilized for the generation of electricity and transportation fuels. High level of biocomponents in microalgae has exhibited the potential feedstock for the generation of ecofriendly biofuels. Microalgae-derived biofuels are suitable as carbon-neutral replacements for petroleum oil/ fuel [80].

Fermentation process for metabolic conversion of microalgal compounds has been used for the synthesis of bioethanol and higher alcohols synthesis. Major biocomponents (carbohydrates, proteins, and lipids) of microalgal biomass are, have been utilized for maximum biofuel generation. Efficient pretreatment methods for algal biomass hydrolysis can help in enhancing the bioavailability of substrates (simple sugar, amino acid, and fatty acid). Biocomponents have been useful for the generation of various biofuels (bioethanol, higher alcohol, and biodiesel) through fermentation and transesterification processes. Maximum utilization of algal biomasses has been successful to get economically feasible biofuel production [81]. The technical potential of producing biofuel has been reported in naturally occurring microalgae spirulina. It was grown in such environment which contained nitrates, phosphates and carbon dioxide from the atmosphere. After drying of algae and grinding, this algal matter was subjected to acid hydrolysis to extract carbohydrates for the formation of an algal sugar solution. Fermentation of this algal sugar solution (15.2 g.L^{-1} of reducing sugar) by *Saccharomyces cerevisiae* was performed to produce ethanol (0.85-1.0%). Varied ethanol percentages were found based on the variation of hydrolysis time, the concentration of sulfuric acid and fermentation time. This process provided the assurance in a defined way for biofuel production from other carbohydrates enriched microalgae [82].

4.3   Biobutanol production

Biobutanol is a potential biofuel, exhibiting similar energy to that of gasoline. Design and selection of a complete industrial scale of biobutanol production plant using microalgae as the feedstock have been reported [83]. *Clostridium acetobutylicum* has produced the butanol titre (3.86g.L^{-1}) with yield (0.13 g/g-carbohydrate) via Acetone n-butanol and ethanol (ABE) fermentation by using microalgae biodiesel residues as a substrate with one-third of carbohydrate residue. Biological butanol production from microalgae biodiesel residues reported for further research on fermentation strategies for improving the production yield of n-butanol. Normally the highest butanol yield of 0.4 g/g-glucose has been reported with 60 g.L^{-1} of glucose and 18 g.L^{-1} of butyrate *by C. acetobutylicum* with approximately one-third of carbohydrate residual [84].

Clostridium saccharoperbutylacetonicum N1- 4 in ABE fermentation has used microalgae biomass (cultivated wastewater) as feedstock by utilization proper pretreatment and enzymatic hydrolysis. Addition of 1% glucose has significantly improved the ABE yield of 0.311 g/g (1.6 fold) and production 0.102 g.L^{-1}h^{-1} [85]. photoautotrophically cultivated microalgae *Chlorella vulgaris* (111g of acid-pretreated biomass) has been utilized for butanol production (titre~3.37g.L^{-1}) with a *Clostridium acetobutylicum* strain in acetone-butanol-ethanol (ABE) fermentation. Microalgae have shown their capability to mitigate CO$_2$ emission via the conversion of CO$_2$ into biomass (as a source of abundant in carbohydrates) and emerged as third-generation feedstock for fermentation [86].

Microalgal biomass has been examined as a biodiesel feedstock due to its ability to accumulate oil bodies, and it has been considered a promising substrate for ABE fermentation because its structure contains starch. Modelling approaches have been utilized under a single-nutrient basis to predict lipid formation. A predictive multi-parameter kinetic model has been used for optimization of the starch formation during microalgae cultivation. The algal growth rate is affected by nitrogen, phosphorus, and the carbon sources with an exhibition of a compartmentalized biomass structure of active biomass with starch and lipids components. Lab-scale cultures of *Chlamydomonas reinhardtii* CCAP 11/32C under various nutrient concentrations have grown mixotrophically. Model fitting parameters have been used through an in-house developed optimization algorithm via linking stochastic and deterministic methods with avoidance of getting trapped in local optima. This model has used the optimal conditions for maximum starch formation. Biobutanol has been produced through renowned ABE fermentation (a microbial process) [87].

4.4   Biohydrogen production

Bio-hydrogen from microalgae (cyanobacteria) has enhanced our commercial awareness as a clean source of biofuel and has shown its potential as an alternative, reliable and renewable energy source. Photosynthetic hydrogen production from microalgae is reported as promising options for clean energy. Advances in hydrogen-fuel-cell technology have provided an eco-friendly way of biofuel production and its use for generating electricity with the release of water as a byproduct. Advances in genetic/metabolic engineering in microalgal species have enhanced the photobiological hydrogen production with manipulation of competing for metabolic pathways by modulating the certain key enzymes (hydrogenase or nitrogenase or both) [88].

Biological route of H_2 production is found at low operating cost-effective with the requisition of economic viability. Utilization of large-scale microalgae biomass sustainable for hydrogen production via photobioreactors has been started with recent technological progress with modified enzymes involvement as well as genetic or metabolic engineering approaches [89]. Two biological methods for hydrogen production from microalgae are reported. In the first method, microalgae have utilized the light energy to produce hydrogen from water whereas in the second one, bacteria with capability for fermenting the carbohydrates (i.e. structural or stored in the microalgal cell wall) are used to produce hydrogen. These approaches are direct and indirect biophotolysis and strategies for using microalgae as feedstock for dark fermentation for hydrogen generation. Further improvement in hydrogen production can be done by identifying technological bottlenecks, detecting the weaknesses and focusing the research efforts [90].

Microalgae with remarkable molecular pathways have generated biohydrogen as an environmentally friendly and sustainable energy source. The effect of environmental factors has simulated outdoor conditions using *Chlamydomonas reinhardtii* D1 mutant strains and CC124 as a control strain. The D1 mutant strains have shown effective hydrogen generation capacity whereas strain D239-40 has been found more attractive in term of hydrogen output due to having a promising D1 protein mutant. Further developments in both strains were done in genetic and bioprocess aspects to scaling up and commercialization for outdoor cultures [91]. Hydrogen has been reported as a zero-carbon and energy dense alternative energy carrier with clean-burning properties. Biohydrogen production by microalgae has reduced the emissions of GHGs to great extent. Biohydrogen is produced through dark fermentation using sugars, starch, or cellulosic materials whereas microalgae-based biohydrogen production is by photolysis [92]. Algal strain *Chlorella vulagaris* is used to produce the H_2 gas in biophotolysis. Initial substrate concentration (5- 40 $g.L^{-1}$, optimum at 10 $g.L^{-1}$), initial pH (6-10, optimum at 8) and total of nitrogen and phosphate contents (10-30% optimum at 10%) for the Bold's Basal culture (BBC) media concentration are found to affect the hydrogen production (10 $g.L^{-1}$) [93].

4.5 Biodiesel production

Microalgal biodiesel is made up of lipid content, often low quantity in a green microalgal species like *Chlorella vulgaris* as a potential feedstock for biodiesel. Identification with evaluation of the relationships between the critical variables in microalgal species can enhance the lipid yield via characterization of various factors for biodiesel production. Multifactor optimization has reported enhancing the lipid pool to 55% dry cell weight compared to 9% control. *C. vulgaris* cells have been pre-grown in glucose (0.7%) supplemented medium and transferred to the optimized conditions at the second stage. It has boosted to lipid yield (20-fold higher) and titre of 1974 mg.L^{-1} than the control. The transesterified *C. vulgaris* oil has shown the presence of ~82% saturated fatty acids with palmitate and stearate as the major components with oxidative stability of biodiesel. The biodiesel properties such as density, viscosity, acid value, iodine value, calorific value, cetane index, ash and water contents have been compared with the international (ASTM and EN) and Indian standards (IS) for biodiesel quality. *C. vulgaris* biomass (potentially a renewable feedstock) with 55% lipid content has been reported as adequate biodiesel source [94].

Microalgae *Chlorella pyrenoidosa* has been grown autotrophically in a vertical bioreactor for its greater efficiency to enhance the biomass concentration and lipid content and it has inoculated to the vertical bioreactor to increase in growth with the effect of different concentrations of nitrogen source (0-0.4 g.L^{-1} KNO$_3$ with optimal at 0.05 g.L^{-1}) and lipid content (15%). With the decrease in nitrate concentration in the medium has found to decrease the biomass production with increased lipid content. Concentration-effect of nitrate source is found to accumulate more lipids in stationary phase compared to exponential phase. Nitrogen starvation has been an effective approach to enhance lipid for biofuel production in *Chlorella pyrenoidosa* [95].

Microalgae have been considered an attractive medium for capturing the excess CO_2 from the atmosphere. Power plants, automobiles, volcanic eruption, decomposition of organic matters and forest fires generate CO_2 as a GHGs component. Capturing of CO_2 through microalgae species has been reported as a potential carbon source to produce lipids for the generation of biofuel. Recent developments in the field of biological carbon capture through microalgae has been found towards the generation of biodiesel with help in selection of significance of certain key parameters such as selection of efficient strain, microalgal metabolism, cultivation systems (open and closed) and biomass production along with the national and international biodiesel specifications and its properties. The potential use of photobioreactors have been reported for biodiesel production under the influence of various factors viz., light intensity, pH, time, temperature, CO_2 concentration and flow with more focus on economic and future overview on biodiesel production from microalgae [96].

Commercial production of microalgal biodiesel is not found economically viable due to low storage lipid yield. Selection of lipid-rich microalgae is an important task for microalgal biodiesel production. Laboratory screening protocols with a comprehensive assessment of microalgae species have been done under the laboratory as well as fields conditions. *Graesiella* sp. WBG-1 has the capability of accumulating a large amount of storage lipid content (33.4 % DW with 90% of storage TAGs) under natural solar irradiance and temperature with pilot-scale raceway pond. It has shown the feasibility of using a low-cost raceway pond for autotrophic cultivation of microalgae for biodiesel production [97].

4.6   Biogas (Methane) production

Microalgae are potential feedstocks biomass for biogas production. Hydrolysis and methanogenesis are two main bioprocesses for biogas production. Low-cost biorefinery approach has been reported for producing additional products (using carbon dioxide and digestates) rather than methane generated during anaerobic digestion (AD). Limitations of commercialization using microalgae for the production of biogas have also discussed with the implementation of a number of research projects on microalgal species improvement with supported worldwide [98].

For conversion of microalgae into biofuels/biogases, anaerobic digestion (AD) of microalgal biomass has been reported as energy efficient technologies. Improved biogas productivity has been found by breaking up the tough and rigid cell wall of microalgae by pretreatment with the utilization of *Bacillus licheniformis*. Pretreatment process for *Chlorella* sp with pure bacterial culture (0 and 8% v/v) has been reported under an anaerobic condition at 37^0C for 60 h. Enhancement in methane production (9.2–22.7%) has been reported in subsequent AD with more dosages of bacteria for pretreating the microalgal biomass (1–8%). Improvement in volatile fatty acids (VFAs) (17.3–44.2%) has also observed with increased soluble chemical oxygen demands (SCOD) content (16.4–43.4%) [99].

The anaerobic co-digestion has been done for microalgal biomass grown in wastewater and wheat straw. Biochemical methane potential (BMP) tests have performed with different substrate proportions (20–80, 50–50 and 80–20%) on a volatile solid basis. For improvements in their mass biodegradability, the co-digestion of both substrates (microalgal biomass and 50% of wheat) has been evaluated after applying a thermo-alkaline pretreatment (10% CaO at 75^0C for 24 h). Co-digestion of 50% microalgae – 50% wheat straw has been done in mesophilic lab-scale reactors with enhanced methane yield (77%) [100]. Ruminal fluid as a source of hydrolytic microorganisms has been utilized for the pretreatment of microalgae (*Senedesmus*). In a bioreactor, use of enriched culture microalgae to the ruminal fluid ratio (S/X: 0.5) has been reported for the hydrolysis (29%) with subsequent production of methane yield for 7 days. The main predominant ruminal hydrolytic bacteria were principally *Clostridium, Proteocatella* and *Pseudomonas* [101].

Conclusion

Microalgae species are available as alternative substrate /source for biofuel synthesis as these have shown an environmentally friendly nature with the capability to capture CO_2 gas for the accumulation of rich oil contents for biofuel (biodiesel, bioethanol / n-butanol, methane or hydrogen gas as production capability). Microalgae are utilized as low cost and promise substrate for third generation biofuels production. Microalgae-MFC has been utilized in electricity generation by using the electrons released to the anode electrode. Microalgae have also been utilized in electricity generation via wastewater treatment with removal of toxic compounds. Lipid degradation processes in microalgae have been reported for their survival and cells grown under fluctuating environmental conditions. Microalgae cultivation and process designing have been reported for algal biodiesel production. Lipids from microalgae have been utilized for the production of biodiesel fuel by transesterification process. Lab-scale photobioreactors (PBRs) have been utilized for cultivation of microalgae via utilization of wastewater or organic compost with sufficient nutrients to sustain its growth. We need to focus more research on the development of microalgae species with more lipid storage capacity via utilization of wastewater treatment and ability to utilize the excess CO_2. As microalgae species are not a competitor for arid / fertilized land (utilize for food cultivation) they shoould be promoted as a source for biofuel production for future energy demand.

Abbreviations

ABE: Acetone–Butanol–Ethanol; **ACCase:** Acetyl-CoA carboxylase; **AD:** Anaerobic digestion; **Algal Res:** Algal research; **Appl Biochem Biotechnol:** Applied biochemistry and biotechnology; **ASTM:** American section of the international association for testing materials; **BBC:** Bold's basal culture; **Biol Evol:** Biology evolution **Biol:** Biology; **Biomass Bioenergy:** Biomass and bioenergy; **Bioresour Technol:** Bioresource technology; **Biotechnol Adv:** Biotechnology advances; **Biotechnol:** Biotechnology; **BMP:** Biochemical methane potential; **BPV:** Biophotovoltaic; **CaO:** Calcium oxide; **Cas:** CRISPR-associated; **CE:** Columbic efficiency **Chin J Chem Eng:** Chinese journal of chemical engineering; **CO_2:** Carbon dioxide; **COD:** Chemical oxygen demand;

Comm: Communications; **CRISPR:** Clustered Regularly Interspaced Short Palindromic Repeats; **Curr Opin Chem Biol:** Current opinion in chemical biology; **DC/L:** Dry cell per litre; **EN:** European Standard; **Ener:** Energy; **Environ Sci Pollut Res Int:** Environmental science and pollution research international; **Environ:** Environment; **FAME:** Fatty acid monoalkyl esters; **FANTOM:** Functional annotation of the mammalian genome; **FAS:** Fatty acid synthase; **Front Environ Sci Eng:** Frontiers of environmental science and engineering; **GHGs:** Green house gases; **GM:** Genetically modified; **g.L^{-1}:** Gram per litre; **HR:** Homologous recombination; **HTL:** Hydrothermal liquefaction; **Int:** International; **IS:** Indian standards; **J Acad Indust Res**: Journal of academia and industrial research; **J Biosci Bioeng:** Journal of bioscience and bioengineering; **Kbp:** Kilo-base pair; **KNO$_3$:** Potassium nitrate; **LC-PUFA:** Long-chain polyunsaturated fatty acids ; **LHC:** Light-harvesting chlorophyll; **mA:** Milli ampere; **Mbp:** mMega base pair; **Mdh:** Malate dehydrogenase (gene), **Metab Eng**: Metabolic Engineering; **MFCs:** Microbial fuel cells; **Microb Cell Fact:** Microbial cell factory; **Microbiol:** Microbiology; **mW/m^2/L:** Milliwatt per square meter per litre; **NADH:** Nicotinamide adenine dinucleotide; **nmoleH$_2$mg Chl^{-1}s^{-1}):** Nanomole hydrogen milligram per chlorophyll per second; **OCND:** One-stage continuous N-deprivation; **OCNL:** One-stage continuous N-limitation; **OCNS:** One-stage continuous N-sufficiency; **ORR:** Oxygen reduction reaction; **PBRs:** Photobioreactors; **pfl:** pyruvate formate-lyase (gene); **Physiol:** Physiology; **PGM:** Platinum group metal; **Prikl Biokhim Mikrobiol:** Prikladnaia biokhimiia i mikrobiologiia; **Proc Natl Acad Sci**: Proceedings of the national academy of sciences; **Process Biochem:** Process biochemistry; **PS:** Photosystem; **PUFAs:** Polyunsaturated fatty acids; **Renew Sust Energ Rev:** Renewable and sustainable energy reviews; **Res J Pharm Biol Chem Sci:** Research journal of pharmaceutical, biological and chemical sciences; **RIKEN** : Rikagaku kenkyūjyo; **RNAi:** RNA interference; **RRDE:** Rotating ring disk electrode; **S/X:** Substrate/ microbial biomass; **SSM:** Sacrificial support method; **SCOD:** Soluble chemical oxygen demands; **SMFC:** Single-chamber MFC; **Sta:** Starch accumulation; **TAG:** Triacylglyceride; **TALEN**: Transcription activator-like effector nucleases; **TBNS:** Two-stage batch N-starvation; **Ter:** Trans-enoyl-CoA reductase; **TG:** Triacylglycerol; **Thaps3:** *Thalassiosira pseudonana;* **TP:** Total phosphate; **Trends Biotechnol:** Trends in biotechnology

References

[1]   A. Singh, P.S. Nigam, J.D. Murphy, Renewable fuels from algae: An answer to debatable land based fuels. Bioresour. Technol. 102 (2011) 10-16. https://doi.org/10.1016/j.biortech.2010.06.032

[2]   S. Pandit, D. Das, Role of Microalgae in Microbial Fuel Cell. D. Das, (Eds.), Algal
        Biorefinery: An Integrated Approach, Capital Publishing Company Springer,
        2015, pp. 375-399.

[3]   R.A. Rozendal, H.V.M. Hamelers, K.Rabaey, J. Keller, C.J.N. Buisman, Towards
        practical implementation of bioelectrochemical wastewater treatment. Trends in
        Biotechnol. 26 (2008) 450–459. https://doi.org/10.1016/j.tibtech.2008.04.008

[4]   S. Kondaveeti, K.S. Choi, R. Kakarla, B. Min, Microalgae *Scenedesmus obliquus*
        as renewable biomass feedstock for electricity generation in microbial fuel cells
        (MFCs), Front. Environ. Sci. Eng. 8 (2013) 784-791.
        https://doi.org/10.1007/s11783-013-0590-4

[5]   I. Gajda, J. Greenman, C. Melhuish, I. Ieropoulos, Self-sustainable electricity
        production from algae grown in a microbial fuel cell system. Biomass Bioenergy,
        82 (2015) 87-93. https://doi.org/10.1016/j.biombioe.2015.05.017

[6]   F. Kong, I.T. Romero, J. Warakanont, Y. Li-Beisson, Lipid catabolism in
        microalgae. New Phytol. 218 (2018) 1340-1348.
        https://doi.org/10.1111/nph.15047

[7]   P. Wensel, G. Helms, B. Hiscox, W.C. Davis, H. Kirchhoff, M. Bule, L. Yu, S.
        Chen, Isolation, characterization, and validation of oleaginous, multitrophic, and
        haloalkaline-tolerant microalgae for two-stage cultivation. Algal Res. 4 (2014) 2-
        11. https://doi.org/10.1016/j.algal.2013.12.005

[8]   M. Faried, M. Samer, E. Abdelsalam, R.S. Yousef, Y.A. Attia, A.S. Ali, Biodiesel
        production from microalgae: Processes, technologies and recent advancements.
        Renew. Sust. Energ. Rev. 79 (2017) 893-913.
        https://doi.org/10.1016/j.rser.2017.05.199

[9]   X.B. Tan, M.K. Lam, Y. Uemura, J.W. Lim, C.Y. Wong, K.T. Lee, Cultivation of
        microalgae for biodiesel production: A review on upstream and downstream
        processing. Chin. J. Chem. Eng. 26 (2018) 17-30.
        https://doi.org/10.1016/j.cjche.2017.08.010

[10]   C.E.F. Silva, A. Bertucco, Bioethanol from microalgae and cyanobacteria: A
        review and technological outlook. Process Biochem. 51 (2016) 1833-1842.
        https://doi.org/10.1016/j.procbio.2016.02.016

[11]   E.I. Lan, J.C. Liao, Metabolic engineering of cyanobacteria for 1-butanol
        production from carbon dioxide. Metab. Eng. 13 (2011) 353-363.
        https://doi.org/10.1016/j.ymben.2011.04.004

[12] J. Milano, H.C. Ong, H.H. Masjuki, W.T. Chong, M.K. Lam, P.K. Loh, V.Vellayan, Microalgae biofuels as an alternative to fossil fuel for power generation. Renew. Sust. Energ. Rev. 58 (2016) 180-197. https://doi.org/10.1016/j.rser.2015.12.150

[13] N. Pragya, K.K. Pandey, P.K. Sahoo, A review on harvesting, oil extraction and biofuels production technologies from microalgae. Renew. Sust. Energ. Rev. 24 (2013) 159-171. https://doi.org/10.1016/j.rser.2013.03.034

[14] B.H. Kiepper, Microalgae Utilization in Wastewater Treatment. UGA Cooperative Extension Bulletin 1419, the University of Georgia and Ft. Valley State University, 2013.

15] I. Khozin-Goldberg, Lipid Metabolism in Microalgae. M.A. Borowitzka, J. Beardall, J.A. Raven, (Eds.), The Physiology of Microalgae, Springer, 2016, pp.413-484.

[16] S. Bellou, M.N. Baeshen, A.M. Elazzazy, D. Aggeli, F. Sayegh, G. Aggelis, Microalgal lipids biochemistry and biotechnological perspectives. Biotechnol. Adv. 32 (2014) 1476-93. https://doi.org/10.1016/j.biotechadv.2014.10.003

[17] C.H. Tan, P.L. Show, J-S. Chang, T.C. Ling, J. C.-W. Lan, Novel approaches of producing bioenergies from microalgae: A recent review. Biotechnol. Adv. 33 (2015) 1219-1227. https://doi.org/10.1016/j.biotechadv.2015.02.013

[18] J.L. Blatti, J. Michaud, M.D. Burkart, Engineering fatty acid biosynthesis in microalgae for sustainable biodiesel. Curr. Opin. Chem. Biol. 17 (2013) 496-505. https://doi.org/10.1016/j.cbpa.2013.04.007

[19] J.A. Gimpel, E.A. Specht, D.R. Georgianna, S.P. Mayfield, Advances in microalgae engineering and synthetic biology applications for biofuel production. Curr. Opin. Chem. Biol. 17 (2013) 489-95. https://doi.org/10.1016/j.cbpa.2013.03.038

[20] A.W.D. Larkum, I.L. Ross, O. Kruse, B. Hankamer, Selection, breeding and engineering of microalgae for bioenergy and biofuel production. Trends in Biotechnol. 30 (2012) 198-205. https://doi.org/10.1016/j.tibtech.2011.11.003

[21] D. Fleury, J. Marshall, A modular photosynthetic microbial fuel cell with interchangeable algae solar compartments. BioRxiv (2017) 166793. https://doi.org/10.1101/166793

Materials Research Forum LLC
doi: http://dx.doi.org/10.21741/9781644900116-2

[22] C. Xu, K. Poon, M.M. Choi, R. Wang, Using live algae at the anode of a microbial fuel cell to generate electricity. Environ. Sci. Pollut. Res. Int. 22 (2015) 15621-15635. https://doi.org/10.1007/s11356-015-4744-8

[23] X.Y. Wu, T.S. Song, X.J. Zhu, P. Wei, C.C. Zhou, Construction and operation of microbial fuel cell with *Chlorella vulgar* is biocathode for electricity generation. Appl. Biochem. Biotechnol. 171 (2013) 2082-2092. https://doi.org/10.1007/s12010-013-0476-8

[24] E. Kojima, K. Zhang, Growth and hydrocarbon production of microalga *Botryococcus braunii* in bubble column photobioreactors. J. Biosci. Bioeng. 87 (1999) 811–815. https://doi.org/10.1016/S1389-1723(99)80158-3

[25] M. Hannon, J. Gimpel, M. Tran, B. Rasala, S. Mayfield, Biofuels from algae: challenges and potential. Biofuels 1 (2010) 763–784. https://doi.org/10.4155/bfs.10.44

[26] J.N. Rosenberg, G. A. Oyler, L. Wilkinson, M. J. Betenbaugh, A green light for engineered algae: redirecting metabolism to fuel a biotechnology revolution. Curr. Opin. Biotechnol. 19 (2008) 430-436. https://doi.org/10.1016/j.copbio.2008.07.008

[27] W. Fang, Y. Si, S. Douglass, D. Casero, S.S. Merchant, M. Pellegrini, I. Ladunga, P. Liu, M.H. Spalding. Transcriptome-wide changes in *Chlamydomonas reinhardtii* gene expression regulated by carbon dioxide and the CO2 concentrating mechanism regulator CIA5/CCM1. Plant Cell 24 (2012) 1876–1893. https://doi.org/10.1105/tpc.112.097949

[28] R. Zhang, W. Patena, U. Armbruster, S.S. Gang, S.R. Blum, M.C. Jonikas, High-throughput genotyping of green algal mutants reveals random distribution of mutagenic insertion sites and endonucleolytic cleavage of transforming DNA. Plant Cell 26 (2014)1398–1409. https://doi.org/10.1105/tpc.114.124099

[29] A. Molnar, F. Schwach, D.J. Studholme, E.C. Thuenemann, D.C Baulcombe, miRNAs control gene expression in the single-cell alga *Chlamydomonas reinhardtii*. Nature 447 (2007) 1126–1129. https://doi.org/10.1038/nature05903

[30] Y.T. Li, D.X. Han, G.R. Hu, M. Sommerfeld, Q.A. Hu, Inhibition of starch synthesis results in overproduction of lipids in *Chlamydomonas reinhardtii*. Biotechnol. Bioeng. 107 (2010) 258–268. https://doi.org/10.1002/bit.22807

[31] B.S. Liu, C. Benning, Lipid metabolism in microalgae distinguishes itself. Curr.
     Opin. Biotechnol. 24 (2013) 300–309.
     https://doi.org/10.1016/j.copbio.2012.08.008

[32] M. La Russa, C. Bogen, A. Uhmeyer, A. Doebbe, E. Filippone, O. Kruse, J.H.
     Mussgnug Functional analysis of three type-2 DGAT homologue genes for
     triacylglycerol production in the green microalga *Chlamydomonas reinhardtii*. J.
     Biotechnol. 162 (2012) 13–20. https://doi.org/10.1016/j.jbiotec.2012.04.006

[33] E.M. Trentacoste, R.P. Shrestha, S.R. Smith, C. Gle, A.C. Hartmann, M.
     Hildebrand, W.H. Gerwick, Metabolic engineering of lipid catabolism increases
     microalgal lipid accumulation without compromising growth. Proc. Natl. Acad.
     Sci. 110 (2013)19748–19753 https://doi.org/10.1073/pnas.1309299110

[34] W.L. Yu, W. Ansari, N.G. Schoepp, M.J. Hannon, S.P. Mayfield, M.D. Burkart,
     Modifications of the metabolic pathways of lipid and triacylglycerol production in
     microalgae. Microb. Cell Fact. 10 (2011) 11-15. https://doi.org/10.1186/1475-
     2859-10-91

[35] H. Teramoto, A.Nakamori, J. Minagawa, T. Ono, Light-intensity-dependent
     expression of Lhc gene family encoding light-harvesting chlorophyll-a/b proteins
     of photosystem II in *Chlamydomonas reinhardtii*. Plant Physiol. 130 (2002) 325–
     333. https://doi.org/10.1104/pp.004622

[36] R.M. Dent, C.M. Haglund, B.L. Chin, M.C. Kobayashi, K.K. Niyogi, Functional
     genomics of eukaryotic photosynthesis using insertional mutagenesis of
     Chlamydomonas reinhardtii. Plant Physiol. 137 (2005) 545–556.
     https://doi.org/10.1104/pp.104.055244

[37] O. Kilian, C.S. Benemann, K.K. Niyogi, B. Vick, High-efficiency homologous
     recombination in the oil-producing alga Nannochloropsis sp. Proc. Natl. Acad. Sci.
     108 (2011) 21265–21269 https://doi.org/10.1073/pnas.1105861108

[38] M.A. Scranton, J.T. Ostrand, F.J. Fields, S.P. Mayfield, Chlamydomonas as a
     model for biofuels and bio-products production. The Plant J. 82 (2015) 523–531.
     https://doi.org/10.1111/tpj.12780

[39] S.S. Merchant, J. Kropat, B.S. Liu, J. Shaw, J. Warakanont TAG, You're it!
     Chlamydomonas as a reference organism for understanding algal triacylglycerol
     accumulation. Curr. Opin. Biotechnol. 23 (2012) 352–363.
     https://doi.org/10.1016/j.copbio.2011.12.001

[40]   P. Sharma, P. Patil, N. Rao, K.V. Swamy, M.B. Khetmalas, G.D. Tandon,
       Mapping Biodiversity of Indigenous Freshwater Chlorophytes. Res. J. Pharm.
       Biol. Chem. Sci. (RJPBCS), 5(3) (2014) 1632-1639.

[41]   I. Haq, A. Muhammad, U. Hameed, Comparative assessment of *Cladophora*,
       Spirogyra and *Oedogonium* biomass for the production of fatty acid methyl esters.
       Prikl. Biokhim. Mikrobiol. 50 (2014) 80-84.
       https://doi.org/10.7868/S0555109913060093

[42]   C. Wang, Z. Wang, F. Luo, Y. Li, The augmented lipid productivity in an
       emerging oleaginous model alga *Coccomyxa subellipsoidea* by nitrogen
       manipulation strategy. World J. Microbiol. Biotechnol. 33 (2017) 160-168.
       https://doi.org/10.1007/s11274-017-2324-4

[43]   G. Blanc, I. Agarkova, J. Grimwood, A. Kuo, A. Brueggeman, D.D. Dunigan, J.
       Gurnon, I. Ladunga, E. Lindquist, S. Lucas, J. Pangilinan, T. Pröschold, A.
       Salamov, J. Schmutz, D. Weeks, T.Yamada, A. Lomsadze, M. Borodovsky, J.M.
       Claverie, I.V. Grigoriev, J.L. Van Etten, The genome of the polar eukaryotic
       microalga *Coccomyxa subellipsoidea* reveals traits of cold adaptation. Genome
       Biol. 13 (2012) R39. https://doi.org/10.1186/gb-2012-13-5-r39

[44]   A.J. Cornish, R.Green, K. Gärtner, S.Mason, E.L. Hegg, Characterization of
       hydrogen metabolism in the multicellular green alga *Volvox carteri*. PloS One 10
       (2015) e0125324. https://doi.org/10.1371/journal.pone.0125324

[45]   S.E. Prochnik, J. Umen, A.M. Nedelcu, A. Hallmann, S.M. Miller, I.Nishii, P.
       Ferris, A. Kuo, T. Mitros, L.K. Fritz-Laylin, U. Hellsten, J. Chapman, O.
       Simakov, S.A. Rensing, A. Terry, J. Pangilinan, V. Kapitonov, J. Jurka, A.
       Salamov, H. Shapiro, J. Schmutz, J. Grimwood, E. Lindquist, S. Lucas, I.V.
       Grigoriev, R. Schmitt, D. Kirk, D.S. Rokhsar. Genomic analysis of organismal
       complexity in the multicellular green alga Volvox carteri. Sci. 329 (2010) 223-
       226. https://doi.org/10.1126/science.1188800

[46]   N. Sumiya, S. Miyagishima, Metabolic engineering of Cyanidioschyzon merolae.
       T. Kuroiwa, S. Miyagishima, S. Matsunaga, N. Sato, H. Nozaki, (Eds.),
       Cyanidioschyzon merolae: A New Model Eukaryote for Cell and Organelle
       Biology (). Springer Singapore, 2018, pp. 343-354

[47]   M. Matsuzaki, O. Misumi, I.T. Shin, S. Maruyama, M. Takahara, S.Y.
       Miyagishima, T. Mori, K. Nishida, F. Yagisawa, K. Nishida, Y. Yoshida, Y.
       Nishimura, S. Nakao, T. Kobayashi, Y. Momoyama, T. Higashiyama, A. Minoda,
       M. Sano, H. Nomoto, K. Oishi, H. Hayashi, F. Ohta, S. Nishizaka, S. Haga, S.

Miura, T. Morishita, Y. Kabeya, K. Terasawa, Y. Suzuki, Y. Ishii, S. Asakawa, H. Takano, N. Ohta, H. Kuroiwa, K. Tanaka, N. Shimizu, S. Sugano, N. Sato, H. Nozaki, N. Ogasawara, Y. Kohara, T. Kuroiwa, Genome sequence of the ultrasmall unicellular red alga *Cyanidioschyzon merolae* 10D. Nature 428 (2004) 653-257. https://doi.org/10.1038/nature02398

[48]   N. Sato, T. Moriyama, N. Mori, M. Toyoshima, Lipid metabolism and potentials of biofuel and high added-value oil production in red algae. World J. Microbiol. Biotechnol. 33 (2017) 74. https://doi.org/10.1007/s11274-017-2236-3

[49]   D. Bhattacharya, D.C. Price, C.X. Chan, H. Qiu, N. Rose, S.Ball, A.P.M. Weber, M. C. Arias, B. Henrissat, P.M. Coutinho, A. Krishnan, S. Zäuner, S. Morath, F. Hilliou, A. Egizi, M.M. Perrineau, H.S. Yoon, Genome of the red alga *Porphyridium purpureum*. Nature Comm. 4 (2013) 1941. https://doi.org/10.1038/ncomms2931

[50]   T. Selvaratnam, H. Reddy, T. Mupaneni, F.O. Holguin, N. Nirmalakhandan, S. Deng, P.J. Lammers, Optimizing energy yields from nutrient recycling using sequential hydrothermal liquefaction with *Galdieria sulphuraria,* Algal Res. 12 (2015) (2015) 74-79.

[51]   G. Schönknecht, W.H. Chen, C.M. Ternes, G.G. Barbier, R.P. Shrestha, M. Stanke, A. Bräutigam, B.J. Baker, J.F. Banfield, R.M. Garavito, K. Carr, C. Wilkerson, S.A. Rensing, D. Gagneul, N.E. Dickenson, C. Oesterhelt, M.J. Lercher, A.P. Weber, Gene transfer from bacteria and archaea facilitated evolution of an extremophilic eukaryote. Sci. 339 (2013) 1207-1210. https://doi.org/10.1126/science.1231707

[52]   K. Yamane, S. Matsuyama, K. Igarashi, M. Utsumi, Y. Shiraiwa, T. Kuwabara, Anaerobic coculture of microalgae with T*hermosipho globiformans* and *Methanocaldococcus jannaschii* at 68°C enhances generation of n-alkane-rich biofuels after pyrolysis. Appl. Environ. Microbiol. 79 (2013) 924-930. https://doi.org/10.1128/AEM.01685-12

[53]   B.A. Read, J. Kegel, M.J. Klute, A. Kuo, S.C. Lefebvre, F. Maumus, C. Mayer, J. Miller, A. Monier, A. Salamov, J. Young, M. Aguilar, J.M. Claverie, S. Frickenhaus, K. Gonzalez, E.K. Herman, Y.C. Lin, J. Napier, H. Ogata, A.F. Sarno, J. Shmutz, D. Schroeder, C. de Vargas, F. Verret, von P. Dassow, K. Valentin, Y. Van de Peer, G. Wheeler, J.B. Dacks, C.F. Delwiche, S.T. Dyhrman, G. Glöckner, U. John, T. Richards, A.Z. Worden, X. Zhang, I.V. Grigoriev, Pan genome of the phytoplankton Emiliania underpins its global distribution. Nature 499 (2013) 209-213. https://doi.org/10.1038/nature12221

[54] E.Y. Artamonova1, T. Vasskog, H.C. Eilertsen, Lipid content and fatty acid composition of *Porosira glacialis* and *Attheya longicornis* inresponse to carbon dioxide (CO2) aeration. PLoS One 12 (2017) e0177703. https://doi.org/10.1371/journal.pone.0177703

[55] J.A. Raymond, H.J. Kim, Possible role of horizontal gene transfer in the colonization of sea ice by algae. PLoS One 7 (2012) e35968. https://doi.org/10.1371/journal.pone.0035968

[56] H. Chtourou, I. Dahmen, I. Dahmen, F. Karray, A. Dhouib, Biodiesel production of amphora sp. and navicula sp. by different cell disruption and lipid extraction methods. J. Biobased Materials Bioener. 9 (2015) 588-595. https://doi.org/10.1166/jbmb.2015.1563

[57] M Hildebrand, R.M. Abbriano, J.E.W. Polle, J.C. Traller, E.M. Trentacoste, S.R Smith, A.K. Davis, Metabolic and cellular organization in evolutionarily diverse microalgae as related to biofuels production. Cur Opin Chem. Biol 17 (2013) 506–514 https://doi.org/10.1016/j.cbpa.2013.02.027

[58] C.E. Moore, B. Curtis, T. Mills, G. Tanifuji, J.M. Archibald, Nucleomorph genome sequence of the cryptophyte alga *Chroomonas mesostigmatica* CCMP1168 reveals lineage-specific gene loss and genome complexity. Genome Biol. Evol. 4 (2012) 1162–1175. https://doi.org/10.1093/gbe/evs090

[59] P. Pandey, V.N. Shinde, R.L. Deopurkar, S.P. Kale, S.A. Patil, D. Pant, Recent advances in the use of different substrates in microbial fuel cells toward wastewater treatment and simultaneous energy recovery. Appl. Ener. 168 (2016) 706-723. https://doi.org/10.1016/j.apenergy.2016.01.056

[60] C. Santoro, C. Arbizzani, B. Erable, I. Ieropoulos, Microbial fuel cells: From fundamentals to applications: A review. J. Power Sour. 356 (2017) 225-244. https://doi.org/10.1016/j.jpowsour.2017.03.109

[61] C. Santoro, S. Rojas-Carbonell, R. Awais, R. Gokhale, M.Kodali, A. Serov, K. Artyushkova, P. Atanassov, Influence of platinum group metal-free catalyst synthesis on microbial fuel cell performance. J. Power Sour. 375 (2018) 11–20. https://doi.org/10.1016/j.jpowsour.2017.11.039

[62] B.E. Logan, M.J. Wallack, K.-Y. Kim, W. He, Y. Feng, P.E. Saikaly, Assessment of microbial fuel cell configurations and power densities. Environ. Sci. Technol. Lett. 2 (2015) 206–214. https://doi.org/10.1021/acs.estlett.5b00180

[63] M. Sawa, A. Fantuzzi, P. Bombelli, C.J. Howe, K. Hellgardt, P.J. Nixon, Electricity generation from digitally printed cyanobacteria Nature Comm. 8 (2017) 1327.

[64] S. Choi, Microscale microbial fuel cells: Advances and challenges. Biosens. Bioelectron. 69 (2015) 8−25. https://doi.org/10.1016/j.bios.2015.02.021

[65] B.E. Logan, Microbial Fuel Cells, John Wiley and Sons, New Jersey, 2008.

[66] R. Timmers, M. Rothballer, D. Strik, M. Engel, S. Schulz, M. Schloter, A. Hartmann, B. Hamelers, C. Buisman, Microbial community structure elucidates performance of *Glyceria maxima* plant microbial fuel cell. Appl. Microbiol. Biotechnol. 94 (2012) 537-548. https://doi.org/10.1007/s00253-012-3894-6

[67] C. Xu, Algae grown anode microbial fuel cell and its application in power generation and biosensor. Open Access Theses and Dissertations. 169, Hong Kong Baptist University, (2016)

[68] R. Huarachi-Olivera, A. Due-as-Gonza, U. Yapo-Pari, P. Vega, M. Romero-Ugarte, J. Tapia, L. Molina, A. Lazarte-Rivera, D.G. Pacheco-Salazar, M. Esparza, Bioelectrogenesis with microbial fuel cells (MFCs) using the microalga Chlorella vulgaris and bacterial communities. Elect. J. Biotechnol. 31 (2018) 34–43. https://doi.org/10.1016/j.ejbt.2017.10.013

[69] A. Gunawardena, S. Fernando, F. To, Performance of a Yeast-mediated Biological Fuel Cell. Int. J. Mol. Sci. 9 (2008) 1893–1907. https://doi.org/10.3390/ijms9101893

[70] A.S. Mathuriya, V.N. Sharma, electricity generation by *Saccharomyces cerevisae* and Clostridium acetobutylicum via microbial fuel cell technology: A comparative study. Adv. Biol. Res. 4 (2010) 217-223.

[71] M. Rosenbaum, U. Schröder, F. Scholz, Utilizing the green alga *Chlamydomonas reinhardtii* for microbial electricity generation: a living solar cell. Appl. Microbiol. Biotechnol. 68 (2005) 753–756. https://doi.org/10.1007/s00253-005-1915-4

[72] S. Angioni, L. Millia, P. Mustarelli, E. Doria, M. E. Temporiti, B. Mannucci, F. Corana, E. Quartarone. Photosynthetic microbial fuel cell with polybenzimidazole membrane: synergy between bacteria and algae for wastewater removal and biorefinery. Heliyon 4 (2018) e00560 https://doi.org/10.1016/j.heliyon.2018.e00560

[73] H. Liu, B.E. Logan, Electricity generation using an air cathode single chamber microbial fuel cell in the presence and absence of a proton exchange membrane. Environ. Sci. Technol. 38 (2004) 4040–4046. https://doi.org/10.1021/es0499344

[74] H. Jiang, Combination of microbial fuel cells with microalgae cultivation for bioelectricity generation and domestic wastewater treatment. Environ. Eng. Sci. 34 (2017) 489–495. https://doi.org/10.1089/ees.2016.0279

[75] S.B. Velasquez-Orta, D. Werner, J.C. Varia, S. Mgana, Microbial fuel cells for inexpensive continuous in-situ monitoring of groundwater quality. Water Res. 117 (2017) 9-17 https://doi.org/10.1016/j.watres.2017.03.040

[76] T.H. Pham, J.K. Jang, I.S. Chang, B.H. Kim, Improvement of cathode reaction of a mediatorless microbial fuel cell. J. Microbiol. Biotechnol. 14 (2004) 324–329.

[77] A.D. Tharali, N. Sain, W.J. Osborne, Microbial fuel cells in bioelectricity production. J. Front. Life Sci. 9 (2016) 252-266. https://doi.org/10.1080/21553769.2016.1230787

[78] K. Li, S. Liu, X. Liu, An overview of algae bioethanol production. Int. J. Ener Res. 38 (2014) 965–977. https://doi.org/10.1002/er.3164

[79] F.S. Eshaq, M.N. Ali, M.K. Mohd, Production of bioethanol from next generation feed-stock alga Spirogyra species. Int. J. Eng. Sci. Technol. 3 (2011) 1749–1755.

[80] O.K. Lee, D.H. Seong, C.G. Lee, E.Y. Lee, Sustainable production of liquid biofuels from renewable microalgae biomass. J. Ind. Eng. Chem. 29 (2015) 24–31. https://doi.org/10.1016/j.jiec.2015.04.016

[81] M.M. El-Dalatony, E.S. Salama, M.B. Kurade, S.H.A. Hassan, S.E. Oh, S. Kim, B.H. Jeon, Utilization of microalgal biofractions for bioethanol, higher alcohols, and biodiesel production: A review. Energies 10 (2017) 2110-2125. https://doi.org/10.3390/en10122110

[82] M.N.B. Hossain, J.K. Basu, M. Mamun, The production of ethanol from micro-algae spirulina. Procedia Eng. 105 (2015) 733-738. https://doi.org/10.1016/j.proeng.2015.05.064

[83] T.K. Yeong, K. Jiao, X. Zeng, L. Lin, S.Pan, M.K. Danquah, Microalgae for biobutanol production–Technology evaluation and value proposition. Algal Res. 31 (2018) 367-376. https://doi.org/10.1016/j.algal.2018.02.029

[84] H.H. Cheng, L.M. Whang, K.C. Chan, M.C. Chung, S.H. Wu, C.P. Liu, S.Y. Tien, S.Y. Chen, J.S. Chang, W.J. Lee, Biological butanol production from microalgae-

based biodiesel residues by *Clostridium acetobutylicum*. Bioresour. Technol. 184 (2015) 379-385. https://doi.org/10.1016/j.biortech.2014.11.017

[85] J. Ellis, N. Hengge, R. Sims, C. Miller, Acetone, butanol, and ethanol production from wastewater algae, Bioresour. Technol. 111 (2012) 491–495. https://doi.org/10.1016/j.biortech.2012.02.002

[86] H. Zhang, D. Jin, X.J. Zhao, Butanol production using carbohydrate-enriched *Chlorella vulgaris* as feedstock. Adv. Materials Res. 830 (2014) 122-125.

[87] G.F. Torres, J. Pittman, C, Theodoropoulos, Optimization of microalgal starch formation for the biochemical production of biobutanol. 27th European Symposium on Computer-Aided Process Engineering - Barcelona, Spain, 1 -5th Oct 2017. https://doi.org/10.1016/B978-0-444-63965-3.50485-2

[88] W. Khetkorn, R.P. Rastogi, A. Incharoensakdi, P. Lindblad, D. Madamwar, A. Pandey, C. Larroche Microalgal hydrogen production–A review. Bioresour. Technol. 243 (2017) 1194-1206. https://doi.org/10.1016/j.biortech.2017.07.085

[89] A. Sharma, S.K. Arya, Hydrogen from algal biomass: A review of production process. Biotechnol. Reports 15 (2017) 63-69. https://doi.org/10.1016/j.btre.2017.06.001

[90] G. Buitrón, J. Carrillo-Reyes, M. Morales, C. Faraloni, G. Torzill, Biohydrogen production from microalgae. C. Gonzalez-Fernandez, R. Mu-oz, (Eds.), Microalgae-Based Biofuels and Bioproducts. From Feedstock Cultivation to End-products, Woodhead Publishing Series in Energy, 2017, pp.209-234.

[91] S.S. Oncel, A.K.C.Faraloni, E. Imamoglu, M. Elibol, G.Torzillo, F.V. Sukan, Biohydrogen production from model microalgae *Chlamydomonas reinhardtii*: A simulation of environmental conditions for outdoor experiments. Int. J. Hydrogen Ener. 40 (2015) 7502-7510. https://doi.org/10.1016/j.ijhydene.2014.12.121

[92] D. Nagarajan, D.J. Lee, A. Kondo, J.S. Chang, Recent insights into biohydrogen production by microalgae - From biophotolysis to dark fermentation. Bioresour. Technol. 227 (2017) 373-387. https://doi.org/10.1016/j.biortech.2016.12.104

[93] W.M. Alalayah, Y.A. Alhamed, A. Al-Zahrani, G. Edris, Experimental investigation parameters of hydrogen production by algae *Chlorella vulgaris*. International Conference on Chemical, Environment & Biological Sciences (CEBS-2014) Sept. 17-18, 2014 Kuala Lumpur (Malaysia).

[94]   N. Mallick, S. Mandal, A.K. Singh, M. Bishai, A. Dash, Green microalga
Chlorella vulgaris as a potential feedstock for biodiesel. J. Chem. Technol.
Biotechnol. 87 (2012) 137-145. https://doi.org/10.1002/jctb.2694

[95]   C. Soumya, H.A.P. Avadhani, R. Vidhya, V. Moses, Production of biofuel from
micro algae (*Chlorella pyrenoidosa*) using vertical reactor system and effect of
nitrogen on growth and lipid content. J. Acad. Indust. Res. 4 (2015) 179-182.

[96]   M. Mondal, S. Goswami, A. Ghosh, G. Oinam, O.N. Tiwari, P. Das, K. Gayen, M.
K. Mandal, G.N. Halder, Production of biodiesel from microalgae through
biological carbon capture: A review. 3 Biotech. 7 (2017) 99.

[97]   X. Wen, K. Du, Z. Wang, X. Peng, L. Luo, H. Tao, Y. Xu, D. Zhang, Y. Geng
Yeguang, Effective cultivation of microalgae for biofuel production: a pilot-scale
evaluation of a novel oleaginous microalga *Graesiella* sp. WBG-1. Biotechnol.
Biofuels 9 (2016) 123. https://doi.org/10.1186/s13068-016-0541-y

[98]   C. Cavinato, A. Ugurlu, I. de Godos, E. Kendir, C. Gonzalez-Fernandez, Biogas
production from microalgae. R. Mu-oz, C. Gonzalez-Fernandez (Eds.),
Microalgae-Based Biofuels and Bioproducts, From Feedstock Cultivation to End-
products, Woodhead Publishing Series in Energy 2017, pp. 155-182.

[99]   S. He, X. Fan, N.R. Katukuri, X.Yuan, F.Wang, R.B. Guo, Enhanced methane
production from microalgal biomass by anaerobic bio-pretreatment. Bioresour.
Technol. 204 (2016) 145-151. https://doi.org/10.1016/j.biortech.2015.12.073

[100] M. Solé-Bundó, C.Eskicioglu, M. Garfí, H.Carrère, I.Ferrer, Anaerobic co-
digestion of microalgal biomass and wheat straw with and without thermo-alkaline
pretreatment. Bioresour. Technol. 237 (2017) 89-98.
https://doi.org/10.1016/j.biortech.2017.03.151

[101] M. Barragán-Trinidad, J. Carrillo-Reyes, G. Buitrón, Hydrolysis of microalgal
biomass using ruminal microorganisms as a pretreatment to increase methane
recovery. Bioresour. Technol. 244 (2017) 100-107.
https://doi.org/10.1016/j.biortech.2017.07.117

Chapter 3

Microbial Fuel Cell Operating Conditions

Naveen Patel[1a*], Dhananjai Rai[2b], MD. Zafar Ali Khan[3b], Shivam Soni[1d], Umesh Mishra[1e] and Biswanath Bhunia[4f]

[1]Department of Civil Engineering, National Institute of Technology, Agartala, Agartala-799046, Tripura, India

[2]Department of Civil Engineering, Bundelkhand Institute of Engineering and Technology, Jhansi, Jhansi-284128, Uttar Pradesh, India

[3]Department of Civil Engineering, Government Polytechnic College, Gonda, Gonda-271001, Uttar Pradesh, India

[4]Department of Bio Engineering, National Institute of Technology Agartala, Agartala-799046, Tripura, India

[a]naveenrbverma30@gmail.com, [b]djrai028@gmail.com, [c]zkhan441@gmail.com, [d]sonishivamsoni@gmail.com, [e]umishra123@rediffmail.com, [f]bbhunia@gmail.com

Abstract

The exploitation of fossil fuels such as coal, oil, and gas have been increasing and is creating a worldwide energy crisis. Because of global environmental concerns about pollution and energy uncertainty, sustainable and clean energy resources are needed to be investigated. Microbial fuel cell (MFC) is such a technology, which uses organic compounds and converts it into electrical energy through catalytic reactions by microorganisms. The electricity generation efficiency of MFC is greatly influenced by various physical and chemical parameters i.e. operating conditions like temperature, pH, type of oxidant, electrode surface area, electrolyte conductivity and substrate, etc.

Keywords

MFC, Operating Condition, Substrate, Oxidant, Electrode, Conductivity

List of Abbreviations

MFC   Microbial fuel cell

PTFE   Polytetra fluoroethylene

CoTMPP Cobalt tetramethoxyphenylporphyrin

AC	Activated carbon
RVC	Reticulated vitreous carbon
CNT	Carbon nanotubes
MEA	Membrane electrode assembly
UMFC	Upflow Microbial fuel cell
PEM	Proton exchange membrane
CEM	Cation exchange membrane
AEM	Anion exchange membrane
BPM	Bipolar membrane
COD	Chemical oxygen demand
CE	Coulombic efficiency
OCP	Open circuit potential

Contents

1.   Introduction

Development and industrialization are obligatory to produce power from available resources to fulfil the world's increasing energy demand. Fossil fuels such as natural gas, coal, and petroleum are being exploited more to fulfil the energy requirements. Dependence on these natural sources is unsustainable due to pollution and limited availabilities which have further created the problem of a global energy crisis. Lots of research is going in the quest for developing highly efficient energy transformations and ways to utilize alternative renewable energy sources [1]. Hence, the development of an alternate long-term energy source is necessary. Bioelectricity generation using microbial fuel cells (MFCs) is one of the areas of green energy productions utilizing renewable resources. MFC is a bio-electrochemical device, which utilizes organic matter and degrades them into smaller molecules through the catalytic activities of microorganisms to generate electricity [2]. The MFC's performance is largely affected by many physical and chemical parameters, such as membrane/separator, types of oxidants, electrode surface area [3], electrode material [4], reactor configuration [5], pH, types of oxidants, electrolyte conductivity, substrates (fuels) and temperature [6]. By optimizing these operating parameters, the performance of MFC can be enhanced by decreasing the polarizations/internal resistance. This chapter encapsulates details related to the effects of these operating parameters on MFC performance

2.   Cathode Electrode

2.1   Cathode Electrode with Catalysts

Cathode catalyst plays an important role in enhancing the cathode performance by lowering the cathodic activation overpotential and reducing the cost of construction. The catalyst cost in air cathode MFCs comprises 47% of the total capital cost [7]. One of the most regularly used cathode catalysts is platinum (Pt). It is used for the reduction of oxygen in both aqueous and air cathode MFCs. However, the Pt catalysts are expensive and can only be used at laboratory scale. To decrease the loading ratio of platinum without further effect on its catalytic performance, various studies have been carried out by researchers. By employing an electron beam evaporation method for Pt deposition on carbon paper, the thickness of platinum layer has been reduced up to 1000 A. In the same study, reduction in loading along with increase in current density (0.42 A m^{-2}) as compared to large scale (commercial) available Pt-black (0.22 A m^{-2}, Vulcan XC-72, E-Tek) was observed [8]. The binding energies of oxygen can be reduced by using Pt in a combination of transition metals like Fe, Co and Ni. Higher power density (1681 mW m^{-2}) was obtained, when carbon electrode doped Pt-Co alloy in the ratio of 2:1 as compared to the power density (1315 mW m^{-2}) obtained by using platinum-coated carbon paper for MFC operation [9]. By doping MnO_2 with cobalt, an increase in the value of current generation was observed from 86 mW m^{-2} (when only MnO_2 was used as a control) to 180 mW m^{-2}, which is close to the value obtained by Pt (198 mW m^{-2}) [10]. Polytetrafluoroethylene (PTFE) and perfluorosulphonic acid (Nafion) are the most commonly used binders for coating the catalyst (Pt or CoTMPP stands for cobalt tetramethoxyphenylporphyrin) on the electrode substrate.

A comparative study was performed by Cheng et al,. for employing two binders namely Nafion and PTFE for coating Pt catalyst in a single chamber air cathode MFCs. During their study, they observed the maximum power of 480 mW m^{-2} when Nafion was used as binder whereas 331 mW m^{-2} was the maximum power density when using PTFE as a binder. Because of the hydrophobicity of PTFE, biofilm formation was found to be low and thin on using PTFE as binder [11]. Coating of CoTMPP catalyst using Nafion as binder was also studied and the generated power was 369 mW m^{-2}. In terms of cost, Nafion binder cost was 500 times more than the PTFE binder. Nowadays, researchers have developed biocathodes for reducing overpotentials. Biocathodes have the ability to perform reduction reaction in MFC electrodes and are microbially catalysed. These biocathodes are being considered as an alternative for replacing expensive catalysts. The biocathodes are getting popularity because of their several advantages such as low cost, reclamation nature, efficiency in wastewater treatment, environmental compatibility, and biogas generation capability.

2.2 Cathode Electrode without Catalysts

Electrodes without catalyst like plain carbon paper can be used in MFCs but such electrodes have the ability to produce power density of smaller quantity i.e. 0.04 mW m^{-3} with respect to the cathode working volume [12]. High power density of 50 mW m^{-3} (cathode working volume) was reported by Frehuia et al. when the air cathode MFC was operated by using granular graphite electrode [13]. Plain cathode electrodes with larger areas have been used by various researchers. Graphite rods, graphite plates, carbon material (like carbon cloth or carbon paper) and activated carbon (AC) with high porosity are essential for higher power generation [13]. For reduction of oxygen, various chemicals such as HNO_3, H_3PO_4 and KOH can be used for pre-treatment of AC cathodes.

Use of AC with HNO_3 for MFC operations, among all these available chemicals for pre-treatment, leads to exhibit higher power density of 170 mW cm^{-2} as compared to 51 mW m^{-2} which was obtained without pre-treatment of AC. But still these cathode electrodes without catalyst due to some reason have relatively smaller power density when compared with Pt-coated AC [14].

3.   Anode electrode Materials and surface modification

Carbon-based materials like graphite rod, graphite plate, carbon cloth, graphite fibre brush, carbon paper and reticulated vitreous carbon (RVC) have been commonly used as anode material for most of the MFC operations. The properties like good electrical conductivity, chemical stability and lower resistance of these carbon materials makes them suitable for using as anode materials in MFCs. Use of high conductivity electrode helps in increasing the power of MFCs by decreasing internal resistance (up to a few ohms). The electrical conductivities for graphite fibre, carbon paper and carbon cloth evaluated in a study conducted by Logan were 1.6, 0.8, and 2.2 Ω cm^{-1} respectively [15]. The anode electrodes made of carbon paper are quite often used in MFC operations as these are quite thin and brittle as compared to other electrodes. This property facilitates them to get easily connected to electrical wire and to be further sealed with non-conductive resin. Graphite sheets, rods and plates can be used for anode electrode either separately or in combination with AC. Use of graphite felt for MFC operations was found to be more efficient as compared to graphite [16]. Because of the spongy structure of the graphite, the power density increased up to 23% by increasing the anode thickness to 1.27 cm from 0.32 cm [17]. The carbon cloths because of their highly flexibility, porosity and firmness provide a higher surface area for bacterial growth and can be used in flat plate MFC operations. Graphite fibre brushes because of their properties like low electrical resistance and high surface area availability for growth of bacteria have been found suitable for application in MFC operations. Graphite fibre brushes are made up of graphite fibres, which have been wounded with titanium or stainless steel wires (electrically conductive and non-corrosive). With the application of graphite fibre brushes as anode electrode material, increase in power density of about 2.4 times has been noticed when compared with the use of carbon paper as anode in MFC operations (from 600 to 1430 mW m^{-2}) [15, 18]. A Comparison study performed using carbon paper and carbon brush anodes for MFC operation (anode electrode materials) in terms of power generation is shown in Fig.1.

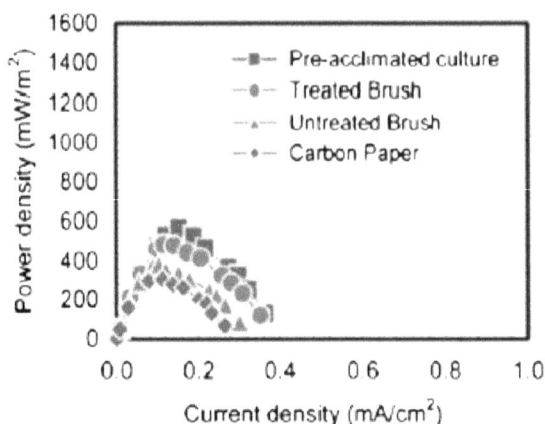

Fig. 1. Power density curves in single chamber bottle type MFCs with carbon paper and carbon brush anodes [18].

Reticulated vitreous carbon (RVC) because of its, high conductivity, porosity and brittle nature has been used by many researchers but its cost is higher than other carbon materials used as an anode electrode. In recent years, stainless steel, gold, ceramic with carbon and titanium are few other non-carbon materials which have been used as anode materials [19, 20]. In a study conducted by Dumas et al., on the use of stainless steel as anode material in MFC operations, low power density was generated (4 mW m^{-2}) [21]. Whereas another study performed by Richter et al., revealed that gold cannot be used by it self as anode electrode [22].

Surface modification of anode electrode has been attempted in order to increase the performance of MFC operations using a particular anode. In one of the successful modifications suggested by Cheng and Logan, the carbon cloth anode was treated with 5% NH$_3$ gas at 700°C using helium as carrier gas. It has been found that the power increased (1640 4 to 1970 4 mW m^{-2}) and lag period reduced up to 50% along with an increase in positive surface charge from 0.38 to 3.9 meq m^{-2}. It has been observed that surface coating with materials like polyaniline, carbon nanotubes (CNT), metals and composites of these materials (polymer/CNTs) resulted in improving power generation. The surface area and adhesion of bacteria on the surface of the material can be increased

Microbial Fuel Cells: Materials and Applications   Materials Research Forum LLC
Materials Research Foundations **46** (2019) 53-74   doi: http://dx.doi.org/10.21741/9781644900116-3

by coating polyaniline on the MFC anodes, which further help, in enhancing the maximum power densities of MFC operations [24]. The coating of the anode with CNT helps in increasing the transfer of an electron to the electrode from bacteria and also helps in decreasing the internal resistance of MFC [25]. There are different types of surface treatments which include; (a) heating of carbon mesh at 450°C for 30 min [26], (b) oxidation of carbon electrodes surface using acid (sulphuric acid) and (c) combination of both heat and acid treatment [27]. These treatments lead to (1) lesser contaminant formation (2) Increase in specific surface area and (3) more positive charge transfer to electrode [27].

Recently, applying of 1.85 V (vs. Ag/AgCl) to graphite plates in electrochemical oxidation method had gained a lot of attention. The power density with pre-treatment can be increased (39 to 58%) [28]. Guo et al., has reported that on increasing the surface roughness of graphite rod/sheets/plates by sandpaper, higher power generation can be achieved [29]. By use of mediators like neutral red and methylene blue (MB), electrons can be transferred from bacterial cell to electrode [30]. Continuous wash out during batch mode operation is one of the major limitations related to the use of these mediators which are required to be supplemented continuously, so that the performance of MFC can be maintained. This problem of washing out can be solved by using immobilized lipases which are sustainable for higher duration.

4.   Reactor configuration

Reactor volume variation, spacing of electrode, hydraulic flow pattern, oxygen supply, and membrane area are the reactors configuration and these are the main operating conditions that affect the performance of MFC. Double chamber "H" type MFCs, among various types of MFCs, have been used most frequently by the researchers. One of the major important aspects of using this type of MFC is that it helps in limiting the substrate and oxygen crossover with increase in diffusion of proton through ion exchange compartment [31]. The performance of MFC was decreased as the distance between the electrodes was increased as it leads to increase in internal resistance. Therefore, clamping of electrodes along with the membranes or separators has been done by various researchers [32]. In double chamber MFCs, by maintaining small spacing between electrodes, the internal resistance was lower (93Ω) [32] compared to the value obtained by other researches who performed the operation by maintaining larger spacing of electrodes for MFC operations (672Ω) [33]. Unknown power densities were developed due to the application of different configurations of the MFCs and this unknown power generated can be calculated by using Eq.1, which is based on the difference in internal and external resistances of the two systems [31, 34].

$$P_a = \frac{P_h \times R^2_{i,h}}{R_{e,h}} \times \frac{R_{e,a}}{R^2_{i,a}}$$

(eq. 1)

Where,

P_a = power density (mW m^{-2}) with advanced configuration of MFC.

P_h = power density with H type MFC.

$R_{i,h}$ = internal resistance of H type MFC.

$R_{i,a}$ = internal resistance of an advanced cell

MFCs configuration with advanced ion transport design leads to increase in power production which was nearly 17% more than the salt bridge MFCs [34]. Just like the previous study, it was observed that bottle type MFCs were less efficient than the cube type MFCs [35]. A comparative study was carried out between single chamber bottle and cube type MFCs by Logan et al. and it was reported that the higher power density was obtained by a cube type MFC than the bottle type MFC (i.e. 2400 mWm^{-2} for cube type and 1200 mWm^{-2} for bottle type). Further, the internal resistance for cube MFC was smaller i.e. (8 Ω) whereas, internal resistance for the bottle type was higher (20 Ω). Sandwich type MFCs with membrane electrode assembly (MEA) has shown higher power density as compared to MFC with spacing between the cathode and anode electrodes of 1.7 cm. However, main problems such as water leakage, evaporation and high oxygen demand are associated with the use of single chamber MFC. These problems can be eliminated by using PTFE diffusion layers on cathode, which will further help in increasing the CE and maximum power densities [36]. He et al. have studied two different configurations of upflow MFCs (UMFCs) and they have found maximum power density of 170 mWm^{-2} from one reactor and 29.2 mWm^{-2} from other reactor with internal resistances of 84 Ω and 17 Ω respectively [26, 37]. The maximum density (309 mWm^{-2}) was obtained by flat plate MFC [12].

5   Membranes/Separators Tested In MFC

5.1   Ion Exchange Membrane

Exchange of cations or protons in MFC operations from anode to cathode chamber is basically done with the help of proton exchange membrane (PEM) or cation exchange membrane (CEM). Nafion-117 with sulphonate (SO$_3$) and hydrophobic fluorocarbon backbone (CF2-CF2) groups is the most common CEM, for diffusion of cations. Whereas

Microbial Fuel Cells: Materials and Applications   Materials Research Forum LLC
Materials Research Foundations **46** (2019) 53-74   doi: http://dx.doi.org/10.21741/9781644900116-3

CMI-700 is another most frequently used CEM/PEM, it consists of strong acidic polymeric membrane with divinylbenzene and gel polystyrene structure. CMI-700 in comparison with Nafion-117 has better mechanical durability, cation conductivity and sustainability but exhibits higher ohmic resistance. Moreover, in recent years several low cost CEMs (SPEEK and Hyflon) have been developed which have shown better performance than the conventional CEMs like Nafion and CMI-700, in terms of power density, ohmic resistance and conductivity [38, 39]. CEMs have certain limitations in MFC operations, which include; (1) proton and cation competition, (2) biofouling and (3) high cost for practical operations. Competition in proton and cation transport results in decrease in pH at anode which further causes increase in internal resistance and variation in power densities [40]. Anion exchange membranes (AEMs) can be used in place of CEMs in MFC operations as these have ability to produce higher current generation. The protons generated at the anode are consumed by carbonate or phosphate anions and without decrease in pH at anode these are efficiently transferred to the cathode and anions thus help in intensifying the performance of MFCs. A comparative study carried out on CEM and AEM in MFCs operation by Kim et al. showed maximum power densities gained by both as 0.48 and 0.61 mW m^{-2} respectively [35]. Bipolar membranes (BPMs) composed of two monopolar membranes of AEM and CEM, have been used in the MFCs operation for transporting both positive and negative ions. Because of their unique composition, BPMs can be used in process of desalination [41].

5.2   Size Selective Separators

Based on pore size, the size separators have been classified into two categories and these are; (1) course porous separators, and (2) microporous separators. Examples of microporous membranes are ultrafiltration and microfiltration membranes whereas examples of course pore separators are nylon mesh, glass fibre, and fabric materials. Various types of the microporous separators have been used for studying the process used in water and wastewater treatments because of their durability and performance. Anode and cathode solutions are physically separated by microporous separators which have the ability to allow the diffusion of charged as well as neutral species because of their effective pore sizes. Zuo et al. conducted a study on Tabular MFCs, which were formed by coating catalytic and conductive layers on UF and higher power density of 17.7 mW m^{-3} was observed by these as compared to 6.6 mW m^{-3} with CEM alone [42]. Fabric, glass and nylon mesh filter (course pore filter) have been reported to show great practical applications because of their cost-effectiveness and higher power density generation [32]. According to a comparative study, the power density was increased when size selective separators for MFC operations was used instead of an ion exchange membrane [31, 43]. However, the major limitations related to the size-selective separators have been higher diffusion of oxygen and substrate cross over. Though, this problem can be solved by increasing the distance between electrodes as suggested by few researchers but it can also lead to increase in internal resistance [44]. Therefore, further research is required to be done in this field so that the MFCs performance with size-selective separators can be optimized.

6. Effect of Temperature on MFC Performance

The temperature effect on MFC operating conditions is very important to be studied for maximum power generation. Temperature variation causes effect on certain parameters of MFC operations which includes Gibbs free energy, kinetics (activation energy, mass transfer coefficient, and solution conductivity), electrode potential, chemical solubility and the distribution as well as the activity of microorganisms on the electrodes [45]. In one of the research, different operational temperatures (30, 35, 40 and 45°C) were used for operating MFC in order to study the effect of temperature on MFCs operating performance. It was observed that power densities increased from 764 to 1065 mW m^{-2} with increase in temperature along with decrease in internal resistance from 177 to 121 Ω [46]. However, on further increase in temperature to 53°C, catalytic activity diminished, whereas if the temperature was reduced to 10°C from 30°C, the potentials of anode and cathode get reduced [46]. Min et al. reported decrease in the lag period and increase in power density at 30°C [47]. In another study by Guerrini et al, it was noted that as the operational temperature increased from 4 to 35°C, the current density also increased to 100 mA m^{-2} from 20 mA m^{-2} [19].

7. Substrates (Fuels) in the MFC Anode Chamber

Substrates immensely help in determining the MFCs operating conditions and hence are considered to be one of the most important factors to be examined. Substrates have been used for determination of maximum electricity generated by MFC operation [48]. Till now, several substrates from pure (simple) to mixed (complex) types have been used in MFCs for generation of electricity. Sources of substrates include agricultural wastewater, domestic wastewater, industrial and animal wastewater and these sources have different organic levels, complexity and microbial community. Complexity, chemical oxygen demand (COD), bacterial community and pH of the wastewater are the most important factors on which generation of electrode depends upon. As compared to complex substrates, simple substrates can be easily utilised as source of energy [49]. Further, the complex substrates require more time to be broken down into simpler molecules and then utilised as source of energy. When simple substrates like acetate and glucose were used in MFC operations in place of the complex ones like starch and cellulose, higher

Microbial Fuel Cells: Materials and Applications   Materials Research Forum LLC
Materials Research Foundations **46** (2019) 53-74   doi: http://dx.doi.org/10.21741/9781644900116-3

Coulombic efficiency (CE) and electricity generation were observed. The double-chambered MFC acetate fed has shown high CE value of 72.3% as compared to 43, 15 and 36% exhibited by butyrate, glucose and propionate respectively [50]. Whereas in term of power density, glucose as substrate has the maximum of 156 as compared to 58, 51.4 and 64.3 mW m^{-2} of acetate, butyrate and propionate substrates respectively.

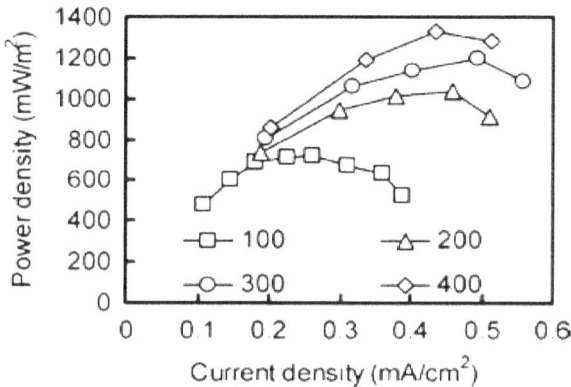

Fig. 2. Effect of solution concentration (conductivity) of 100 to 400 mM on MFC power generation [6].

8. Electrolyte Conductivity

There are two categories in which the electrolytes of MFC have been classified. These are; Anodic solution (anolyte) and cathodic solution (catholyte). Both of these play a very important role in preventing imbalances and hence lead to better MFC performance. According to the study conducted by Fan et al., the level of internal resistance of electrolyte was increased (47.3 to 82.2%) as the phosphate buffer reduced to 50 mM from 200 mM [51]. In another study of membrane-free single chambered MFC, the internal resistance was decreased from 161 to 91 Ω as NaCl (100mM) was added into 100mM phosphate buffer as a supplement [6]. However, no significant change was observed on further increasing the electrolyte concentration to 400 mM (200mM NaCl+200mM phosphate) from 200 mM (100mM NaCl+100mM phosphate). Electrolyte concentration greatly affect the MFCs internal resistance of anode, cathode and solution. In another study, it has been reported that 36-78% of the total internal resistance was contributed by the electrolyte resistance, whereas 2-35% of the resistance was contributed by the anode [51]. Whereas in term of power density, a special observation was reported by Cheng and Logan, that with increase of the electrolyte conductivity from 1.7 Ms cm^{-1} to 7.8 mS cm^{-1}, the power density increased to 33 mW m^{-3} from 16 mW m^{-3} [52]. This increase in power density was observed up to 60 mW m^{-3} when the conductivity was increased to 20 mS cm^{-1}. In another study, it has been found that the addition of buffer in the primary sludge led to increase in power generation of up to 1.5 times [53]. This increase in power density was observed up to 60 mW m^{-3} till the conductivity was increased to 20 mS cm^{-1}. The power generated in a single chambered air cathode MFC can be increased by 85%, as the buffer concentration (conductivity) of solution level increased to 400 mM from 100 mM as shown in Fig. 2 [6].

9. Electrolyte pH in Governing MFC Performances

The pH of the electrolyte plays a very crucial role in maintaining the proton transfer in the MFC, bacterial growth along with its metabolism on the anode, and oxygen reduction at the cathode. Electron and proton are generated at the anode and eventually transferred to the cathode for reduction process, leading to the production of electricity [54, 55]. Till now, in most of the studies pH changes and potential imbalances between the anode and the cathode have been maintained by adding a buffer [56].

As the pH decreases, the acidic condition develops on the anode because of the production of H$^+$ ions due to the oxidation of organic substrates by the bacteria present on it, whereas the cathode electrode is alkaline in nature due to the production of OH$^-$ ions. Optimum pH plays a very important role in generating electrons and electrogenic bacteria which exhibit a positive effect on MFCs current generation. Different bacterial species in the anodic chamber are mainly responsible for variation in optimum pH condition [57]. In a study by He et al. it was observed that with an increase in the pH from 5 to 7, the anodic polarisation resistance was decreased whereas, with an increase in pH from 7 to 10, the polarisation resistance increased [57]. Maximum power densities can be obtained by using primary sludge from a wastewater treatment plant at neutral pH (rather than acidic) [11]. While in another study, the maximum power generated in a fed-batch MFC was obtained at pH of 6, by using anaerobic sewage sludge [58]. With using the pure culture of *Bacillus strain*, maximum MFC performance was achieved at pH of 10.5 (more alkaline pH) [59]. Operational pH is the main factor responsible due to which Coulombic efficiencies of MFC get influenced. The following Nernst equations (Eqs. 2 and 3) can be used for calculation of potential of oxygen reduction on cathode at pH 7.

$$E = E^O - \frac{RT}{nF} \ln \frac{1}{[o_2]^{1/2}[H^+]^2}$$

eq. (2)

$$E = E^O - \frac{(8.314J / molK)(298.15K)}{(2)(96,500C / mol)} \ln \frac{1}{[0.2]^{1/2}[10^{-7}]^2} = 0.805V$$

eq. (3)

Where,

E = half-cell reduction potential

E^O = standard potential

R = universal gas Constant

T = temperature in Kelvin

n = number of electrons transferred, and

F = Faraday's constant (the number of Coulombs per mole of electrons).

Oxygen redox potentials of the cathode, previously reported by Logan et al. at various pHs were 0.627 V (pH 10) and 0.805 V (pH 7) [2]. In another report, it was found that when pH value decreased from 7 to 3.5, the value of open circuit potential (OCP) increased to 1.38 from 0.90 V, whereas an increase in basic nature led to drop in OCP [60].

10.   Oxidants in an MFC Cathode

Various types of oxidants have been used in MFC operations for improving the overall MFCs performance and determination of half-cell potential. The various types of oxidants such as O_2 (dissolved and gaseous), $Fe(CN)_6^{3-}$, fumarate, NO_3^-, SO_4^{2-}, MnO_4^{2-} and cytochromes have been used in MFC cathode as electron acceptors. Fig. 3 shows, some of the common oxidants and their reduced forms, which are being used in the MFC cathode. As per one of the research by Logan, it was found that oxygen because of its availability and high redox potential (0.82 V), is the most commonly used oxidant in MFCs [15]. But the theoretical value of cathode reduction potential with oxygen has been lower than the practical value. The Eq. 4 shows that four electrons are required by oxygen to get reduced to water but this process gets hindered due to the formation of hydrogen peroxide with the use of two electrons.

$$O_2 + 4H^+ + 4e^- \rightarrow 2H_2O \qquad \text{(eq. 4)}$$

$$O_2 + 2H^+ + 2e^- \rightarrow H_2O_2 \qquad \text{(eq. 5)}$$

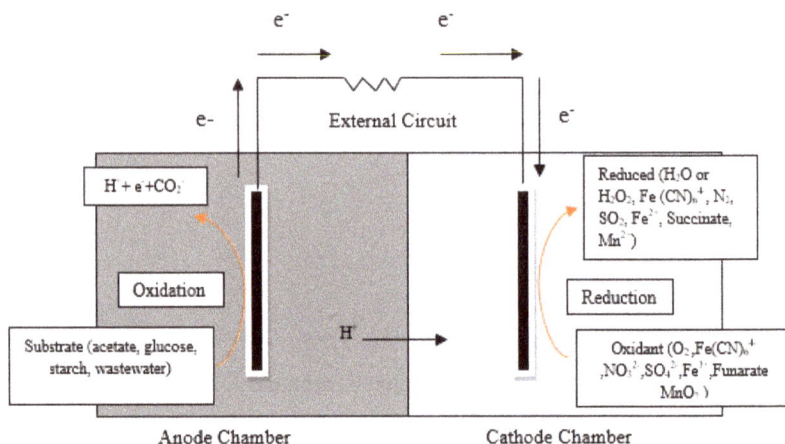

Fig. 3. Possible electron acceptors in MFC cathode chambers.

The formation of water is faster than hydrogen peroxide. However, hydroxide formation is primarily responsible for the degradation of cathode materials and MFCs membrane, but it also acts as a disinfectant for microbial contamination [2]. Some researchers have tried to use Fe $(CN)_6$ $^{3-}$ as an oxidant in place oxygen as it has high reduction potential as well as the ability to replace the use of precious catalyst like platinum [60]. With the use of ferricyanide catholyte, the voltage generation was increased to 11% (0.65 V) in comparison to the MFC with aeration (0.578 V) [61]. In one of the similar study like above, MFC catholytes with aerated ferricyanide have higher power generation of 155% than the MFC catholytes with ferricyanide or only with aeration where power generation was about 111% [62]. Permanganate's open circuit potential (OCP) of the cathode for MFC operation was found to be higher as compared to that of oxygen (i.e. 1.7 V at pH 3.5 for permanganate whereas 0.8V at pH 7 for oxygen) [60]. Nitrate and sulphate with biocathode (microbial interaction) along with a significant amount of oxygen concentration can be used as final electron acceptors [63]. The performance of MFC gets affected according to the concentration of reactant i.e. partial pressure of oxidisers (for example use of oxygen in place of normal air) in the cathode according to Nernst Eq.3. There are three possible effects due to the increase in partial pressures of oxygen [64]. These are (1) increase in OCP; (2) decrease in activation losses; and (3) increase in limiting current due to a reduction in mass transport. Similar results were reported in one of the studies by Kakarla et al. where MFCs performance (air cathode) was evaluated by increasing the oxygen concentrations, supplied from externally connected algae bioreactor [54].

Conclusions

MFCs architecture, temperature, electrode and catalyst materials, pH, membrane/ separator, conductivity and substrate concentration are the physical and chemical factors which are very crucial for efficient, economic and sustainable MFC power generation and hence optimization of these parameters is essential. Optimization becomes more essential especially at the time of real field application of these MFCs, therefore the main physicochemical parameters affecting the MFC operations need to be thoroughly determined and optimized.

References

[1]   K. Chandrasekhar, K. Amulya, S.V. Mohan, Solid phase bio-electrofermentation of food waste to harvest value-added products associated with waste remediation, Waste Manag. 45 (2015) 57-65. https://doi.org/10.1016/j.wasman.2015.06.001

[2]   B.E. Logan, B. Hamelers, R. Rozendal, U. Schröder, J. Keller, S. Freguia, P. Aelterman, W. Verstraete, K. Rabaey, Microbial fuel cells: methodology and technology, Environ Sci. Technol. 40 (2006) 5181-5192. https://doi.org/10.1021/es0605016

[3]   S.E. Oh, B.E. Logan, Proton exchange membrane and electrode surface areas as factors that affect power generation in microbial fuel cells, Appl. Microbiol. Biotechnol. 70 (2006) 162-169. https://doi.org/10.1007/s00253-005-0066-y

[4]   D.H. Park, J.G. Zeikus, Improved fuel cell and electrode designs for producing electricity from microbial degradation, Biotechnol. Bioeng. 81 (2003) 348-355. https://doi.org/10.1002/bit.10501

[5]   Y. Fan, H. Hu, H. Liu, Enhanced Coulombic efficiency and power density of air-cathode microbial fuel cells with an improved cell configuration, J. Power Sources. 171 (2007) 348-354. https://doi.org/10.1016/j.jpowsour.2007.06.220

[6]   H. Liu, S. Cheng, B.E. Logan, Power generation in fed-batch microbial fuel cells as a function of ionic strength, temperature, and reactor configuration, Environ. Sci. Technol. 39 (2005) 5488-5493. https://doi.org/10.1021/es050316c

[7]   R.A. Rozendal, H.V. Hamelers, K. Rabaey, J. Keller, C.J. Buisman, owards practical implementation of bioelectrochemical wastewater treatment, Trends Biotechnol. 26 (2008) 450-459. https://doi.org/10.1016/j.tibtech.2008.04.008

[8]   H.I. Park, U. Mushtaq, D. Perello, I. Lee, S.K. Cho, A. Star, M. Yun, Effective and low-cost platinum electrodes for microbial fuel cells deposited by electron beam evaporation, Energy Fuels. 21 (2007) 2984-2990. https://doi.org/10.1021/ef070160x

[9]   Z. Yan, M. Wang, J. Liu, R. Liu, J. Zhao, Glycerol-stabilized NaBH4 reduction at room-temperature for the synthesis of a carbon-supported PtxFe alloy with superior oxygen reduction activity for a microbial fuel cell, Electrochim. Acta. 141 (2014) 331-339. https://doi.org/10.1016/j.electacta.2014.06.137

[10]   X. Li, B. Hu, S. Suib, Y. Lei, B. Li, Manganese dioxide as a new cathode catalyst in microbial fuel cells, J. Power Sources. 195 (2010) 2586-2591. https://doi.org/10.1016/j.jpowsour.2009.10.084

[11]   S. Cheng, H. Liu, B.E. Logan, Power densities using different cathode catalysts (Pt and CoTMPP) and polymer binders (Nafion and PTFE) in single chamber microbial fuel cells, technology, Environ. Sci. Technol. 40 (2006) 364-369. https://doi.org/10.1021/es0512071

[12]   B. Min, B.E. Logan, Continuous electricity generation from domestic wastewater and organic substrates in a flat plate microbial fuel cell, Environ. Sci. Technol. 38 (2004) 5809-5814. https://doi.org/10.1021/es0491026

[13]   S. Freguia, K. Rabaey, Z. Yuan, J. Keller, Non-catalyzed cathodic oxygen reduction at graphite granules in microbial fuel cells, Electrochim. Acta. 53 (2007) 598-603. https://doi.org/10.1016/j.electacta.2007.07.037

[14]   M. Dutoanu, D. Erable, S.S. Kumar, M.M. Ghangrekar, K. Scott, Effect of chemically modified Vulcan XC-72R on the performance of air-breathing cathode in a single-chamber microbial fuel cell, Bioresour. Technol. 101 (2010) 5250-5255. https://doi.org/10.1016/j.biortech.2010.01.120

[15]   B.E. Logan, Microbial fuel cells, John Wiley & Sons. Inc., New Jersey. 2007. https://doi.org/10.1002/9780470258590

[16] S.K. Chaudhuri, D.R. Lovley, Electricity generation by direct oxidation of glucose in mediatorless microbial fuel cells, Nat. Biotechnol. 21 (2003) 1229. https://doi.org/10.1038/nbt867

[17] Y. Ahn, B.E. Logan, Altering anode thickness to improve power production in microbial fuel cells with different electrode distances, Energy Fuels. 27 (2012) 271-276. https://doi.org/10.1021/ef3015553

[18] B. Logan, S. Cheng, V. Watson, G. Estadt, Graphite fiber brush anodes for increased power production in air-cathode microbial fuel cells, Environ. Sci. Technol. 41 (2007) 3341-3346. https://doi.org/10.1021/es062644y

[19] E. Guerrini, P. Cristiani, M. Grattieri, C. Santoro, B. Li, S. Trasatti, Electrochemical behavior of stainless steel anodes in membraneless microbial fuel cells, J. Electrochem. Soc. 161 (2014) H62-H67. https://doi.org/10.1149/2.096401jes

[20] R. Thorne, H. Hu, K. Schneider, P. Bombelli, A. Fisher, L.M. Peter, A. Dent, P.J. Cameron, Porous ceramic anode materials for photo-microbial fuel cells, J. Mater. Chem. 21 (2011) 18055-18060. https://doi.org/10.1039/c1jm13058g

[21] C. Dumas, A. Mollica, D. Féron, R. Basséguy, L. Etcheverry, A. Bergel, Marine microbial fuel cell: use of stainless steel electrodes as anode and cathode materials, Electrochim. Acta. 53 (2007) 468-473. https://doi.org/10.1016/j.electacta.2007.06.069

[22] H. Richter, K. McCarthy, K.P. Nevin, J.P. Johnson, V.M. Rotello, D.R. Lovley, Electricity generation by Geobacter sulfurreducens attached to gold electrodes, Langmuir. 24 (2008) 4376-4379. https://doi.org/10.1021/la703469y

[23] S. Cheng, B.E. Logan, Ammonia treatment of carbon cloth anodes to enhance power generation of microbial fuel cells, Electrochem. Commun. 9 (2007) 492-496. https://doi.org/10.1016/j.elecom.2006.10.023

[24] H.-Y. Tsai, C.-C. Wu, C.-Y. Lee, E.P. Shih, Microbial fuel cell performance of multiwall carbon nanotubes on carbon cloth as electrodes, J. Power Sources. 194 (2009) 199-205. https://doi.org/10.1016/j.jpowsour.2009.05.018

[25] Y. Qiao, C.M. Li, S.J. Bao, Q.L. Bao, Carbon nanotube/polyaniline composite as anode material for microbial fuel cells, J. Power Sources. 170 (2007) 79-84. https://doi.org/10.1016/j.jpowsour.2007.03.048

[26] Z. He, S.D. Minteer, L.T. Angenent, Electricity generation from artificial wastewater using an upflow microbial fuel cell, Environ. Sci. Technol. 39 (2005) 5262-5267. https://doi.org/10.1021/es0502876

[27] Y. Feng, Q. Yang, X. Wang, B.E. Logan, Treatment of carbon fiber brush anodes for improving power generation in air–cathode microbial fuel cells, J. Power Sources. 195 (2010) 1841-1844. https://doi.org/10.1016/j.jpowsour.2009.10.030

[28] D.A. Lowy, L.M. Tender, Harvesting energy from the marine sediment–water interface: III. Kinetic activity of quinone-and antimony-based anode materials, J. Power Sources. 185 (2008) 70-75. https://doi.org/10.1016/j.jpowsour.2008.06.079

[29] K. Guo, A. Prévoteau, S.A. Patil, K. Rabaey, Engineering electrodes for microbial electrocatalysis, Curr. Opin. Biotechnol. 33 (2015) 149-156. https://doi.org/10.1016/j.copbio.2015.02.014

[30] K. Wang, Y. Liu, S. Chen, Improved microbial electrocatalysis with neutral red immobilized electrode, J. Power Sources. 196 (2011) 164-168. https://doi.org/10.1016/j.jpowsour.2010.06.056

[31] S. Kondaveeti, J. Lee, R. Kakarla, H.S. Kim, B. Min, Low-cost separators for enhanced power production and field application of microbial fuel cells (MFCs), Electrochim. Acta. 132 (2014) 434-440. https://doi.org/10.1016/j.electacta.2014.03.046

[32] S. Choi, J.R. Kim, J. Cha, Y. Kim, G.C. Premier, C. Kim, Enhanced power production of a membrane electrode assembly microbial fuel cell (MFC) using a cost effective poly [2, 5-benzimidazole](ABPBI) impregnated non-woven fabric filter, Bioresour. Technol. 128 (2013) 14-21. https://doi.org/10.1016/j.biortech.2012.10.013

[33] J. Sun, Y. Hu, Z. Bi, Y. Cao, Improved performance of air-cathode single-chamber microbial fuel cell for wastewater treatment using microfiltration membranes and multiple sludge inoculation, J. Power Sources. 187 (2009) 471-479. https://doi.org/10.1016/j.jpowsour.2008.11.022

[34] B. Min, J. Kim, S. Oh, J.M. Regan, B.E. Logan, Electricity generation from swine wastewater using microbial fuel cells, Water Res. 39 (2005) 4961-4968. https://doi.org/10.1016/j.watres.2005.09.039

Microbial Fuel Cells: Materials and Applications　　　　　　　　Materials Research Forum LLC
Materials Research Foundations **46** (2019) 53-74　　　　　doi: http://dx.doi.org/10.21741/9781644900116-3

[35]　J.R. Kim, S. Cheng, S.-E. Oh, B.E. Logan, Power generation using different cation, anion, and ultrafiltration membranes in microbial fuel cells, Environ. Sci. Technol. 41 (2007) 1004-1009. https://doi.org/10.1021/es062202m

[36]　S. Cheng, H. Liu, B.E. Logan, Increased performance of single-chamber microbial fuel cells using an improved cathode structure, Electrochem. Commun. 8 (2006) 489-494. https://doi.org/10.1016/j.elecom.2006.01.010

[37]　Z. He, N. Wagner, S.D. Minteer, L.T. Angenent, An upflow microbial fuel cell with an interior cathode: assessment of the internal resistance by impedance spectroscopy, Environ. Sci. Technol. 40 (2006) 5212-5217. https://doi.org/10.1021/es060394f

[38]　S. Ayyaru, S. Dharmalingam, Improved performance of microbial fuel cells using sulfonated polyether ether ketone (SPEEK) TiO2–SO3H nanocomposite membrane, RSC Adv. 3 (2013) 25243-25251. https://doi.org/10.1039/c3ra44212h

[39]　I. Ieropoulos, J. Greenman, C. Melhuish, Improved energy output levels from small-scale microbial fuel cells, Bioelectrochemistry. 78 (2010) 44-50. https://doi.org/10.1016/j.bioelechem.2009.05.009

[40]　W.W. Li, G.P. Sheng, X.W. Liu, H.Q. Yu, Recent advances in the separators for microbial fuel cells, Bioresour. Technol. 102 (2011) 244-252. https://doi.org/10.1016/j.biortech.2010.03.090

[41]　F. Harnisch, U. Schröder, F. Scholz, The suitability of monopolar and bipolar ion exchange membranes as separators for biological fuel cells, Environ. Sci. Technol. 42 (2008) 1740-1746. https://doi.org/10.1021/es702224a

[42]　Y. Zuo, S. Cheng, D. Call, B.E. Logan, Tubular membrane cathodes for scalable power generation in microbial fuel cells, Environ Sci Technol. 41 (2007) 3347-3353. https://doi.org/10.1021/es0627601

[43]　J. Moon, S. Kondaveeti, B. Min, Evaluation of lowcost separators for increased power generation in single chamber microbial fuel cells with membrane electrode assembly, Fuel Cells. 15 (2015) 230-238. https://doi.org/10.1002/fuce.201400036

[44]　J.M. Moon, S. Kondaveeti, T.H. Lee, Y.C. Song, B. Min, Minimum interspatial electrode spacing to optimize air-cathode microbial fuel cell operation with a membrane electrode assembly, Bioelectrochem. 106 (2015) 263-267. https://doi.org/10.1016/j.bioelechem.2015.07.011

[45]   A. Larrosa-Guerrero, K. Scott, I. Head, F. Mateo, A. Ginesta, C. Godinez, Effect of temperature on the performance of microbial fuel cells, Fuel. 89 (2010) 3985-3994. https://doi.org/10.1016/j.fuel.2010.06.025

[46]   Y. Liu, V. Climent, A. Berna, J.M. Feliu, Effect of temperature on the catalytic ability of electrochemically active biofilm as anode catalyst in microbial fuel cells, Electroanal. 23 (2011) 387-394. https://doi.org/10.1002/elan.201000499

[47]   B. Min, Ó.B. Román, I. Angelidaki, Importance of temperature and anodic medium composition on microbial fuel cell (MFC) performance, Biotechnol. Lett. 30 (2008) 1213-1218. https://doi.org/10.1007/s10529-008-9687-4

[48]   Z. Li, X. Zhang, Y. Zeng, L. Lei, Electricity production by an overflow-type wetted-wall microbial fuel cell, Bioresour. Technol. 100 (2009) 2551-2555. https://doi.org/10.1016/j.biortech.2008.12.018

[49]   D. Pant, G. Van Bogaert, L. Diels, K. Vanbroekhoven, A review of the substrates used in microbial fuel cells (MFCs) for sustainable energy production, Bioresour. Technol. 101 (2010) 1533-1543. https://doi.org/10.1016/j.biortech.2009.10.017

[50]   K.-J. Chae, M.-J. Choi, J.-W. Lee, K.-Y. Kim, I.S. Kim, Effect of different substrates on the performance, bacterial diversity, and bacterial viability in microbial fuel cells, Bioresour. Technol. 100 (2009) 3518-3525. https://doi.org/10.1016/j.biortech.2009.02.065

[51]   Y. Fan, E. Sharbrough, H. Liu, Quantification of the internal resistance distribution of microbial fuel cells, Environ. Sci. Technol. 42 (2008) 8101-8107. https://doi.org/10.1021/es801229j

[52]   S. Cheng, B.E. Logan, Increasing power generation for scaling up single-chamber air cathode microbial fuel cells, Bioresour. Technol. 102 (2011) 4468-4473. https://doi.org/10.1016/j.biortech.2010.12.104

[53]   V. Vologni, R. Kakarla, I. Angelidaki, B. Min, Increased power generation from primary sludge by a submersible microbial fuel cell and optimum operational conditions, Bioprocess Biosyst. Eng. 36 (2013) 635-642. https://doi.org/10.1007/s00449-013-0918-2

[54]   R. Kakarla, J.R. Kim, B.-H. Jeon, B. Min, Enhanced performance of an air–cathode microbial fuel cell with oxygen supply from an externally connected algal bioreactor, Bioresour. Technol. 195 (2015) 210-216. https://doi.org/10.1016/j.biortech.2015.06.062

[55]   R. Kakarla, B. Min, Evaluation of microbial fuel cell operation using algae as an oxygen supplier: carbon paper cathode vs. carbon brush cathode, Bioprocess Biosyst. Eng. 37 (2014) 2453-2461. https://doi.org/10.1007/s00449-014-1223-4

[56]   R. Kakarla, B. Min, Photoautotrophic microalgae Scenedesmus obliquus attached on a cathode as oxygen producers for microbial fuel cell (MFC) operation, Int. J. Hydrog. Energy. 39 (2014) 10275-10283. https://doi.org/10.1016/j.ijhydene.2014.04.158

[57]   Z. He, Y. Huang, A.K. Manohar, F. Mansfeld, Effect of electrolyte pH on the rate of the anodic and cathodic reactions in an air-cathode microbial fuel cell, Bioelectrochem. 74 (2008) 78-82. https://doi.org/10.1016/j.bioelechem.2008.07.007

[58]   Y. Zhang, B. Min, L. Huang, I.J.A. Angelidaki, Generation of electricity and analysis of microbial communities in wheat straw biomass-powered microbial fuel cells, Appl. Environ. Microbiol. 75 (2009) 3389-3395. https://doi.org/10.1128/AEM.02240-08

[59]   T. Akiba, H. Bennetto, J. Stirling, K. Tanaka, Electricity production from alkalophilic organisms, Biotechnol. Lett. 9 (1987) 611-616. https://doi.org/10.1007/BF01033196

[60]   S. You, Q. Zhao, J. Zhang, J. Jiang, S. Zhao, A microbial fuel cell using permanganate as the cathodic electron acceptor, J. Power Sources. 162 (2006) 1409-1415. https://doi.org/10.1016/j.jpowsour.2006.07.063

[61]   S.V. Mohan, S.V. Raghavulu, D. Peri, P.N. Sarma, Bioelectronics, Integrated function of microbial fuel cell (MFC) as bio-electrochemical treatment system associated with bioelectricity generation under higher substrate load, Biosen. Bioelect. 24 (2009) 2021-2027. https://doi.org/10.1016/j.bios.2008.10.011

[62]   L. Wei, H. Han, J. Shen, Effects of cathodic electron acceptors and potassium ferricyanide concentrations on the performance of microbial fuel cell, Int. J. Hydrog. Energy. 37 (2012) 12980-12986. https://doi.org/10.1016/j.ijhydene.2012.05.068

[63]   H. L. Song, Y. Zhu, J. Li, Electron transfer mechanisms, characteristics and applications of biological cathode microbial fuel cells–A mini review, Arab. J. Chem (2015). https://doi.org/10.1016/j.arabjc.2015.01.008

[64] J. Jiang, Q. Zhao, J. Zhang, G. Zhang, D.J. Lee, Electricity generation from bio-treatment of sewage sludge with microbial fuel cell, Bioresour. Technol. 100 (2009) 5808-5812. https://doi.org/10.1016/j.biortech.2009.06.076

Microbial Fuel Cells: Materials and Applications   Materials Research Forum LLC
Materials Research Foundations **46** (2019) 75-100   doi: http://dx.doi.org/10.21741/9781644900116-4

Chapter 4

Microbial Fuel Cells Characterization

Fatma Aydin Unal[1,2], Mehmet Harbi Calimli[1,3], Hakan Burhan[1], Fatma Sismanoglu[1], Busra Yalcın[1], Fatih Şen[1]*

[1]Sen Research Group, Biochemistry Department, Faculty of Arts and Science, Dumlupınar University, Evliya Çelebi Campus, 43100 Kütahya, Turkey

[2]Metallurgical and Materials Engineering Department, Faculty of Engineering, Alanya Alaaddin Keykubat University, 07450 Alanya/Antalya, Turkey

[3]Tuzluca Vocational High School, Igdir University, Igdir, Turkey

fatihsen1980@gmail.com

Abstract

Microbial fuel cells (MFCs) received considerable interest in recent years and represent a promising technology based on the conversion of chemical energy into electrical energy by microbial catalysis. However, their power output and stability is low due to the influence of the cathode, electrolyte, membrane, microorganism of the anode, and cell configuration. The choice of appropriate characterization techniques and methods to examine these problems, which limit the development of MFCs, is extremely important. For this reason, this chapter focuses on biochemical and electrochemical characterization techniques in order to examine a variety of surface morphology, analytical and spectroscopic details.

Keywords

Fuel Cell, Microbial Fuel Cell, Characterization, Electrochemistry, SEM

Contents

1.   Introduction

Microbial fuel cells (MFCs) contain inorganic-organic substances and generate electrical current by using bacteria as a catalyst. The electrons generated by the bacteria in these materials are transmitted from the cathode to the anode. Under normal circumstances, a positive current causes the reverse electron current from the positive point to the negative point. For this purpose, different chemical mediators are used to generate the current using bacteria from the electrode-free systems. The resulting electrons reach the cathode to merge with protons thanks to the oxygen in the air and a separating material on the anode, and so water is formed as a result of this process [1]. To characterize the contents of the microbial fuel cell, some different techniques, such as CLSM (confocal laser scanning microscopy), SEM (scanning electron microscopy), and fluorescence microscopy are performed. These techniques detect growth patterns, heterogeneity, and biofilm thickness-density on the anode surface in microbial fuel cells [2–5]. Besides, the porosity of membrane materials and electrode are also characterized by using these techniques. Other techniques, such as spectroscopic, electrochemical, and biochemical are carried out to evaluate the structure of the product and the usage of the substrate and the mediators in the microbial fuel cells. Among these, the spectroscopic techniques, such as High-Pressure Liquid Chromatography (HPLC) and Ultraviolet-Visible Spectroscopy

(UV-VIS) [6] are used to detect the types of chemicals. The use of redox mediator as a biochemical method is a quantitative measurement technique [7]. Cyclic voltammetry (CV) [8], differential pulse voltammetry (DPV) [9], chronoamperometry (CA) [10], chronopotentiometry (CP) [11], polarization curves [12], power curves [13], coulombic efficiency [14], resistance [15], and energy efficiency [16] are electrochemical methods which allow qualitative analysis. Among these techniques, cyclic voltammetry is the most widely used method in biological fuel cells, and it is the basic structure to provide electron transfer in the system [7].

2.   Characterization techniques of microbial fuel cells

Lately, microbial fuel cells have been paid growing attention [17]. However, their power output and stability is low due to the effect of the cathode, electrolyte, membrane, anodic microorganism, and cell configuration, etc. The selection of appropriate characterization techniques and evaluation methods are crucial to solve these problems. To effectively use microbial fuel cells, it is necessary to take advantage of developments in different science and engineering fields. Microbiology, molecular biology, electrochemistry, material science, and environmental engineering provide information and technology for the effective use of biological fuel cells. Looking at recent studies, it seems that the research and characterization techniques for microbial fuel cells have increased steadily. In summary, analysis methods, such as biochemical, electrochemical, and spectroscopic methods, should be used to evaluate structurally and functionally the materials used in microbial fuel cells [7–18]. In this chapter, these techniques are discussed in detail.

2.1   Electrochemical techniques

Electrochemistry gives the fundamental information of corrosion, battery coating, electrode coating, biomedical applications, and electrosynthesis. The following reaction depicts the primary reaction for electrochemistry.

$$O_x + ne^- \leftrightarrow R_{ed} \tag{1}$$

Where O_x and R_{ed} are the oxidized and reduced species, and n is the number of electrons involved. The activity of the species presents the redox reaction activity and the standard potential effect the electrode surface potential in the fuel cells. This is expressed by the Nernst equation as follows [19]:

$$E = E^0 + \frac{RT}{nF} \ln \frac{a_{ox}}{a_{red}} \tag{2}$$

Electrochemical/analytical techniques are very important to analyze mechanisms (microbial, physiological, chemical and electron transport). Thus, these techniques are useful for optimizing the operation of microbial fuel cells, promoting the electron transfer process, and continuous innovation. Unlike conventional electrochemical systems, the operation of microbial fuel cells is limited by optimal microbial growth and sustainability requirements. Typical conditions contain ambient temperature and pressure, electrolytes with a low concentration of ions, near neutral pH, etc. The following sections present the electrochemical techniques that can be used in complex microbial fuel cells systems [20].

2.1.1 Voltammetric measurements

2.1.1.1 Cyclic voltammetry (CV)

In biochemistry, voltammetry is widely used to determine the catalytic flow and redox conversion which occur on the surface of the electrodes in the presence of the enzyme substrate. The linear sweep voltammetry (LSV) is another measure of linear waveform versus potential time at the electrodes. The linear sweep voltammetry can measure the scan rate depending on the slope of potential versus time graph [21]. A set of an electrolytic microbial fuel cell consists of a reference electrode (RE), a counter electrode (CE), and a working electrode (WE) [22]. The working electrode and the reference electrode allow the cell potential to be measured [20]. In other words, the current measurement in the cell is provided by these two electrodes. The representation of cyclic voltammetry is given in Fig.1 and 2. The oxidation of the analyte present in the electrolyte increases with the forward scan by increasing the potential. The surface of the electrode is closed when the concentration of the analyte is decreased, and the potential starts to decrease at the end of the process. A peak occurs for an analyte which can be oxides in any range of the determined potential range. With the reversal of the previous potential, the potential level reaches a quantity that can reduce the initial oxidation product and the occurrence of an inverted polarity current from the forward scanning. Cyclic voltammetry in microbial fuel cells is used to obtain information about the redox reaction potentials in the analyte and the electrochemical ratios of the compounds in these reactions.

The significant parameters related to the reaction in biological fuel cells are obtained by anionic and cationic peak data, such as potentials and currents (E_{pa}, E_{pc}, I_{pa}, and I_{pc}) taken from cyclic voltammogram. $\Delta E = E_{pa} - E_{pc}$ equation is used to evaluate the level of electrochemical reversibility for redox reaction using the peak potentials [23].

Microbial Fuel Cells: Materials and Applications — Materials Research Forum LLC
Materials Research Foundations **46** (2019) 75-100 — doi: http://dx.doi.org/10.21741/9781644900116-4

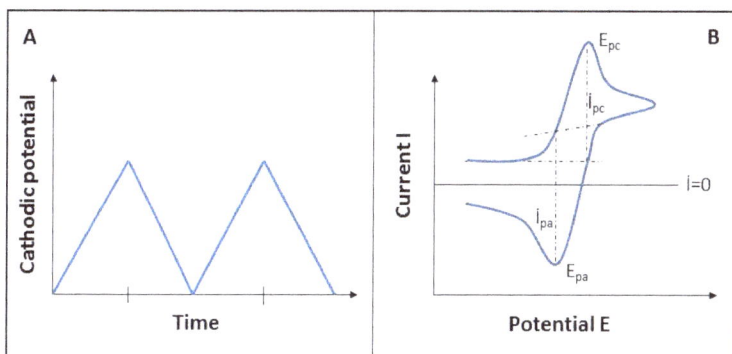

Figure 1. A. Excitation signal: potential vs. time in CV; B. Cyclic voltammogram: current vs. Potential.

The following equation is used to determine the potential of redox reaction occurring in a microbial fuel cell.

$$E^0 = \frac{E_{pa} + E_{pc}}{2} \tag{3}$$

Generally, the analyte concentration determines the peak height in a reaction having electron transfer and diffusion control. The following equation is called Randles-Sevcik (Eq. 4) which is used to calculate analyte current.

$$I_p = 0.4463 \, nFAC \left(\frac{nFvD}{RT}\right)^{1/2} \tag{4}$$

where C is the concentration of analyte, A is the working electrode area (cm^2), D is a constant of diffusion for analyte (cm^2/s), v is the scan rate (V/s), D is the diffusion coefficient of the analyte (cm^2/s), and n, F, R, T have their usual significance as defined above for the Nernst Eq. (2). Using the Randle-Sevcik Equation, it is shown that the graph between the square root of the scan rate and the peak current is linear [21]. In the simplest case of Langmuir adsorption [21], the shape of the voltammogram is entirely symmetric with identical areas under the two waves, and the peak potentials satisfy $E_{pa} = E_{pc} = E^0$. In this case, the peak current is proportional to v rather than to $v^{1/2}$.

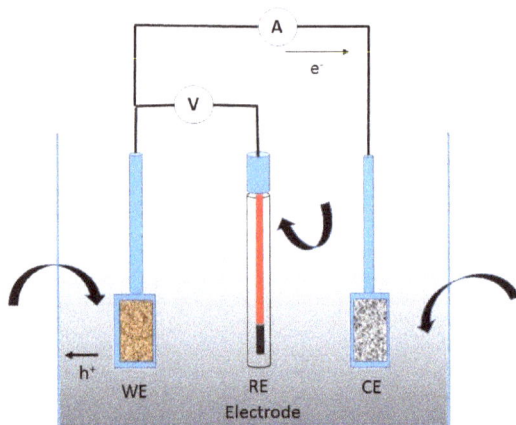

Figure 2. The principle of a three-electrode system.

It is relatively easy to determine whether a reaction is a reversible couple, with the help of the cyclic voltammetry method. The required anion/cation ratio for an effective reversible electrochemical system is 1, and the electrode potential separation rate in the room conditions is 59.2 mV/n (n is the number of transferred electrons in electrochemical reaction). The non–reversible systems have more differences with lower peak numbers and peak potentials compared to reversible ones. The reason for this is the extension of the time required for electron transfer in non-returnable systems [20]. In microbial fuel cell researches, cyclic voltammetry methods were applied for (1) examining the direct or indirect electron transfer mechanism between the biological membrane and the electrode, (2) determining the anodic oxidation potential and the cathodic reduction potential, including biological or chemical potential; (3) testing the performance of the catalyst system [19]. The use of cyclic voltammetry technology is easy, but it takes time. However, background experiments are necessary to guarantee the accuracy of the mechanical work. The anode may contain peak current and peak potential direct or indirect electron transfer from electrode interface reactions when the chamber is filled with microbial assemblies. There are many unknown factors for complex systems, such as wastewater [20]. The cyclic voltammetry curves are affected by the following factors: pre-treatment of the electrode surface, microorganism types and their thermodynamic properties, the rate of electron transfer reactions concentration of electroactive species and their diffusion rates and sweep rate.

It should be noted that many electrode materials used in microbial fuel cells can not fabricate reversible electrochemical reactions even for classical reversible redox couple $Fe(CN)_6^{3-}$ /$Fe(CN)_6^{4}$ [20–24]. This is mainly due to the fact that the heterogeneous processes of electrode reactions can be considerably infected by the microstructure, roughness, and functional groups present on the electrode surface. For example, an electrode will fabricate different electrode reactions and different cyclic voltammetry curves before and after polishing. For this reason, the electrode and operating terms must be suitable for the proper definition of kinetic parameters [20].

2.1.2 Electrochemical spectroscopy technique (EST)

The electrochemical spectroscopy technique (EST) is applied to make the technical examination. The displacement of the voltammetry electrodes applied with this technique is accomplished by applying sweeps and potential steps. A small voltage is applied with alternating current in the system, and then, the system response is investigated in the case of equilibrium. In the electrochemical spectroscopy technique, there is no interference during the application, so no damage occurs to the material. In other words, measurements made in this system can be done without interruption. However, in other characterization techniques, microbial fuel cells are degraded and require reanalysis of the samples. In electrochemical spectroscopy technique, microbial fuel is also referred to as a solid electrochemical material, since the voltage-current in the cell can be kept at the same level. Electrochemical spectroscopy technique is widely used in corrosion and various electrochemical fields. Electrochemical spectroscopy technique is also handy in examining the effects of different internal resistances on the impedance of the fuel cells [7]. Microbial fuel cells attract so much attention in producing alternative energy from wastewater and organic wastes. However, excessive voltage loss and reduced electrical power limit the use of microbial fuel cells. To overcome these factors which define the use of microbial fuel cells, researchers have focused on avoiding cell voltage loss and increasing electrical power density. For this purpose, in techniques such as electrochemical spectroscopy technique, impedance is applied to determine the components forming an internal resistance. Electrochemical spectroscopy technique is an advanced technique for characterizing internal resistance components such as ohm resistance, diffusion resistance, charge transfer resistance in fuel cells. In addition to these, the electrochemical spectroscopy technique is recently applied to determine the performance of anodes and used in applications to determine the electrochemical behaviour and resistance of redox regulators. During the implementation of the electrochemical spectroscopy technique, there were no visible effects that would cease the components and activities of the microbial fuel cells [25–27].

2.1.3 Chronoamperometric measurements

2.1.3.1 Chronoamperometry (CA)

As a simple electrochemical technique, the chronoamperometry (CA) is widely used in electroanalysis and electropolymerization [28]. Chronoamperometry records the current generated on the working electrode against time. While the cyclic voltammetry measures the current response in a dynamic state, in chronoamperometry, a specific potential measures a transient current response to a particular state of equilibrium. This current measured with chronoamperometry fully demonstrates all the oxidation and reduction occurred on the surface electrode.

2.1.4 Potentiometric measurements

2.1.4.1 Chronopotentiometry (CP)

The chronopotentiometry (CP) is a potentiometric measuring device measure the half cell potential of the electrodes in the microbial fuel cells. chronopotentiometry method measures the current by monitoring polarization curves with a potentiostat [29]. The potentiostat in chronopotentiometry as a measuring device can measure either the generated current or the potential correctly. The potentiostat also consists of three electrode systems which are a reference, counter, and working electrodes [1]. Generally, when the cathode is a counter electrode, the biocatalysis anode is set as the working electrode (Fig. 3). The reference electrode should be placed as close as possible to the working electrode immersed into the electrolyte. This avoids significant potential loss between the working electrode and a reference electrode which is responsible for the false potential measurement [30–31].

To record a polarization curve with the potentiostat, the potential of the working electrode is varied at a certain scan-rate from open circuit voltage (OCV) to short circuit condition (null R_{ext}) [1–32]. The potential scan of the reverse current continues until coming back to the starting point, and a slow scan rate of 1 mV s^{-1} provides a good polarization curve that is not relevant to oxidation/reduction reactions [32–33]. After recording the polarization curve, the power curve can be plotted by calculating the power generated (P) by the microbial fuel cells for each set R_{ext} through the second Ohm law (equation 5) [1–31].

$$P = E.I = E.(E / R_{ext}) = E^2/R_{ext} \tag{5}$$

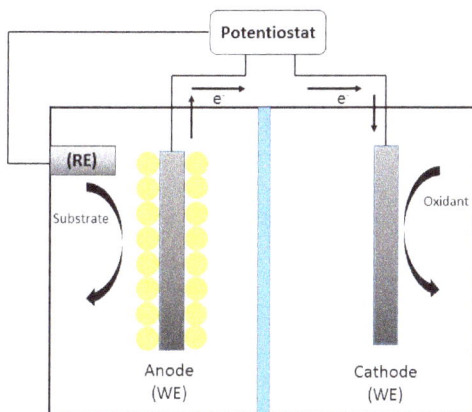

Figure 3. Three-electrode-setup for potentiostatic polarization curve recording.

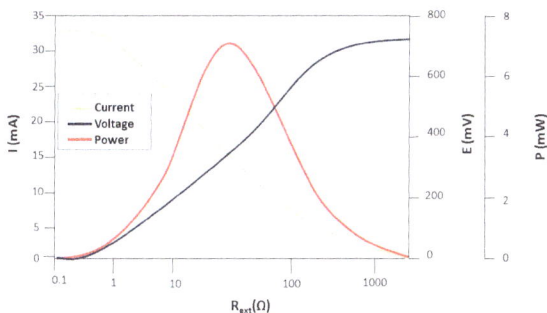

Figure 4. MFC current (I), voltage (E) and power (P) as function of applied external resistance (R_{ext}).

Since there is no current for the open circuit condition (infinite $_{Rext}$), no energy is generated. The power rises with the current until Maximum Power Point (MPP) is reached. Symmetric power curves are recorded for microbial fuel cells with high R_{int}. When maximum power point is applied, R_{ext} equals R_{int} (i.e., maximum power transfer theorem). With low R_{int}, power generation increases in microbial fuel cells [1]. For this reason, one of the primary objectives of microbial fuel cells is to reduce the internal resistance of the system [34–35]. An example of how the microbial fuel cells electrical parameters (I, E and P) change when R_{ext} is changed is given in Fig. 4 [1–31].

2.1.5 Polarization curves

Polarization curves (P versus I) which are used to detect current as a voltage indication are effective tools to characterize and analyze fuel cells. A potentiostat of microbial fuel cell recording polarization curves gets data for electrodes. In the absence of potentiometers, variable resistance boxes are used. The potential of electrodes is measured periodically with decreasing the load. Ohm law gives the current carried. A reference electrode is used to compare the performance of electrochemical set as mentioned in the above sections. An appropriate scan rate (for example 1 mVs^{-1}) must be chosen to record data of the polarization curve in measurements with a potentiostat. In cases where an external resistance is given to the electrochemical microbial fuel cell system, current and potential data should be recorded when steady-state conditions are met and depending on external conditions or the set of the electrochemical fuel cell, the steady-state conditions are established within a few minutes. The state of the system is temporary and this time varies depending on the substrate concentration in the anode. This affects substrate-product mass transfer for current and voltage. Multiple batch cycles provide the data of the polarization curves which allow the calculation of coulombic data for each resistance [1]. Long-term recording can cause the formation of slips in electrochemical fuel cells. The regions for Polarization curves are given as follow.

(i) Beginning with the open circuit potentials having zero current where the activation losses predominate, the voltage initially drops steeply.

(ii) In this region where the ohmic losses predominate, the voltage drops occur slowly, and the obtained graph of voltage drop versus the current is linear.

(iii) A rapid drop in voltage at high currents is seen in this region (solid line, Figure 5A), and as seen in Figure 5A, the concentration losses are dominated. A linear polarization curve is present in Figure 5A. The internal resistance (R_{int}) for a microbial fuel cell is calculated from the polarization curves (for example, Rint = -ΔE/ΔI=26 Ω; Figure 5A, dashed line) [1–36]. These curves are used to find potential and current for each electrode and all of microbial fuel cells (Figure 5).

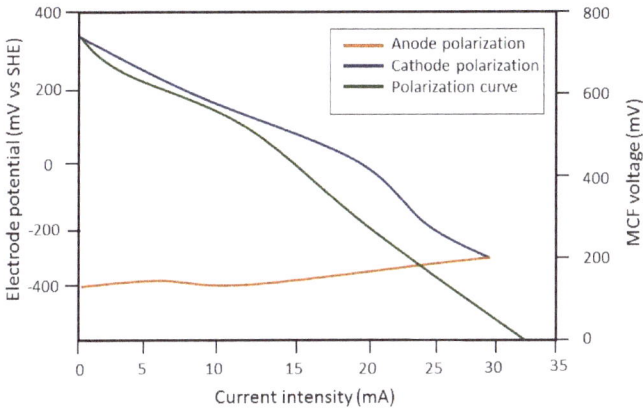

Figure 5. Anode (in brown) and cathode (in green) polarization curves addition gives the total MFC polarization curve (in blue).

There are different methods to track the polarization curve of microbial fuel cells [35]. The simplest one is called the "variable resistance method." This consists of connecting different external resistances (R_{ext}) to the microbial fuel cells and measuring the response regarding output voltage. Using first Ohm's law, current intensity can be calculated for each set resistance as Eq. 6 [37]:

$$I = E / R_{ext} \tag{6}$$

Before obtaining the polarization curve, it is suggested that the microbial fuel cells are entirely set to the open circuit potential in the open circuit mode for several hours. Then, R_{ext} must be changed in descending order, starting from the highest value (in the order of 10^4 Ω) to conclude at the lowest one (1 Ω is sufficient). To achieve a meaningful result, once the system reaches steady-state conditions again, the voltage on each setting resistance must be recorded. Steady state can take a few minutes or more to set up. This depends on the system and the amplitude of the R_{ext} variation. It is recommended to wait at least 2 minutes between obtaining each voltage. The polarization curve should be

Microbial Fuel Cells: Materials and Applications   Materials Research Forum LLC
Materials Research Foundations **46** (2019) 75-100   doi: http://dx.doi.org/10.21741/9781644900116-4

recorded in both senses (in other words, down from high to R_{ext}, and vice versa). Also, the entire polarization curve measurement should not last more than 2-3 hours since it can negatively affect the exoelectrogenic metabolism. There must be a compromise between the number of R_{ext} being tested (giving the curve resolution) and the time required for the measurement (the final result). There is no general rule for this, and there must be work arounds for each tested microbial fuel cells [38].

2.1.6 Power curves

Power is a significant parameter that can be used to reflect microbial fuel cells performance. Polarization curves are acquired from the power curve [1]. Also, the microbial fuel cells describe the power output as a function of the current. Generally, there is a single maximum (called Maximum Power Point or MPP) parabolic shape that arises when the outer resistance is equal to the inner resistance of the cell [35]. Internal resistance (R_{int}) expresses a general measure of microbial fuel cell's internal voltage losses. It is also geometrically described by the slope of the linear region of the polarization curve. A polarization curve's linear approach based on open circuit potential and R_{int} values is reported in Equation 7.

$$E = OCV - IR_{int} \tag{7}$$

An example of measuring polarization and power curves in microbial fuel cells is shown in Figure 6.

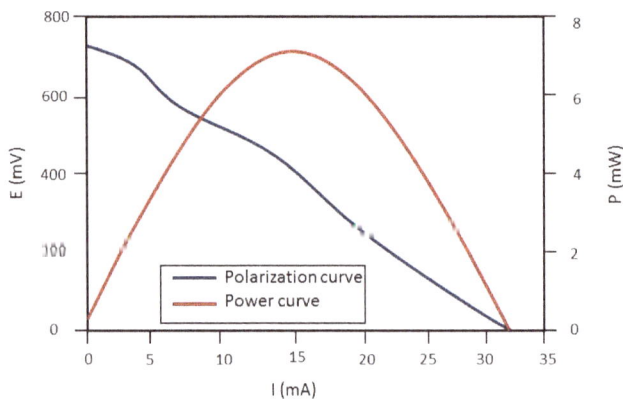

Figure 6. Example of MFC polarization curve (in blue) and the power curve (in red).

Ohmic resistances are the basic building blocks of microbial fuel cells, and that provide the primary functions for the formation of the central features of these cells, partly due to the low ionic conductivity of the substrate solutions. However, usually due to the low level of optimization in fuel cell design, internal resistance dominates ohmic resistance. By the time the linear polarization curve is obtained, the slope becomes equal to the internal resistance. E is the electromotive force, and R_{int} is the internal resistance of the microbial fuel cells [20]. When the internal resistance is equal to the external resistance, the highest power value is obtained. However, when the polarization curves are not linear, the power output is closely related to the concentration losses.

2.2   Coulombic efficiency (CE)

ε_c represents the coulombic efficiency which can be described as the ratio of charge of microbial fuel cells output conducted from the substrate to the anode, and in the case of conversion of all the substrates to current, the maximum possible coulombic yield is obtained [1]. When acetate is used as a substrate, it can be converted to electrons or methane. The type of these products depends on the type of microorganisms. All of the electrons that can not be transferred to the anodes are briefly defined as losses. Possible sinks of electrons are; the formation of methane outside of the bacterial growth with oxygen, sulphate or nitrate at the anode, the production of hydrogen by fermentation, and the oxidation carried out by electron receivers. The coulombic efficiency is reduced in case of the presence of oxygen as an alternative electron acceptor for in the bacteria, media, wastewater, or via membrane diffusion [39–40].

If the microbial fuel cells runs in fed-batch mode, ε_{cb} can be calculated from the following formula:

$$\varepsilon_{Cb} = \frac{M \int_0^{tb} i\, dt}{FbV_{An}\Delta COD} \qquad .$$

(9)

Where M is the molecular weight of oxygen 32 g/mol, F *is* the Faraday constant,= 96,485 C/mol, b=4 corresponds to the number of electrons exchanged per mole of oxygen, V_{An} is the volume of liquid in the anode compartment, and ΔCOD is the change in COD (chemical oxygen demand) over time t_b (g/L) [38–40]. For continuous flow through the system, microbial fuel cells can achieve its continuous steady operation state when the load resistor (external resistor) remains constant, and the current does not change with time. Therefore, calculation of ε_{cb} can be further simplified and calculated as follows [34–41]:

$$\varepsilon_{Cb} = \frac{Mi}{FbV\Delta COD} \qquad (10)$$

where $VCOD$ is the difference in the effluent and influent COD (g/L).

2.3   Resistance

In a real microbial fuel cells system, energy loss is accompanied by internal resistance from microorganism to electron transfer to the last electron receiver. The internal resistance can diminish the yield by reducing the output voltage [42].

2.4   Degradation efficiency

A microbial fuel cell is not only used to generate electricity, but it can also be proposed as a method for treating wastewater. For this reason, it is significant to appreciate the overall performance ranges for removal of BOD, COD, or TOC. Other factors, such as solubility of the particles and removal of nutrients may also be necessary. COD removal is an essential process to get energy and coulombic measurements in wastewater treatments, so much attention is given to the COD removal performance. The COD removal efficiency (COD) can be calculated as the ratio between the removed and the fluid COD, and the equation can be expressed as [1–43]:

$$\varepsilon_{COD} = \frac{\Delta COD}{COD}. \qquad (11)$$

2.5   Energy efficiency

Energy efficiency is described as the energy recovery for power production and the theoretical energy release of microbial fuel cells. The comparison of electricity generation of microbial fuel cells with sewage treatment systems is based on the amount of energy recovered. The total energy efficiency in biological fuel cells is defined as the ratio of the energy yield obtained over a certain period to the combustion heat obtained from the added substrate. For batch-type feeding, the equation is shown as follows [1];

$$\eta MFC = \frac{\int^{th} Ui\,dt}{m_{addes}\,\Delta H} \qquad (12)$$

where H is the heat of combustion (J / mol) and m is the amount (mol) of substrate added. This is usually only calculated for substances that are known when H is not known for actual wastewater. In microbial fuel cells, the energy efficiency ranges from 2% to 50% or more when biodegradable substrates are used [1].

2.6   Other characterization techniques

Biological fuel cells can also be examined using different analytical characterization methods, such as spectroscopic, electrochemical, surface, biochemical, etc. The structural characterization methods, such as transmission electron microscopy (TEM), scanning electron microscopy (SEM), high-resolution transmission electron microscopy (HRTEM), and confocal laser scanning microscopy (CLSM) are simple and useful techniques for the definition of properties of the materials on the anode of electrochemical systems (to reveal their heterogeneity, thickness, and density of biofilm on the anode). Biochemical analyzes in microbial fuel cells are used to assess the redox mediators used in electrolysis systems quantitatively. However, the spectroscopic techniques, such as liquid chromatography-mass spectrometry (LC-MS), high-performance liquid chromatography (HPLC), and ultraviolet-visible spectroscopy (UV-VIS) are used effectively in the characterization of materials in microbial fuel cell's systems. Due to the identification of qualitative properties of materials used in microbial fuel cells electrochemical methods, such as differential pulse (DPV) and cyclic voltammetry (CV) are widely used [7–44]. The structures providing electron shuttling according to the electrochemical activity are determined based on the pixel in the analysis. One of the useful biological fuel cells characterization techniques is cyclic voltammetry, and it is commonly used to determine compounds in microbial fuel cells [7–14].

2.6.1 Scanning electron microscopy (SEM)

To determine the biofilm morphology, samples from biofilm are examined with the help of scanning electron microscopy (SEM) [42]. Electrode material analysis is supplemented with high-resolution imaging and elemental analysis using scanning electron microscopy (SEM). The probe in this technique is a focused electron beam, usually with a diameter of 10 to 20 nm that results in higher resolution images of surface features than achievable with light microscopy. Contrary to atomic force microscopy (AFM), the probe penetrates into the sample and therefore allows information to be acquired on its elemental and phase composition which can also be used to enhance image contrast.

The main experimental factors crucial to the image quality are sample preparation, electron acceleration voltage, spot size, and detector position. The acceleration voltage is proportional to the penetration depth of the electrons into the sample where they are scattered due to elastic and inelastic collisions. The fraction of elastic to inelastic collisions depends on the atomic number Z of the sample material and determines the shape as well as the size of the electron beam interaction volume. Collisions with atoms that have low Z values result in higher penetration depths and are mainly inelastic. The loss in energy causes the electrons to form a pear-shaped interaction volume in the size region of micrometres. Increasing the acceleration voltage would only increase the penetration depth for a given Z and not the interaction volume. The elasticity of collisions increases with Z resulting in larger scattering angles and smaller, more spherical interaction volumes. Elemental and phase composition of a sample, therefore, have a strong influence on the energy and scattering angle of the reflected electrons, and compositional analysis becomes more reliable and more significant for the differences in Z [1]. Two types of reflected electrons are distinguished and require separate detectors, the backscattered electrons with kinetic energy higher than 50 eV and the secondary electrons with lower energy. The high energy backscattered electrons result from elastic collisions with deflection angles larger than 90° and are measured for the compositional analysis while secondary electrons lost their kinetic energy as a result of the inelastic collision and are detected at lower deflection angles for imaging.

Consequently, the fraction and energy of backscattered electrons increase with the atomic number of a sample which aids the compositional analysis. In practice, high acceleration voltage and large spot size are employed to improve the count of backscattered electrons and to reduce error in the compositional analysis. However, these settings are detrimental for the image resolution since they increase the probe size. Backscattered electrons also originate from a larger interaction volume than the secondary electrons measured with the same acceleration voltage and spot size which further diminishes the usefulness of backscattered electrons for imaging. High-resolution image acquisition is therefore performed separately from the compositional analysis at low acceleration voltage and small spot size in order to minimize the probe size and the interaction volume. In some instances where phase contrast is preferred to resolution, backscattered electrons are used for imaging as well. Phase contrast is also called Z-contrast since the atomic number determines the energy difference between electrons reflected from different phases. The phase contrast is improved with the same methods as the compositional analysis. Thus, a compromise between resolution and phase visualization is necessary when such measurements are performed.

Interference from charging and radiation damage is an additional problem for the imaging of less conductive samples, especially when the biological specimen is investigated. The sputter coating of several nanometers of gold, palladium, chromium, or platinum becomes necessary if the sample is not sufficiently conductive to discharge into the sample. Biological samples also have to be chemically fixed with cross-linking agents and dehydrated before the metal coating since otherwise the water evaporates under the high vacuum conditions of the measurement and scatter the electron beam. These requirements reduce the data representativeness of the sample in its natural form and

Microbial Fuel Cells: Materials and Applications   Materials Research Forum LLC
Materials Research Foundations **46** (2019) 75-100   doi: http://dx.doi.org/10.21741/9781644900116-4

make the in-situ observation of biological processes impossible. However, environmental scanning electron microscopy (ESEM) has been developed as an SEM adaption without the need for most sample preparation steps to overcome these problems. The high vacuum is only maintained around the electron beam for most of the beam path through a series of pressure limiting apertures and differential pumping while the sample is kept in low vacuum. Specimens in environmental scanning electron microscopy are therefore imaged closer to the electron source to reduce the scattering from gas molecules. Such a technology presents a promising combination of the benefits from electron microscopy and in-situ biofilm imaging and can help in future investigations of exoelectrogenic activity of biofilms on electrodes.

2.6.2 Atomic force microscopy (AFM)

Atomic force microscopy (AFM) is a scanning probe technique for the investigation of topography and electrostatic charge on the surface. In this technique, most of the surface quality calculations are made on the topographic map, and the image quality has a critical precaution for data evaluation. The minimization of sources for imaging artefacts is therefore even more essential to the reliability of the results than the tip quality. Most artefacts can be avoided with regular calibration and adjustment of scan speed, setpoint, and signal amplification according to the sample properties [45].

Last, but not least, there are also many other spectroscopic and microscopic techniques, such as X-ray diffraction, X-ray photoelectron spectroscopy, nuclear magnetic resonance spectroscopy, transmission electron microscopy, Raman spectroscopy, energy dispersive analysis, etc [46–54]. All of these techniques have been used for identification of the materials which have been used in most of the applications [55–63]. They are all complementary techniques in order to describe the full properties of the materials [60–64–72].

Conclusions

In conclusion, microbial fuel cells (MFCs) have received considerable interest due to their promising technology based on the conversion of chemical energy into electrical energy by microbial catalysis. There are many types of electrochemical characterization techniques such as cyclic voltammetry (CV), differential pulse voltammetry (DPV), chronoamperometry (CA), chronopotentiometry (CP), etc. There are also other techniques such as atomic force microscopy, scanning electron microscopy, etc to identify the related materials in microbial fuel cells. However, the choice of appropriate characterization techniques and methods to solve the problems related to microbial fuel cells is extremely important as mentioned in this chapter. In this regard, biochemical and electrochemical characterization techniques, a variety of surface morphology, analytical, spectroscopic, microscopic techniques will be used very commonly for identification of materials in microbial fuel cells in the near future.

References

[1]   B. E. Logan, B. Hamelers, R. Rozendal, U. Schröder, J. Keller, S. Freguia, P. Aelterman, W. Verstraete, & K. Rabaey, Microbial fuel cells: Methodology and technology. Environmental Science and Technology, 40 (2006) 5181–5192. https://doi.org/10.1021/es0605016.

[2]   N. K. Dhami, M. S. Reddy, & A. Mukherjee, We Are Intechopen, The World's Leading Publisher of Open Access Books Built by Scientists, For Scientists TOP 1 % Control of A Proportional Hydraulic System. Waste Water - Evaluation and Management, 2 (2012) 137–164. https://doi.org/10.5772/32009.

[3]   D. V. P. Sanchez, D. Jacobs, K. Gregory, J. Huang, Y. Hu, R. Vidic, & M. Yun, Changes in Carbon Electrode Morphology Affect Microbial Fuel Cell Performance with Shewanella Oneidensis MR-1. Energies, 8 (2015) 1817–1829. https://doi.org/10.3390/en8031817.

[4]   N. Uria, I. Ferrera, & J. Mas, Electrochemical Performance and Microbial Community Profiles in Microbial Fuel Cells in Relation to Electron Transfer Mechanisms. BMC Microbiology, 17 (2017) 1–12. https://doi.org/10.1186/s12866-017-1115-2.

[5]   T. T. Nguyen, T. T. T. Luong, P. H. N. Tran, H. T. V. Bui, H. Q. Nguyen, H. T. Dinh, B. H. Kim, & H. T. Pham, A lithotrophic Microbial Fuel Cell Operated with Pseudomonads-Dominated Iron-Oxidizing Bacteria Enriched at The Anode. Microbial Biotechnology, 8 (2015) 579–589. https://doi.org/10.1111/1751-7915.12267.

[6]   S. Ishii, S. Suzuki, T. M. Norden-Krichmar, K. H. Nealson, Y. Sekiguchi, Y. A. Gorby, & O. Bretschger, Functionally Stable and Phylogenetically Diverse Microbial Enrichments from Microbial Fuel Cells during Wastewater Treatment. PLoS ONE, 7 (2012) e30495. https://doi.org/10.1371/journal.pone.0030495.

[7]   N. S. Ramaraja, P. Ramasamy, Electrochemical Impedance Spectroscopy for Microbial Fuel Cell Characterization. Journal of Microbial & Biochemical Technology, S6 (2013) 1–14. https://doi.org/10.4172/1948-5948.S6-004.

[8]   A. E. Franks & K. P. Nevin, Microbial Fuel Cells, A Current Review. Energies, 3 (2010) 899–919. https://doi.org/10.3390/en3050899.

[9] A. R. Khaskheli, J. Fischer, J. Barek, V. Vyskočil, Sirajuddin, & M. I. Bhanger, Differential pulse Voltammetric Determination of Paracetamol in Tablet and Urine Samples at A Micro-Crystalline Natural Graphite-Polystyrene Composite Film Modified Electrode. Electrochimica Acta, 101 (2013) 238–242. https://doi.org/10.1016/j.electacta.2012.09.102.

[10] P. S. Patil, S. H. Mujawar, A. I. Inamdar, & S. B. Sadale, Electrochromic Properties of Spray Deposited Tio2-Doped WO 3 Thin Films. Applied Surface Science, 250 (2005) 117–123. https://doi.org/10.1016/j.apsusc.2004.12.042.

[11] M. Hou, L. Chen, Z. Guo, X. Dong, Y. Wang, & Y. Xia, A clean and membrane-free chlor-alkali process with decoupled Cl_2 and H_2/NaOH production. Nature Communications, 9 (2018) 438. https://doi.org/10.1038/s41467-018-02877-x.

[12] M. Vuković, B. Pesic, N. Štrbac, I. Mihajlović, & M. Sokić, Linear Polarization Study of The Corrosion of Iron in The Presence of Thiobacillus Ferrooxidans Bacteria. International Journal of Electrochemical Science, 7 (2012) 2487–2503.

[13] I. Ieropoulos, J. Greenman, D. Lewis, & O. Knoop, Energy Production and Sanitation Improvement Using Microbial Fuel Cells. Journal of Water, Sanitation and Hygiene for Development, 3 (2013) 383-396. https://doi.org/10.2166/washdev.2013.117.

[14] J. R. Kim, G. C. Premier, F. R. Hawkes, R. M. Dinsdale, & A. J. Guwy, Development of A Tubular Microbial Fuel Cell (MFC) Employing A Membrane Electrode Assembly Cathode. Journal of Power Sources, 187 (2009) 393–399. https://doi.org/10.1016/j.jpowsour.2008.11.020.

[15] S. Cheng & H. Liu, Rapid Deblurring for Spiral fMRI. In Vivo, 13 (2005) 2426–2432. https://doi.org/10.1021/es051652w.

[16] Z. Ge, Energy-efficient Wastewater Treatment by Microbial Fuel Cells: Scaling Up and Optimization, Virginia Polytechnic Institute and State University, 2015.

[17] J. Zhang, E. Zhang, K. Scott, & J. G. Burgess, Enhanced Electricity Production by Use of Reconstituted Artificial Consortia of Estuarine Bacteria Grown As Biofilms. Environmental Science and Technology, 46 (2012) 2984–2992. https://doi.org/10.1021/es2020007.

[18] R. P. Ramasamy, V. Gadhamshetty, L. J. Nadeau, & G. R. Johnson, Impedance Spectroscopy as A Tool for Non-Intrusive Detection of Extracellular Mediators in Microbial Fuel Cells. Biotechnology and Bioengineering, 104 (2009) 882–891. https://doi.org/10.1002/bit.22469.

[19] L. Hussein, Dissertation zur Erlangung des Doktorgrades Decorated
      Nanostructured Carbon Materials for Abiotic and Enzymatic Biofuel Cell
      Applications, 2016.

[20] F. Zhao, R. C. T. Slade, & J. R. Varcoe, Techniques for The Study and
      Development of Microbial Fuel Cells: An Electrochemical Perspective. Chemical
      Society Reviews, 38 (2009) 1926–1939. https://doi.org/10.1039/b819866g.

[21] A. Mittal, L. Kurup, & J. Mittal, Freundlich and Langmuir Adsorption Isotherms
      and Kinetics for The Removal of Tartrazine from Aqueous Solutions Using Hen
      Feathers. Journal of Hazardous Materials, 146 (2007) 243–248.
      https://doi.org/10.1016/j.jhazmat.2006.12.012.

[22] R. Huarachi-Olivera, A. Dueñas-Gonza, U. Yapo-Pari, P. Vega, M. Romero-
      Ugarte, J. Tapia, L. Molina, A. Lazarte-Rivera, D. G. Pacheco-Salazar, & M.
      Esparza, Bioelectrogenesis with Microbial Fuel Cells (MFCs) Using The
      Microalga Chlorella Vulgaris and Bacterial Communities. Electronic Journal of
      Biotechnology, 31 (2018) 34–43. https://doi.org/10.1016/j.ejbt.2017.10.013.

[23] F. Haque, M. Rahman, E. Ahmed, P. Bakshi, & A. Shaikh, A Cyclic Voltammetric
      Study of the Redox Reaction of Cu(II) in Presence of Ascorbic Acid in Different
      pH Media. Dhaka University Journal of Science, 61 (2013) 161–166.
      https://doi.org/10.3329/dujs.v61i2.17064.

[24] V. Horvat-radošević, K. Kvastek, & D. Križekar, Kinetics of the [Fe (CN) 6] 3-
      /[Fe (CN) 6] 4- Redox Couple Reaction on Anodically Passivated FesoB20. 70
      (1997) 537–561.

[25] M. Sindhuja, N. S. Kumar, V. Sudha, & S. Haripriya, Equivalent Circuit
      Modeling of Microbial Fuel Cells Using Impedance Spectroscopy. Journal of
      Energy Storage, 7 (2016) 136–146. https://doi.org/10.1016/j.est.2016.06.005.

[26] Z. Lu, P. Girguis, P. Liang, H. Shi, G. Huang, L. Cai, & L. Zhang, Biological
      Capacitance Studies of Anodes In Microbial Fuel Cells Using Electrochemical
      Impedance Spectroscopy. Bioprocess and Biosystems Engineering, 38 (2015)
      1325–1333. https://doi.org/10.1007/s00449-015-1373-z.

[27] K. C. Honeychurch, The voltammetric Behaviour of Lead at A Hand Drawn Pencil
      Electrode and Its Trace Determination in Water by Stripping Voltammetry.
      Analytical Methods, 7 (2015) 2437–2443. https://doi.org/10.1039/c4ay02987a.

[28]   L. M. B. and A. C. Michael., An Introduction to Electrochemical Methods in Neuroscience. In B.L. Michael AC,ed., Electrochemical Methods for Neuroscience. (Boca Raton (FL): CRC Press/Taylor & Francis, 2007).

[29]   M. Tachibana, K. Ishida, Y. Wada, M. Aizawa, & M. Fuse, Study of Polarization Curve Measurement Method for Type 304 Stainless Steel in BWR High Temperature-High Purity Water. Journal of Nuclear Science and Technology, 46 (2009) 132–141. https://doi.org/10.1080/18811248.2007.9711514.

[30]   R. Awasthi, Madhu, & R. N. Singh, Application of Graphene in Electrochemical Devices. Handb. Funct. Nanomater., 3 (2014) 239–262.

[31]   D. Molognoni, Microbial Fuel Cells Application to Wastewater Treatment : Laboratory Experience And Controlling Strategies PhD Thesis of, 2014.

[32]   K. Kalle, Kalle Koivuniemi Bioelectricity Production From Simulated Mining and Forest Industry Wastewaters in Microbial Fuel, 2016.

[33]   D. Hidalgo, Politecnico di Torino Politecnico Di Torino Doctor of Philosophy in Chemical Engineering Department of Applied Science and Technology Development of Innovative Materials Used in Electrochemical Devices for The Renewable Production of Hydrogen And Electric, 2015. https://doi.org/10.6092/polito/porto/2588827.

[34]   H. O. Mohamed, M. A. Abdelkareem, M. Park, J. Lee, T. Kim, G. Prasad Ojha, B. Pant, S. J. Park, H. Y. Kim, & N. A. M. Barakat, Investigating The Effect of Membrane Layers on The Cathode Potential of Air-Cathode Microbial Fuel Cells. International Journal of Hydrogen Energy, 42 (2017) 24308–24318. https://doi.org/10.1016/j.ijhydene.2017.07.218.

[35]   G. Massaglia, Politecnico di Torino Development of New Nanostructured Electrodes in Microbial Fuel Cells (MFCs), 2017. https://doi.org/10.6092/polito/porto/2676549.

[36]   S. K. Kamaraj, S. M. Romano, V. C. Moreno, H. M. Poggi-Varaldo, & O. Solorza-Feria, Use of Novel Reinforced Cation Exchange Membranes for Microbial Fuel Cells. Electrochimica Acta, 176 (2015) 555–566. https://doi.org/10.1016/j.electacta.2015.07.042.

[37]   V. M. Ortiz-Martínez, M. J. Salar-García, F. J. Hernández-Fernández, & A. P. de los Ríos, Development and characterization of A New Embedded Ionic Liquid Based Membrane-Cathode Assembly for Its Application in Single Chamber

Microbial Fuel Cells. Energy, 93 (2015) 1748–1757.
https://doi.org/10.1016/j.energy.2015.10.027.

[38]   D. S. Rodrigues, Microbial Community Optimization for Electricity Generation in
       Microbial Fuel Cells, 2017.

[39]   N. T. Trinh, J. H. Park, S. S. Kim, J. C. Lee, B. Y. Lee, & B. W. Kim, Generation
       Behavior of Elctricity in A Microbial Fuel Cell. Korean Journal of Chemical
       Engineering, 27 (2010) 546–550. https://doi.org/10.2478/s11814-010-0066-1.

[40]   D. Majumder, J. P. Maity, M. J. Tseng, V. R. Nimje, H. R. Chen, C. C. Chen, Y.
       F. Chang, T. C. Yang, & C. Y. Chen, Electricity Generation and Wastewater
       Treatment of Oil Refinery in Microbial Fuel Cells Using Pseudomonas Putida.
       International Journal of Molecular Sciences, 15 (2014) 16772–16786.
       https://doi.org/10.3390/ijms150916772.

[41]   I. Satar, W. R. W. Daud, B. H. Kim, M. R. Somalu, M. Ghasemi, M. H. A. Bakar,
       T. Jafary, & S. N. Timmiati, Performance of Titanium–Nickel (Ti/Ni) and
       Graphite Felt-Nickel (GF/Ni) Electrodeposited by Ni as Alternative Cathodes for
       Microbial Fuel Cells. Journal of the Taiwan Institute of Chemical Engineers, 89
       (2018) 67–76. https://doi.org/10.1016/j.jtice.2018.04.010.

[42]   G. Li & P. Miao, Electrochemical Analysis of Proteins and Cells (Berlin,
       Heidelberg: Springer Berlin Heidelberg, 2013). https://doi.org/10.1007/978-3-642-
       34252-3.

[43]   S. Van Denhouwe, Combining Constructed Wetlands and Microbial Fuel Cells for
       Enhanced Wastewater Treatment, 2013.

[44]   A. Sotres, L. Tey, A. Bonmatí, & M. Viñas, Microbial Community Dynamics in
       Continuous Microbial Fuel Cells Fed with Synthetic Wastewater and Pig Slurry.
       Bioelectrochemistry, 111 (2016) 70–82.
       https://doi.org/10.1016/j.bioelechem.2016.04.007.

[45]   N. Ngadi, Mechanisms of Molecular Brush Inhibition of Protein Adsorption onto
       Steel Surface. 2009.

[46]   F. Sen, Y. Karatas, M. Gulcan, & M. Zahmakiran, Amylamine Stabilized
       Platinum(0) Nanoparticles: Active and Reusable Nanocatalyst in The Room
       Temperature Dehydrogenation of Dimethylamine-Borane. RSC Advances, 4
       (2014) 1526–1531. https://doi.org/10.1039/c3ra43701a.

[47]   S. Eris, Z. Daşdelen, Y. Yıldız, & F. Sen, Nanostructured Polyaniline-rGO
       Decorated Platinum Catalyst with Enhanced Activity and Durability for Methanol

Oxidation. International Journal of Hydrogen Energy, 43 (2018) 1337–1343. https://doi.org/10.1016/j.ijhydene.2017.11.051.

[48] Y. Yıldız, S. Kuzu, B. Sen, A. Savk, S. Akocak, & F. Şen, Different Ligand Based Monodispersed Pt Nanoparticles Decorated With rGO As Highly Active and Reusable Catalysts for The Methanol Oxidation. International Journal of Hydrogen Energy, 42 (2017) 13061–13069. https://doi.org/10.1016/j.ijhydene.2017.03.230.

[49] Y. Yildiz, H. Pamuk, Ö. Karatepe, Z. Dasdelen, & F. Sen, Carbon Black Hybrid Material Furnished Monodisperse Platinum Nanoparticles as Highly Efficient and Reusable Electrocatalysts for Formic Acid Electro-Oxidation. RSC Advances, 6 (2016) 32858–32862. https://doi.org/10.1039/c6ra00232c.

[50] E. Erken, Y. Yıldız, B. Kilbaş, & F. Şen, Synthesis and Characterization of Nearly Monodisperse Pt Nanoparticles for C_1 to C_3 Alcohol Oxidation and Dehydrogenation of Dimethylamine-borane (DMAB). Journal of Nanoscience and Nanotechnology, 16 (2016) 5944–5950. https://doi.org/10.1166/jnn.2016.11683.

[51] B. Çelik, E. Erken, S. Eriş, Y. Yildiz, B. Şahin, H. Pamuk, & F. Sen, Highly Monodisperse Pt(0)@AC Nps As Highly Efficient and Reusable Catalysts: The Effect of The Surfactant on Their Catalytic Activities in Room Temperature Dehydrocoupling of DMAB. Catalysis Science and Technology, 6 (2016) 1685–1692. https://doi.org/10.1039/c5cy01371b.

[52] B. Çelik, S. Kuzu, E. Erken, H. Sert, Y. Koşkun, & F. Şen, Nearly Monodisperse Carbon Nanotube Furnished Nanocatalysts as Highly Efficient and Reusable Catalyst for Dehydrocoupling of DMAB and C1 to C3 Alcohol Oxidation. International Journal of Hydrogen Energy, 41 (2016) 3093–3101. https://doi.org/10.1016/j.ijhydene.2015.12.138.

[53] G. Baskaya, İ. Esirden, E. Erken, F. Sen, & M. Kaya, Synthesis of 5-Substituted-1H-Tetrazole Derivatives Using Monodisperse Carbon Black Decorated Pt Nanoparticles as Heterogeneous Nanocatalysts. Journal of Nanoscience and Nanotechnology, 17 (2017) 1992–1999. https://doi.org/10.1166/jnn.2017.12867.

[54] B. Sen, S. Kuzu, E. Demir, S. Akocak, & F. Sen, Polymer-Graphene Hybride Decorated Pt Nanoparticles as Highly Efficient and Reusable Catalyst for The Dehydrogenation of Dimethylamine–Borane at Room Temperature. International Journal of Hydrogen Energy, 42 (2017) 23284–23291. https://doi.org/10.1016/j.ijhydene.2017.05.112.

[55] E. Demir, B. Sen, & F. Sen, Highly efficient Pt Nanoparticles and f-MWCNT Nanocomposites Based Counter Electrodes for Dye-Sensitized Solar Cells. Nano-

Structures & Nano-Objects, 11 (2017) 39–45.
https://doi.org/10.1016/j.nanoso.2017.06.003.

[56]   S. Eris, Z. Daşdelen, & F. Sen, Investigation of Electrocatalytic Activity and
       Stability of Pt@F-VC Catalyst Prepared by In-Situ Synthesis for Methanol
       Electrooxidation. International Journal of Hydrogen Energy, 43 (2018) 385–390.
       https://doi.org/10.1016/j.ijhydene.2017.11.063.

[57]   S. Eris, Z. Daşdelen, & F. Sen, Enhanced Electrocatalytic Activity and Stability of
       Monodisperse Pt Nanocomposites for Direct Methanol Fuel Cells. Journal of
       Colloid and Interface Science, 513 (2018) 767–773.
       https://doi.org/10.1016/j.jcis.2017.11.085.

[58]   B. Sen, S. Kuzu, E. Demir, T. Onal Okyay, & F. Sen, Hydrogen Liberation from
       The Dehydrocoupling of Dimethylamine–Borane at Room Temperature by Using
       Novel and Highly Monodispersed RuPtNi Nanocatalysts Decorated with Graphene
       Oxide. International Journal of Hydrogen Energy, 42 (2017) 23299–23306.
       https://doi.org/10.1016/j.ijhydene.2017.04.213.

[59]   B. Çelik, Y. Yildiz, H. Sert, E. Erken, Y. Koşkun, & F. Şen, Monodispersed
       Palladium-Cobalt Alloy Nanoparticles Assembled on Poly(N-Vinyl-Pyrrolidone)
       (PVP) as A Highly Effective Catalyst for Dimethylamine Borane (DMAB)
       Dehydrocoupling. RSC Advances, 6 (2016) 24097–24102.
       https://doi.org/10.1039/c6ra00536e.

[60]   B. Sen, S. Kuzu, E. Demir, E. Yıldırır, & F. Sen, Highly Efficient Catalytic
       Dehydrogenation of Dimethyl Ammonia Borane via Monodisperse Palladium–
       Nickel Alloy Nanoparticles Assembled on PEDOT. International Journal of
       Hydrogen Energy, 42 (2017) 23307–23314.
       https://doi.org/10.1016/j.ijhydene.2017.05.115.

[61]   H. Göksu, B. Çelik, Y. Yıldız, F. Şen, & B. Kılbaş, Superior Monodisperse CNT-
       Supported CoPd (CoPd@CNT) Nanoparticles for Selective Reduction of Nitro
       Compounds to Primary Amines with NaBH4 in Aqueous Medium.
       ChemistrySelect, 1 (2016) 2366–2372. https://doi.org/10.1002/slct.201600509

[62]   Y. Yıldız, I. Esirden, E. Erken, E. Demir, M. Kaya, & F. Şen, Microwave (Mw)-
       assisted Synthesis of 5-Substituted 1H-Tetrazoles via [3+2] Cycloaddition
       Catalyzed by Mw-Pd/Co Nanoparticles Decorated on Multi-Walled Carbon
       Nanotubes. ChemistrySelect, 1 (2016) 1695–1701.
       https://doi.org/10.1002/slct.201600265.

[63]   G. Başkaya, Y. Yıldız, A. Savk, T. O. Okyay, S. Eriş, H. Sert, & F. Şen, Rapid,
       Sensitive, and Reusable Detection of Glucose by Highly Monodisperse Nickel
       Nanoparticles Decorated Functionalized Multi-Walled Carbon Nanotubes.
       Biosensors and Bioelectronics, 91 (2017) 728–733.
       https://doi.org/10.1016/j.bios.2017.01.045.

[64]   B. Şen, A. Aygün, T. O. Okyay, A. Şavk, R. Kartop, & F. Şen, Monodisperse
       Palladium Nanoparticles Assembled on Graphene Oxide with The High Catalytic
       Activity and Reusability in The Dehydrogenation of Dimethylamine-Borane.
       International Journal of Hydrogen Energy, 3 (2018) 2–8.
       https://doi.org/10.1016/j.ijhydene.2018.03.175.

[65]   B. Sen, S. Kuzu, E. Demir, S. Akocak, & F. Sen, Highly Monodisperse RuCo
       Nanoparticles Decorated on Functionalized Multiwalled Carbon Nanotube with
       The Highest Observed Catalytic Activity in The Dehydrogenation of
       Dimethylamine−Borane. International Journal of Hydrogen Energy, 42 (2017)
       23292–23298. https://doi.org/10.1016/j.ijhydene.2017.06.032.

[66]   Y. Yıldız, E. Erken, H. Pamuk, H. Sert, & F. Şen, Monodisperse Pt Nanoparticles
       Assembled on Reduced Graphene Oxide: Highly Efficient and Reusable Catalyst
       for Methanol Oxidation and Dehydrocoupling of Dimethylamine-Borane
       (DMAB). Journal of Nanoscience and Nanotechnology, 16 (2016) 5951–5958.
       https://doi.org/10.1166/jnn.2016.11710.

[67]   S. Akocak, B. Şen, N. Lolak, A. Şavk, M. Koca, S. Kuzu, & F. Şen, One-Pot
       Three-Component Synthesis of 2-Amino-4H-Chromene Derivatives by Using
       Monodisperse Pd Nanomaterials Anchored Graphene Oxide as Highly Efficient
       and Recyclable Catalyst. Nano-Structures and Nano-Objects, 11 (2017) 25–31.
       https://doi.org/10.1016/j.nanoso.2017.06.002.

[68]   Z. Daşdelen, Y. Yıldız, S. Eriş, & F. Şen, Enhanced Electrocatalytic Activity and
       Durability of Pt Nanoparticles Decorated on GO-PVP Hybride Material for
       Methanol Oxidation Reaction. Applied Catalysis B: Environmental, 219 (2017)
       511–516. https://doi.org/10.1016/j.apcatb.2017.08.014.

[69]   H. Goksu, Y. Yıldız, B. Çelik, M. Yazici, B. Kilbas, & F. Sen, Eco-Friendly
       Hydrogenation of Aromatic Aldehyde Compounds by Tandem Dehydrogenation
       of Dimethylamine-Borane in The Presence of A Reduced Graphene Oxide
       Furnished Platinum Nanocatalyst. Catalysis Science & Technology, 6 (2016)
       2318–2324. https://doi.org/10.1039/C5CY01462J.

[70]   H. Göksu, Y. Yıldız, B. Çelik, M. Yazıcı, B. Kılbaş, & F. Şen, Highly Efficient and Monodisperse Graphene Oxide Furnished Ru/Pd Nanoparticles for The Dehalogenation of Aryl Halides via Ammonia Borane. ChemistrySelect, 1 (2016) 953–958. https://doi.org/10.1002/slct.201600207.

[71]   B. Aday, Y. Yildiz, R. Ulus, S. Eris, F. Sen, & M. Kaya, One-Pot, Efficient and Green Synthesis of Acridinedione Derivatives Using Highly Monodisperse Platinum Nanoparticles Supported with Reduced Graphene Oxide. New Journal of Chemistry, 40 (2016) 748–754. https://doi.org/10.1039/c5nj02098k.

[72]   S. Bozkurt, B. Tosun, B. Sen, S. Akocak, A. Savk, M. F. Ebeoğlugil, & F. Sen, A Hydrogen Peroxide Sensor Based on TNM Functionalized Reduced Graphene Oxide Grafted with Highly Monodisperse Pd Nanoparticles. Analytica Chimica Acta, 989 (2017) 88–94. https://doi.org/10.1016/j.aca.2017.07.051.

Materials Research Forum LLC
doi: http://dx.doi.org/10.21741/9781644900116-5

Chapter 5

Paper-Based Microbial Fuel Cell

Suruchee Samparnna Mishra, Swaraj Mohanty and Sonali Mohapatra[*]

Department of Biotechnology, College of Engineering & Technology, Bhubaneswar, India

sonalimohapatra85@gmail.com

Abstract

Microbial fuel cells (MFCs) are devices that convert organic substances into electrical energy with the help of microorganisms. The MFCs are advantageous in an industrial perspective as these can utilize a wide range of substrates. Hence, paper-based MFCs are a promising alternative for bioelectricity generation which can act ideally even without much effort by the operator and can produce efficient power even after just being plummeted onto the liquid solution containing organic matter. Further, the wide-scale applicability of paper-based MFCs along with their cost-effective features has made the device an important ration in the present industrial based energy sector.

Keywords

Microbial Fuel Cell (MFC), Exoelectrogens, Membranes, Electricity

List of Abbreviations

EET- Extracellular electron transfer

FMN- Flavin mononucleotide MFCs -Microbial fuel cells

OM c-Cysts- Outer-membrane c-type cytochromes

PEM -Proton exchange membrane

RF- Riboflavin

SMFCs- Sediment microbial fuel cells

Contents

1.   Introduction

The quality of human life in society directly influences the environmental condition by interfering with the use of energy in different sectors. To keep pace with our extravagant lifestyle, there arises an escalating need not only to produce an adequate amount of energy but also to discover sustainable and environmentally friendly energy sources. Though current industrial sectors rely more or less on renewable energy resources, it will not be sufficient to encounter the augmented energy requirements in the future [1]. Renewable sources such as bioethanol and biodiesel generated from a range of organic or inorganic wastes by anaerobic fermentation combined with other downstream techniques can provide a competitive solution to energy and environmental dilemma concurrently. Nevertheless, an assortment of future renewable and sustainable energy must enclose diverse carbon-neutral based technologies such as microbial fuel cell in order to attain future energy exigency.

Microbial fuel cells (MFCs) came into light due to their ability to exploit the chemical energy of recyclable organic matter to form electrical energy [2] though the whole reaction which is an association of bacterial metabolism with electrochemical reactions on the anode [3]. Moreover, attribution to electricity generation, wastewater treatment, bioremediation of toxic metals, etc. make this particular technology an attractive alternative for future energy demand [4]. MFCs are bio-electrochemical devices that generally consist of an aerobic/anaerobic cathode chamber(electrode, electron acceptor and catalyst) and an anode chamber(anaerobic; includes an electrode, microorganisms and anolyte) alienated by a proton exchange membrane(PEM) [5]. This membrane allows cations precisely H^+ ions to pass the anode chamber and reach the cathode chamber. The whole external circuit is completed by means of a conductive load that connects both anode and cathode within the chambers (Fig. 1).

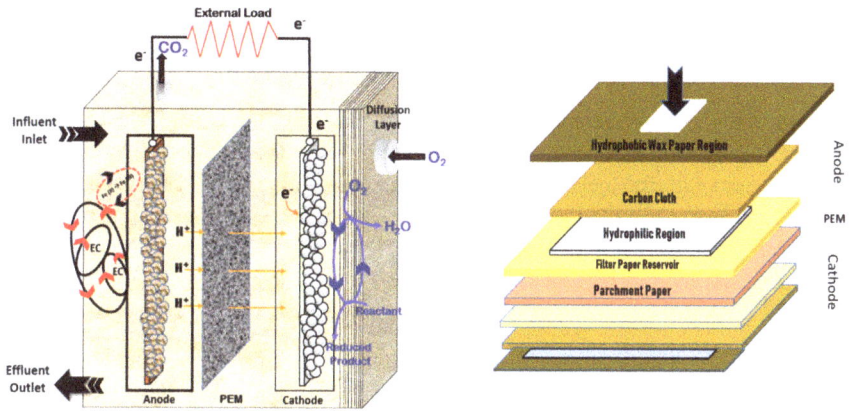

Figure 1. Depicts the construction and working of the paper-based microbial fuel cell.

In-depth study about MFC reveals that on the anode surface, a biofilm develops, which contains electro-active microorganisms capable of transferring electrons generated from the decomposition of organic matter to the electrode. The electrons and protons are then transferred from the organics to the cathodic surface by extracellular electron transfer (EET) and proton exchange membrane (PEM) respectively thereby generating energy for respiration of microorganisms. The electrons are then transferred to the cathode through external electrical connection and merge with oxygen and protons to form H_2O in the presence of a catalyst such as platinum [6, 7]. Rahimnejad et al. [8] illustrated the chemical reactions taking place in both the chambers by equations 1 and 2.

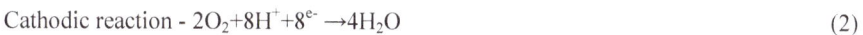

Anodic reaction - $C_2H_4O_2 + 2H_2O \rightarrow 2CO_2 + 8H^+ + 8e^-$ 　　　　　　　　　　 (1)

Cathodic reaction - $2O_2 + 8H^+ + 8e^- \rightarrow 4H_2O$ 　　　　　　　　　　　　　　 (2)

The electrons generated in the bacterial cells transferred directly to the anode or facilitated by a mediator present in the solution [9]. Depending on the mode of electron transfer from the cell to the electrode, microbial fuel cells are categorized as direct electron transfer (DET), mediated electron transfer (MET), electron transfer through nanowire and electron transfer by oxidizing excreted catabolites by microorganisms in the fuel cells [10]. While METs require mediator precisely a chemical to transfer electrons from the bacterial cells to the anode, DETs involve those bacteria that have the capacity to transfer electrons and protons directly into the anode as part of its normal metabolic process such bacteria are precisely called exoelectrogens.

Thus keeping in mind the necessity of microbial fuel cells, the present chapter discusses (a) the role of exoelectrogens in microbial fuel cells, (b) the mechanism and efficacy of paper-based microbial fuel cell and (c) applications of paper-based microbial fuel cells. The future perspective of microbial fuel cells is also stated and this chapter is expected to elucidate the importance of paper-based microbial fuel cells in various domains of biological and engineering aspects.

2.   Exoelectrogens in microbial fuel cell

Exoelectrogens make MFCs the most promising technology in the contemporary energy market due to their capacity to transfer electrons derived from the metabolism of organic matters to the anode. These electrochemically active bacteria possess a unique molecular machinery to transfer the electrons exogenously using a strong oxidizing agent or solid conductor as a final electron acceptor unlike other aerobic and anaerobic bacteria which require oxygen or other chemical compounds for the same. Exoelectrogens exploit an oxidoreductase pathway for direct transfer of electrons through a specialized protein channel onto the anodic surface [11]. This distinctive transport mechanism is carried out by conductive pili, direct contact through a conductive biofilm, or transferring by secreted mediator enzymes. Electrons are generated subsequent to ATP production in electron transport chain (ETC) at the time of oxidative phosphorylation [12]. Further, strong oxidizing agents present in the solution such as iron oxides and manganese oxides act as final electron acceptors to receive those electrons instead of molecular oxygen (O_2) in conventional respiration.

Generally, three types of microbial fuel cells have been identified depending on the types of microorganisms used such as photoautotrophic-type biofuel cells [13], heterotrophic-type biofuel cells [14] and sediment biofuel cells [15].

Photoautotrophic microbial fuel cells involve microorganisms such as photosynthetic bacteria (PSB) or algae/cyanobacteria which utilize light as their source of energy rather than a fuel for the production of electrical energy [16]. Electron transfer process in this type of cells is being facilitated by a photostable mediator. The operation of photoautotrophic microbial fuel cells based on their photosynthetic mechanism (oxygenic or anoxygenic) [12]. These MFCs can function in two operating modes in which firstly, energy can be produced and stored by the cells during illumination followed by the release and processing of energy with a non-photosynthetic biofuel cell or secondly,

Microbial Fuel Cells: Materials and Applications   Materials Research Forum LLC
Materials Research Foundations **46** (2019) 101-124   doi: http://dx.doi.org/10.21741/9781644900116-5

energy can be extracted by a mediator that aids transportation of electrons during illumination.

Tsujimura et al. generated a photoautotrophic microbial fuel cell in which oxidation of water occurs in the anodic compartment to produce dioxygen and protons whereas bilirubin oxidase (BOD) was used as a biocatalyst in the cathodic compartment along with 2,2'-azino-bis(3-ethylbenzothiazoline-6-sulfonate) [ABTS] as a mediator. This MFC generated 0.13 mW power at a potential of 0.26 V and a 500 Ω external resistance [13].

Heterotrophic microbial fuel cells involve those microorganisms which utilize different chemicals such as lactate, pyruvate, sulphate, etc. as their source of energy to generate an electron. These cells are being assisted by a mediator such as iron oxides or manganese oxides. Cooney et al. studied the growth kinetics of *Desulfovibrio desulfuricans Essex* 6 in a microbial fuel cell that produces electrons by utilizing a different source of energy [14]. The researchers involved in the use of both sulphate and lactate as the source of energy for this microbe have observed almost equal growth conditions for both of the energy sources. Hydrogen sulphide produced by the reduction of sulphate was observed not only to inhibit the growth of the microbes but also to narrow down the growth yield and sulphate-specific reduction rate mutually. Likewise, high initial lactate concentration also witnessed to stunt the bacterial growth as well as specific sulphate reduction rates and eventually conferred variable biomass growth yields which might arise due to some tailback in lactate oxidation pathway leading to the production of butanol as a by-product. Apart from these two chemicals, pyruvate was also employed as a carbon source. The authors found that the use of lactate favoured the microbial growth rate and biomass growth yields. A marginally higher current density was produced in case of lactate as compared to pyruvate for equal biomass concentration.

Sediment microbial fuel cells (SMFCs) involve the application of anaerobic microbe sediment in organic matters to which an anode is being inserted. It is then connected to the cathode in the overlying aerobic water through an external circuit. SMFCs produce electrical energy from the electro-potential difference between aerobic water and anaerobic sediments present in the cell [17]. The authors proposed a distinctive prospect to investigate the energy flow pattern through the microbial communities together with their efficiency of collecting energy and the potential for power generation in the natural environment. SMFCs have become favourable as the energy generation source to be readily utilized in different sectors.

A number of microorganisms such as *Geobacter sulfurreducens* [18], *Shewanella oneidensis* [19], *Shewanella putrefaciens* [20, 21], *Escherichia coli* [22], *Rhodoferax ferrireducens* [23], *Pseudomonas aeruginosa, Alcaligenes faecalis, Enterococcus gallinarum, Pseudomonas aeruginosa* and other *Pseudomonas* species, *C. reinhardtii, Synechocystis sp. PCC 6803, Saccharomyces cerevisiae, Rhodopseudomonas palustris* DX-1 [24], and *Ochrobactrum anthropi YZ*-1 [25] etc. have been used as biocatalysts for the production of electrical energy.

G. sulfurreducens is one of the most studied exoelectrogens employed in MFCs. This microbe utilizes various carbon sources for electricity generation and develops a thick biofilm on the anode surface. Primarily, *G. sulfurreducens* secretes riboflavin in a mono-layered biofilm which later associates with outer-membrane c-type cytochromes (OM c-Cysts) to form a complex which initiates electron transfer to the anode surface [26, 27]. As the biofilm develops, this microbe utilizes OM c-Cysts (essentially OmcZ) for electron transport next to the anode surface, while the bacteria respiring away from the electrode generate conductive nanowires (type IV pili) which contribute to the transfer of electrons inside the biofilm and lastly onto the electrode surface [26]. Bond et al. employed *Geobacter sulfurreducens* as biocatalyst in a microbial fuel cell and observed that this particular microbe was capable to transfer electron and proton gained form acetate to cathode efficiently thereby generating electrical energy from bacterial energy [28].

Shewanella oneidensis is also a highly studied exoelectrogen as it flaunts the potential to decrease a variety of electron acceptors [29, 30]. This specific microbe secretes two types of flavin molecules i.e. riboflavin (RF) and flavin mononucleotide (FMN) that combine with cytochromes like OmcA and MtrC respectively as a cofactor to promote electron transfer to the anode surfaces [31]. Hashemi et al. found that *Shewanella oneidensis* MR-1 excreted soluble redox molecules that served as extracellular electron transporter forming almost 70% of the transport from individual bacterial cells to the electrode.

A number of researches have been carried out to identify exoelectrogens and to study their bio-electrochemical activity for decades. Kim et al. performed cyclic voltammetry analysis on the pure strain of *Shewanella putrefaciens* and found that this microorganism retained its electrochemical activity in the presence of nitrate rather than O_2 at an applied potential of +200 mV [20]. This finding led to the conclusion that the performance of a mediator-less microbial fuel cell can be enhanced by incorporating electrochemically active bacteria in presence of nitrate instead of O_2. The same authors in 2002 performed cyclic voltammetry along with a fuel cell type electrochemical cell to monitor the electron transfer from different *Shewanella putrefaciens* and *Escherichia coli* strains to the anode and found both of the methods equally efficient in determining the electrochemical activity of the microorganism without any electrochemical mediators [21]. High concentration of bacteria yielded a high amount of electron transfer thereby generating higher electrical power. The further cyclic volumetric study revealed that electrochemical activity could only be observed in case of anaerobically grown *Shewanella putrefaciens* MR-1, IR-1, and SR-21 strain while aerobically grown *Shewanella putrefaciens* cells as well as aerobically and anaerobically grown *E. coli* cells exhibited no such accomplishments. Likewise, comparable bioelectrochemical activity was also observed in the case of *Escherichia coli* [32].

Park et al. [22] compared electric power production, substrate consumption and growth of E. coli cell in biofuel cells as a biocatalyst. They found that *Escherichia coli* produced 3.1 $\mu A\ cm^{-2}$ electricity by consuming 39 mM of acetate with neutral red electrode whereas, with the native electrode this microbe produced 1.41 $\mu A\ cm^{-2}$ electric power.

Chaudhuri et al. [23] studied another efficient exoelectrogens *Rhodoferax ferrireducens* which are capable of transferring the majority of the electrons gained from the carbon sources such as glucose to the electrode. Though a high amount of electrons was being transferred from the anodic surface to the cathodic surface, there was little possibility to see whether that also affected the energy transfer efficiency as it solely dependent on the current and potential.

Xing et al. [24] also employed phototrophic purple nonsulfur bacteria *Rhodopseudomonas palustris* DX-1 for production of electricity, which is able to generate electricity in microbial fuel cells in MFC and obtained 60 mW/m2 electricity power density. The researchers also compared the electricity densities by using pure culture as well as the mixed culture of bacteria and found enhanced power density in case of the single culture of microorganisms. The authors acquired gigantic achievement by introducing the bio-electrochemical activity of this microbe as it utilizes a wide range of substrates such as volatile acids, yeast extract, and thiosulfate.

Zuo et al. [25] developed a U-tube-shaped MFC to grow a pure culture of exoelectrogenic bacteria *Ochrobactrum anthropi* YZ-1. The researchers observed that this strain was unable to respire using hydrous Fe(III) oxide but could produce 89 mW/m^2 while acetate was employed as the electron donor in the U-tube MFC.

3.   Paper-based microbial fuel cells

Construction of a microbial fuel cell also plays a vital role in the sustenance of this technique in the recent market. As MFCs are frequently fabricated using hazardous materials such as plastic which it may threaten our ecosystem and thus should be avoided. To avoid such problems, recently, paper-based microbial fuel cells made from paper have become a defensible choice [33]. Moreover, due to their light weight and portability, these paper-based MFCs have become a promising backup plan for bioelectricity generation, especially in inaccessible areas. These fuel cells can act ideally even without much effort of the operator and can produce efficient power even after only being plummeted onto the liquid solution containing organic matter.

Moreover, the use of paper can lead the energy sector towards the emergence of light-weighted and eco-friendly electronics along with the auxiliary remunerations of cost-effectiveness and widespread obtainability [34]. Researchers have experimented with different designs of paper MFCs to monitor the amount of generated bioelectricity. Some of these paper-based MFCs such as 2D MFCs, 3D MFCs, microfabricated paper-based MFCs, screen-printed paper microbial fuel cell, stackable, two-chambered, paper-based microbial fuel cell and origami paper-based fuel cell is described in following paragraphs.

Winfield et al. [35] have constructed 2D paper-based MFCs and 3D paper-based MFCs and compared the electricity generating capacity of both the MFCs and found that though both the cells generated current simply by dripping onto the liquid solution. The 3D tetrahedron MFCs generated power for over 2 weeks with the output sufficient for any basic useful applications.

Chouler et al. [36] reported single-component paper MFC that intends at evolving simple, light-weighted single-use sensing device. The membrane-less device was constructed by screen-printing carbon-based electrodes onto a single sheet of paper that also played a role of a separator between the two electrodes. The capillary force of action created a continuous flow. However, the flow pattern and force of flow differed with the range of paper structure. The authors used 1 kΩ resistor to which the electrodes were connected. The exoelectrogens in the biofilm generated electrons (e) and protons (H) that were either transported across the external circuit or diffused through the paper to the cathode. The authors also compared the effect of stacking in this device by folding two pMFCs back-to-back (fpMFC) and found that stacking of two devices together (fpMFC) resulted into a maximum power density and power output of 0.07 μW and 3.0 cm μA cm which were almost 1.7 and 4 times the value obtained with a single cell respectively.

Hashemi et al. [37] constructed a 3D paper-based microbial fuel cell by employing *S. oneidensis* MR-1 as a biocatalyst. The authors allowed the fuel cell to lead the liquid solution through three-dimensional papers which curtailed the need for external force. The cell permitted the flow of the microbial stream and potassium ferricyanide into the chambers where both the chambers were separated by PEM. The positively charged ions were released during the biocatalytic breakdown of the anolyte and reached the cathode. The authors observed the production of 1.3 μW of power and 52.25 μA of current along with the formation of a biofilm on the anode.

Hashemi et al. [37] designed a self-pumped paper-based microbial fuel cell operating under continuous flow condition by the action of capillary motion. The authors employed *Shewanella oneidensis* MR-1 as biocatalyst and conducted this experiment for approximately 5 days without any intrusion and external power source for the duration of the operation. They observed a production of 52.25 μA current and a power density of approximately 25 W/m3 along with biofilm formation on carbon cloth electrode.

Fraiwan et al. [38] presented a micro-fabricated paper-based microbial fuel cell (MFC) generating a maximum power of 5.5 μW/cm. The authors constructed an MFC that involved anodic and cathodic compartments separated by a paper-based proton exchange membrane by intruding sulfonated sodium polystyrene sulfonate and micro-fabricated paper chambers stacked to form multilayer cell by modelling hydrophobic barriers of photoresist. A current of 74 μA was generated immediately after the addition of biomass and catholyte to the MFC. The authors observed that the voltage produced during the electricity production was raised by 1.9 fold when two MFC devices were stacked in series.

Another modification of paper-based microbial fuel cell involves origami based paper microbial fuel cell that blends the skill of origami and the knowledge of MFCs. This device holds the perspective to shift the prototype from traditional microbial fuel cells to flexible paper-based cells. This three-dimensional fuel cell comprises of modular and foldable battery stack generated by 2D paper sheets over high degrees of folding. The multi-layered MFC is constructed on the concept of ninja star-shaped origami design shaped by eight MFC modular blades and is flexible from sharp shuriken to round Frisbee form. Fraiwan et al. [39] reported an origami-based paper microbial fuel cell in which the biomass is added into an inlet of the closed cell stack which was being transported into each MFC module through patterned fluidic concurrently by displaying all the air-cathodes to the air for their cathodic reactions. The device was observed to generate electrical current and potential to lighten an LED for more than 20 min.

In the similar context, Lee et al. [40] designed a fully-integrated stackable 3-D paper-based bacteria-powered fuel cell system that was capable of generating power from microbial metabolism. This particular device was able to deliver on-board energy with one drop of bacteria-containing liquid from wastewater, municipal wastes, biomass and watershed. This device was associated with an air-breathing cathode constructed of paper with activated carbon on the sprayed nickel electrode. The authors gave this device a compact and stackable structure by employing origami techniques.

Fraiwan et al. [41] also developed a stackable and integrable paper-based microbial fuel cell (MFC) for potentially powering on-chip paper-based devices. A cell stack was designed by connecting four MFCs in series by folding them three times. These MFCs were prepared on a T shaped filter paper. Each MFC was constructed by sandwiching multifunctional paper layers for two chambered fuel cell configuration. The device was designed in the form that allowed bacteria-containing anolyte and potassium ferricyanide to flow through predesigned separated fluidic pathways within the paper matrix, filling the reservoir of each device. The MFC stack in series generated a power density of 1.2 μW/cm, which was almost two-fold higher than the earlier reported values on the paper-based MFC stack [37].

4. Different membranes used in paper-based MFCs

The use of microbial fuel cells has created a new era in the field of fuel cell technology and power generation. Due to the unique characteristics (transparency, two ionic oppositely charged diode and flexibility), the paper-based membranes generate an efficient power which can be helpful to operate the low power consuming biosensors and chips with biomedical and healthcare applications.

The transfer of electrons towards the anode and protons towards the cathode across specific membranes generate power due to high reduction potential. The electrons at the end react with oxygen which carries out the reduction process and releases some energy [42, 43].

There are various types of membranes used in paper-based microbial fuel cells(MFCs) for the purpose of power and electricity generation (Table 1).

Table 1: Different types of membranes used in a microbial fuel cell(MFC)

SI No.	Types of Membranes	Composition	Uses	Citation
1	Ceramic Membranes	Ceramic and its composite	Widely use as proton exchanger in low-cost MFCs	[86-88]
2	Polymer Electrolyte Membranes	Metal-Carbon Composite, Metal-Activated Carbon Composite, Metal-Carbon Nanotube(CMT) Composite and Polymer-Carbon Composite etc.	Use as a hybrid membrane for the paper-based microbial fuel cells(MFCs)	[89-91]

3	Graphene and Graphene Oxide-Based Membrane	Composite of Iron and Nitrogen Graphene(Fe-N-G), Graphene Oxide in sulfonated polyether ether ketone(GO-SPEEK) etc.	Serve as the alternative to Nafion membrane used in MFC, Helps to increase the ion exchange property in MFCs	[92-94]
4	Polypyrrole Nanotube Membrane	Polymer of polypyrrole	Use as an anode in MFCs	[95-97]
5	Phenolphthalein Based Membrane	These are multiblock of phenolphthalein based polyarylene ether sulfone nitriles(PESN)	Alkaline fuel cells, Anion exchanger membranes(AEMs)	[98, 99]
6	Polybenzimidazole Membrane	Composite of SBA-15 mesostructured silica functionalized with sulphonic moieties and Polybenzimidazole(PBI)	Use as a separator(Proton exchanger) in wastewater remediation and MFCs	[100, 101]

5. Applications of paper-based MFCs

There are various applications of the microbial fuel cells that have been developed in the past few years due to cutting-edge research in fuel cell technology. Scientists are always trying to make a significant change for the power generation capacity in paper-based microbial fuel cells by certain modifications in the catholyte and anolyte. The generated power helps to trigger the small electric devices, chips and sensors for a defined time period.

The crucial applications of paper-based microbial fuel cells can be broadly classified as below.

Generation of bioelectricity

The paper-based microbial fuel cell (MFC) has been developed by the help of nutrients and food materials for the viability of the bacterial culture which can generate a small amount of power for a prolonged time period [44]. The power can be used to supply energy into small electrical devices with a high thoughtful safety issue and health perspective [45]. The research in the field of bioelectricity generation tells that the cardiac stimulation is also possible by MFC if we can generate a power of 25mWatt.

Rahimnejad and et al. successfully generated the power by using fabricated paper-based MFC and operated one digital clock and ten LED lamps for the duration of 48 hours [46, 47].

Biohydrogen production

The paper-based MFCs can also be designed to produce biohydrogen apart from electricity. These can also be used in the hydrogen economy as a renewable source of hydrogen production to meet the hydrogen demand [45, 48].

The useful hydrocarbons can also be produced from wastewater streams by using paper-based MFCs with the help of electrodes which can generate methane gas by the conversion of carbon present in the effluent stream [49]. During this process, the solid particles are separated by a screening process from the stream and then transferred into a large tank, inside which the anode is coated with a strain of bacteria which purify the water by oxidation and produces electricity [50]. The flow of electrons take place towards the cathode which is coated with a specific group of bacteria hence the carbon dioxide (CO_2), hydrogen (H_2) and electricity are converted into methane gas by the process of electro-methanogenesis [51, 52].

Wastewater treatment

The paper-based MFCs play a vital role in the process of treatment of wastewater as it contains biodegradable organic matters [53, 54]. The wastewaters coming from the household effluents, food and beverage industry, medical and pharmaceutical work fields are allowed to undergo the various biochemical processes in order to remove the metals and metal ions from the leachate [55-58]. The cost-effective biological treatment processes of wastewater with the help of paper-based MFCs are also used for the removal of hydrocarbons, salinity and sulphate and purification of the wastewater [59, 60]. The anaerobic degradation process of wastewater is also helpful for the energy generation and methane gas production after a long-term treatment process [44].

Biosensors

The batteries used in electrochemical sensors are not a continuous source of energy, hence the low power-generating microbial fuel cell (MFC) due to the reaction at the cathode and anode for powering the small elements can be used in sensors [61]. These sensors are used for the detection of toxicity in the environment [62], biological oxygen demand (BOD) [63], bacterial contamination in water [64, 65], hazardous compounds in water [36] and biomedical sensors for the detection of glucose, enzymes and vitamins [66, 67]. These biosensors show a satisfactory performance upon use and can be used for a period of 5 years [45, 68].

MFCs in space

The sources of energy production are very limited and the regular replacement processes of the power sources are not feasible in the space stations so the need of a continuous power source or a source of power generation with a longer period compared to others is essential. The Paper-based MFCs are also used to power spaceships operated remotely which can be triggered by the bioelectricity production due to the wastes [39, 69]. It has been observed that hybrid MFCs are more effective than the lithium-ion power batteries and can supply power to steadily operate low load devices [70]. The small satellites can be operated by the combination of capacitors and microbial fuel cells and can provide real-time information to the data centre [71, 72].

6.   **Future prospective of paper-based MFCs**

The research and development in the field of biotechnology and electronics are going on to overcome the limitations of paper-based microbial fuel cells (MFCs). The principal problem is the low power generation, hence technocrats are much concerned about the enhancement of the generation of a higher power from these processes [73]. The use of nano-electrodes which can increase the surface area and the ultra-capacitor for the storage of energy has also been the main challenge for the scientists [74, 75]. The biotechnologists are working to develop genetically modified microbial strains which can tolerate a wide range of temperature so that these microbial fuel cells can be operated under extreme temperature conditions without slowing down the microbial reactions [76].

The graphene-based nanomaterial plays a great role in the manufacture of small electronics due to its unique properties hence can be used as a solution for microbial fuel cells (MFCs) and the power generation can be enhanced [77]. The paper-based electrodes can be congregated from functionalized 3-dimensional graphene [78]. The graphene nanoparticle plays a vital role in the medicine, biomaterial, biosensors and small biomedical instrumentation hence research for the self-power source by using microbial fuel cell is the challenge for the scientific community [68, 79]. The generation of power for vehicles and automobiles by using fuel cells to overcome the fuel scarcity as well as environmental management is a new key challenge for the technologist [80].

The advancement in microfluidics can help the researchers to produce small paper-based MFC devices with high surface to volume ratio and better performance [81, 82]. The electron transport mechanism is limited to the bacterial membrane so scientists are looking for other permeable membranes with better strength which can be used in MFCs [83-85].

Conclusion

The alternative sources of power generation to massive and expensive batteries are stackable paper-based microbial fuel cells which can produce a power density up to 1.6 Volt by the metabolism of the microbial biomass used in it. The 2-dimensional paper sheets are helpful to produce the 3-dimensional microbial fuel cells stack which has a multifunctional layer forming ability due to its high degree of folding. This paper-based microbial fuel cell has made a revolutionary change in the field of power generation by fuel cell technology as it is very simple in design, the production cost is low and easy to use as a power source for low power chips and sensors.

Reference

[1]   S.P. Sukhatme, Meeting India's future needs of electricity through renewable energy sources, Curr. Sci. 101 (2011) 624-630.

[2]   N. Thepsuparungsikul, N. Phonthamachai, H.Y. Ng, Multi-walled carbon nanotubes as electrode material for microbial fuel cells, Water Sci Technol. 65 (2012) 1208-1214. https://doi.org/10.2166/wst.2012.956

[3]   B.E. Logan, Scaling up microbial fuel cells and other bioelectrochemical systems, Appl. Microbiol. Biotechnol. 85 (2010) 1665-1671. https://doi.org/10.1007/s00253-009-2378-9

[4]   J. Chouler, M. Di Lorenzo, Water quality monitoring in developing countries; can microbial fuel cells be the answer? Biosensors (Basel). 5 (2015) 450-470. https://doi.org/10.3390/bios5030450

[5]   M. Di Lorenzo, A.R. Thomson, K. Schneider, P.J. Cameron, I. Ieropoulos, A small-scale air-cathode microbial fuel cell for on-line monitoring of water quality, Biosens Bioelectron. 62 (2014) 182-188. https://doi.org/10.1016/j.bios.2014.06.050

[6]   Y.C. Yong, Y.Y. Yu, X. Zhang, H. Song, Highly active bidirectional electron transfer by a self-assembled electroactive reduced-graphene oxide-hybridized biofilm, Angew. Chem. Int. Ed. Engl. 53 (2014) 4480-4483. https://doi.org/10.1002/anie.201400463

[7]   C.I. Torres, A. Kato Marcus, B.E. Rittmann, Proton transport inside the biofilm limits electrical current generation by anoderespiring bacteria, Biotechnol. Bioeng. 100 (2008) 872-881. https://doi.org/10.1002/bit.21821

[8]   M. Rahimnejad, A. Adhami, S. Darvari, A. Zirepour, S.E. Oh, Microbial fuel cell
        as new technology for bioelectricity generation: A review, Alexandria Eng. J. 54
        (2015) 745-756. https://doi.org/10.1016/j.aej.2015.03.031

[9]   R.A. Bullen, T.C. Arnot, J.B. Lakeman, F.C. Walsh, Biofuel cells and their
        development, Biosens. Bioelectron. 21 (2006) 2015-2045.
        https://doi.org/10.1016/j.bios.2006.01.030

[10]   M. Mustakeem, Electrode materials for microbial fuel cells: nanomaterial
        approach, Mater. Renew. Sustain Energy. 4 (2015) 22.
        https://doi.org/10.1007/s40243-015-0063-8

[11]   R. Chandra, S.V. Mohan, P.S. Roberto, B.E. Ritmann, R.A.S. Cornejo,
        Biophotovoltaics: conversion of light energy to bioelectricity through
        photosynthetic microbial fuel cell technology, Microbial Fuel Cell, Springer, 2018,
        373-387.

[12]   M. Rosenbaum, F. Aulenta, M. Villano, L.T. Angenent, Cathodes as electron
        donors for microbial metabolism: which extracellular electron transfer
        mechanisms are involved?, Bioresour Technol. 102 (2011) 324-333.
        https://doi.org/10.1016/j.biortech.2010.07.008

[13]   S. Tsujimura, A. Wadano, K. Kano, T. Ikeda, Photosynthetic bioelectrochemical
        cell utilizing cyanobacteria and water-generating oxidase, Enzyme Microb.
        Technol. 29 (2001) 225-231. https://doi.org/10.1016/S0141-0229(01)00374-X

[14]   M.J. Cooney, E. Roschi, I.W. Marison, C. Comminellis, U. Von Stockar,
        Physiologic studies with the sulfate-reducing bacterium Desulfovibrio
        desulfuricans: evaluation for use in a biofuel cell, Enzyme Microb. Technol. 18
        (1996) 358-365. https://doi.org/10.1016/0141-0229(95)00132-8

[15]   K. Rabaey, G. Lissens, S.D. Siciliano, W. Verstraete, A microbial fuel cell capable
        of converting glucose to electricity at high rate and efficiency, Biotechnol Lett. 25
        (2003) 1531-1535. https://doi.org/10.1023/A:1025484009367

[16]   L. Mao, W.S. Verwoerd, Selection of organisms for systems biology study of
        microbial electricity generation: a review, Int. J. Energy Environ. Eng. 4 (2013)
        17. https://doi.org/10.1186/2251-6832-4-17

[17]   T.K. Sajana, M.M. Ghangrekar, A. Mitra, Effect of pH and distance between
        electrodes on the performance of a sediment microbial fuel cell, Water Sci
        Technol. 68 (2013) 537-543. https://doi.org/10.2166/wst.2013.271

[18]   D.R. Bond, D.E. Holmes, L.M. Tender, D.R. Lovley, Electrode-reducing microorganisms that harvest energy from marine sediments, Science. 295 (2002) 483-485. https://doi.org/10.1126/science.1066771

[19]   D.R. Bond, D.R. Lovley, Electricity production by Geobacter sulfurreducens attached to electrodes, Appl. Environ. Microbiol. 69 (2003) 1548-1555. https://doi.org/10.1128/AEM.69.3.1548-1555.2003

[20]   B.H. Kim, T. Ikeda, H.S. Park, H.J. Kim, M.S. Hyun, K. Kano, K. Takagi, H. Tatsumi, Electrochemical activity of an Fe (III)-reducing bacterium, Shewanella putrefaciens IR-1, in the presence of alternative electron acceptors, Biotechnol. Tech. 13 (1999) 475-478. https://doi.org/10.1023/A:1008993029309

[21]   H.J. Kim, H.S. Park, M.S. Hyun, I.S. Chang, M. Kim, B.H. Kim, A mediator-less microbial fuel cell using a metal reducing bacterium, Shewanella putrefaciens, Enzyme Microb. Technol. 30 (2002) 145-152. https://doi.org/10.1016/S0141-0229(01)00478-1

[22]   D.H. Park, S.K. Kim, I.H. Shin, Y.J. Jeong, Electricity production in biofuel cell using modified graphite electrode with neutral red, Biotechnol. Lett. 22 (2000) 1301-1304. https://doi.org/10.1023/A:1005674107841

[23]   S.K. Chaudhuri, D.R. Lovley, Electricity generation by direct oxidation of glucose in mediatorless microbial fuel cells, Nat. Biotechnol. 21 (2003) 1229-1232. https://doi.org/10.1038/nbt867

[24]   D. Xing, Y. Zuo, S. Cheng, J.M. Regan, B.E. Logan, Electricity generation by Rhodopseudomonas palustris DX-1, Environ. Sci. Technol. 42 (2008) 4146-4151. https://doi.org/10.1021/es800312v

[25]   Y. Zuo, D. Xing, J.M. Regan, B.E. Logan, Isolation of the exoelectrogenic bacterium Ochrobactrum anthropi YZ-1 by using a U-tube microbial fuel cell, Appl. Environ. Microbiol. 74 (2008) 3130-3137. https://doi.org/10.1128/AEM.02732-07

[26]   R. Kumar, L. Singh, A.W. Zularisam, Exoelectrogens: recent advances in molecular drivers involved in extracellular electron transfer and strategies used to improve it for microbial fuel cell applications, Renew. Sust. Energ. Rev. 56 (2016) 1322-1336. https://doi.org/10.1016/j.rser.2015.12.029

[27]   Y. Yang, Y. Ding, Y. Hu, B. Cao, S.A. Rice, S. Kjelleberg, H. Song, Enhancing bidirectional electron transfer of Shewanella oneidensis by a synthetic flavin pathway, ACS Synth. Biol. 4 (2015) 815-823. https://doi.org/10.1021/sb500331x

[28] D.R. Bond, D.R. Lovley, Electricity production by Geobacter sulfurreducens attached to electrodes, Appl. Environ. Microbiol. 69 (2003) 1548-1555. https://doi.org/10.1128/AEM.69.3.1548-1555.2003

[29] F. Kracke, I. Vassilev, J.O. Krömer, Microbial electron transport and energy conservation–the foundation for optimizing bioelectrochemical systems, Front Microbiol. 6 (2015) 575. https://doi.org/10.3389/fmicb.2015.00575

[30] P. Parameswaran, T. Bry, S.C. Popat, B.G. Lusk, B.E. Rittmann, C.I. Torres, Kinetic, electrochemical, and microscopic characterization of the thermophilic, anode-respiring bacterium Thermincola ferriacetica, Environ. Sci. Technol. 47 (2013) 4934-4940. https://doi.org/10.1021/es400321c

[31] A. Okamoto, S. Kalathil, X. Deng, K. Hashimoto, R. Nakamura, K.H. Nealson, Cell-secreted flavins bound to membrane cytochromes dictate electron transfer reactions to surfaces with diverse charge and pH, Scientific Reports. 4 (2014) 5628. https://doi.org/10.1038/srep05628

[32] D.A. Gradskov, I.A. Kazarinov, V.V. Ignatov, Bioelectrochemical oxidation of glucose with bacteria Escherichia coli, Russ. J. Electrochem. 37 (2001) 1216-1219. https://doi.org/10.1023/A:1012727918599

[33] S. Choi, Microscale microbial fuel cells: advanceds and challenges, Biosens. Bioelectron. 69 (2015) 8-25. https://doi.org/10.1016/j.bios.2015.02.021

[34] A.K. Yetisen, M.S. Akram, C.R. Lowe, Paper-based microfluidic point-of-care diagnostic devices, Lab Chip. 13 (2013) 2210-2251. https://doi.org/10.1039/c3lc50169h

[35] J. Winfield, P. Milani, J. Greenman, I. Ieropoulos, Passive feeding in paper-based microbial fuel cells, ECS Trans. 85 (2018) 1193-1200. https://doi.org/10.1149/08513.1193ecst

[36] J. Chouler, Á. Cruz-Izquierdo, S. Rengaraj, J.L. Scott, M. Di Lorenzo, A screen-printed paper microbial fuel cell biosensor for detection of toxic compounds in water, Biosens. Bioelectron.102 (2018) 49-56. https://doi.org/10.1016/j.bios.2017.11.018

[37] N. Hashemi, J.M. Lackore, F. Sharifi, P.J. Goodrich, M.L. Winchell, N. Hashemi, A paper-based microbial fuel cell operating under continuous flow condition, Technology. 4 (2016) 98-103. https://doi.org/10.1142/S2339547816400124

[38] A. Fraiwan, S. Mukherjee, S. Sundermier, H.-S. Lee, S. Choi, A paper-based microbial fuel cell: Instant battery for disposable diagnostic devices, Biosens. Bioelectron. 49 (2013) 410-414. https://doi.org/10.1016/j.bios.2013.06.001

[39] A. Fraiwan, S. Choi, A stackable, two-chambered, paper-based microbial fuel cell, Biosens. Bioelectron. 83 (2016) 27-32. https://doi.org/10.1016/j.bios.2016.04.025

[40] H. Lee, S. Choi, An origami paper-based bacteria-powered battery, Nano Energy. 15 (2015) 549-557. https://doi.org/10.1016/j.nanoen.2015.05.019

[41] A. Fraiwan, S. Choi, A stackable, two-chambered, paper-based microbial fuel cell, Biosens. Bioelectron. 83 (2016) 27-32. https://doi.org/10.1016/j.bios.2016.04.025

[42] S.K. Chaudhuri, D.R. Lovley, Electricity generation by direct oxidation of glucose in mediatorless microbial fuel cells, Nat. Biotechnol. 21 (2003) 1229-1232. https://doi.org/10.1038/nbt867

[43] B.E. Logan, Exoelectrogenic bacteria that power microbial fuel cells, Nat. Rev. Microbiol. 7 (2009) 375-381. https://doi.org/10.1038/nrmicro2113

[44] A.S. Mathuriya, V.N. Sharma, Bioelectricity production from paper industry waste using a microbial fuel cell by Clostridium species, J. Biochem. Tech. 1 (2009) 49-52.

[45] M. Rahimnejad, A. Adhami, S. Darvari, A. Zirepour, S.E. Oh, Microbial fuel cell as new technology for bioelectricity generation: A review, Alexandria Eng. J. 54 (2015) 745-756. https://doi.org/10.1016/j.aej.2015.03.031

[46] H.M. Singh, A.K. Pathak, K. Chopra, V.V. Tyagi, S. Anand, R. Kothari, Microbial fuel cells: a sustainable solution for bioelectricity generation and wastewater treatment, Biofuels. (2018) 1-21. https://doi.org/10.1080/17597269.2017.1413860

[47] A.S. Mathuriya, J.V. Yakhmi, Microbial fuel cells–Applications for generation of electrical power and beyond, Crit. Rev. Microbiol. 42 (2016) 127-143. https://doi.org/10.3109/1040841X.2014.905513

[48] A. Ghimire, L. Frunzo, F. Pirozzi, E. Trably, R. Escudie, P.N.L. Lens, G. Esposito, A review on dark fermentative biohydrogen production from organic biomass: process parameters and use of by-products, Appl. Energy. 144 (2015) 73-95. https://doi.org/10.1016/j.apenergy.2015.01.045

[49] W.M. Budzianowski, A review of potential innovations for production, conditioning and utilization of biogas with multiple-criteria assessment, Renew. Sust. Energ. Rev. 54 (2016) 1148-1171. https://doi.org/10.1016/j.rser.2015.10.054

[50] C. Sakdaronnarong, A. Ittitanakam, W. Tanubumrungsuk, S. Chaithong, S. Thanosawan, N. Sinbuathong, C. Jeraputra, Potential of lignin as a mediator in combined systems for biomethane and electricity production from ethanol stillage wastewater, Renew. Energ. 76 (2015) 242-248. https://doi.org/10.1016/j.renene.2014.11.009

[51] A. Kadier, Y. Simayi, P. Abdeshahian, N.F. Azman, K. Chandrasekhar, M.S. Kalil, A comprehensive review of microbial electrolysis cells (MEC) reactor designs and configurations for sustainable hydrogen gas production, Alexandria Eng. J. 55 (2016) 427-443. https://doi.org/10.1016/j.aej.2015.10.008

[52] D.R. Lovley, K.P. Nevin, Microbial production of multi-carbon chemicals and fuels from water and carbon dioxide using electric current, US Patent, 9856449, January 2, 2018.

[53] Z. Xu, Y. Liu, I. Williams, Y. Li, F. Qian, H. Zhang, D. Cai, L. Wang, B. Li, Disposable self-support paper-based multi-anode microbial fuel cell (PMMFC) integrated with power management system (PMS) as the real time "shock" biosensor for wastewater, Biosens. Bioelectron. 85 (2016) 232-239. https://doi.org/10.1016/j.bios.2016.05.018

[54] A. Escapa, M.I. San-Martín, R. Mateos, A. Morán, Scaling-up of membraneless microbial electrolysis cells (MECs) for domestic wastewater treatment: Bottlenecks and limitations, Bioresour. Technol. 180 (2015) 72-78. https://doi.org/10.1016/j.biortech.2014.12.096

[55] H. Boghani, J.R. Kim, R.M. Dinsdale, A.J. Guwy, G.C. Premier, Reducing the burden of food processing washdown wastewaters using microbial fuel cells, Biochem. Eng. J. 117 (2017) 210-217. https://doi.org/10.1016/j.bej.2016.10.017

[56] W. Yang, J. Li, Q. Fu, L. Zhang, X. Zhu, Q. Liao, A simple method for preparing a binder-free paper-based air cathode for microbial fuel cells, Bioresour. Technol. 241 (2017) 325-331. https://doi.org/10.1016/j.biortech.2017.05.063

[57] F. M. Ramírez, H. Addi, F.J. H. Fernández, C. Godínez, A. Pérez de los Ríos, E.M. Lotfi, M. El Mahi, L.J. Lozano Blanco, Air breathing cathode-microbial fuel cell with separator based on ionic liquid applied to slaughterhouse wastewater treatment and bio-energy production, J. Chem. Technol. Biotechnol. 92 (2017) 642-648. https://doi.org/10.1002/jctb.5045

[58] J.M. Sonawane, S.B. Adeloju, P.C. Ghosh, Landfill leachate: a promising substrate for microbial fuel cells, Int. J. Hydrogen Energy. 42 (2017) 23794-23798. https://doi.org/10.1016/j.ijhydene.2017.03.137

[59]   P. Jain, M. Sharma, P. Dureja, P.M. Sarma, B. Lal, Bioelectrochemical approaches for removal of sulfate, hydrocarbon and salinity from produced water, Chemosphere 166 (2017) 96-108. https://doi.org/10.1016/j.chemosphere.2016.09.081

[60]   B. Matturro, C. Cruz Viggi, F. Aulenta, S. Rossetti, Cable bacteria and the bioelectrochemical snorkel: the natural and engineered facets playing a role in hydrocarbons degradation in marine sediments, Front. Microbiol. 8 (2017) 952. https://doi.org/10.3389/fmicb.2017.00952

[61]   A.T. Vicente, A. Araújo, D. Gaspar, L. Santos, A.C. Marques, M.J. Mendes, L. Pereira, E. Fortunato, R. Martins, Optoelectronics and bio devices on paper powered by solar cells, Nanostructured Solar Cells, InTech. 2017. https://doi.org/10.5772/66695

[62]   J.Z. Sun, G.P. Kingori, R.W. Si, D.D. Zhai, Z.H. Liao, D.Z. Sun, T. Zheng, Y.C. Yong, Microbial fuel cell-based biosensors for environmental monitoring: a review, Water Sci. Technol.71 (2015) 801-809. https://doi.org/10.2166/wst.2015.035

[63]   Y. Li, J. Sun, J. Wang, C. Bian, J. Tong, Y. Li, S. Xia, A microbial electrode based on the co-electrodeposition of carboxyl graphene and Au nanoparticles for BOD rapid detection, Biochem. Eng. J. 123 (2017) 86-94. https://doi.org/10.1016/j.bej.2017.03.015

[64]   S. Rengaraj, Á.C. Izquierdo, J.L. Scott, M. Di Lorenzo, Impedimetric paper-based biosensor for the detection of bacterial contamination in water, Sens. Actuators B Chem. 265 (2018) 50-58. https://doi.org/10.1016/j.snb.2018.03.020

[65]   A. Elmekawy, H.M. Hegab, D. Pant, C.P. Saint, Bio-analytical applications of microbial fuel cell–based biosensors for onsite water quality monitoring, J. Appl. Microbiol. 124 (2018) 302-313. https://doi.org/10.1111/jam.13631

[66]   H. Guo, M.H. Yeh, Y. Zi, Z. Wen, J. Chen, G. Liu, C. Hu, Z.L. Wang, Ultralight cut-paper-based self-charging power unit for self-powered portable electronic and medical systems, ACS Nano. 11 (2017) 4475-4482. https://doi.org/10.1021/acsnano.7b00866

[67]   F. Arduini, S. Cinti, V. Scognamiglio, D. Moscone, Paper-based electrochemical devices in biomedical field: recent advances and perspectives, 77 (2017). ISSN 0166-526X. http://dx.doi.org/10.1016/bs.coac.2017.06.005. https://doi.org/10.1016/bs.coac.2017.06.005

[68]   J.P. Esquivel, J. Buser, C.W. Lim, C. Dominguez, S. Rojas, P. Yager, N. Sabate, Single-use paper-based hydrogen fuel cells for point-of-care diagnostic applications, J. Power Sources 342 (2017) 442-451. https://doi.org/10.1016/j.jpowsour.2016.12.085

[69]   H. Wang, J.D. Park, Z.J. Ren, Practical energy harvesting for microbial fuel cells: a review, Environ. Sci. Technol. 49 (2015) 3267-3277. https://doi.org/10.1021/es5047765

[70]   A.S. Commault, O. Laczka, N. Siboni, B. Tamburic, J.R. Crosswell, J.R. Seymour, P.J. Ralph, Electricity and biomass production in a bacteria-Chlorella based microbial fuel cell treating wastewater, J. Power Sources. 356 (2017) 299-309. https://doi.org/10.1016/j.jpowsour.2017.03.097

[71]   R.C. Tyce, J.W. Book, L.M. Tender, Microbial fuel cell power systems, US Patent, 20100081014, April 2 2010.

[72]   P.S. Schrader, C.E. Reimers, P. Girguis, J. Delaney, C. Doolan, M. Wolf, D. Green, Independent benthic microbial fuel cells powering sensors and acoustic communications with the MARS underwater observatory, J. Atmospheric Ocean. Technol. 33 (2016) 607-617. https://doi.org/10.1175/JTECH-D-15-0102.1

[73]   J.M. Sonawane, A. Yadav, P.C. Ghosh, S.B. Adeloju, Recent advances in the development and utilization of modern anode materials for high performance microbial fuel cells, Biosens. Bioelectron. 90 (2017) 558-576. https://doi.org/10.1016/j.bios.2016.10.014

[74]   M.D. Stoller, S. Park, Y. Zhu, J. An, R.S. Ruoff, Graphene-based ultracapacitors, Nano Lett. 8 (2008) 3498-3502. https://doi.org/10.1021/nl802558y

[75]   F. Yu, C. Wang, J. Ma, Capacitance-enhanced 3D graphene anode for microbial fuel cell with long-time electricity generation stability, Electrochim. Acta. 259 (2018) 1059-1067. https://doi.org/10.1016/j.electacta.2017.11.038

[76]   E. Heidrich, J. Dolfing, M.J. Wade, W.T. Sloan, C. Quince, T.P. Curtis, Temperature, inocula and substrate: contrasting electroactive consortia, diversity and performance in microbial fuel cells, Bioelectrochem. 119 (2018) 43-50. https://doi.org/10.1016/j.bioelechem.2017.07.006

[77]   M. Mohammadifar, J. Zhang, I. Yazgan, O. Sadik, S. Choi, Power-on-paper: origami-inspired fabrication of 3-D microbial fuel cells, Renew. Energ. 118 (2018) 695-700. https://doi.org/10.1016/j.renene.2017.11.059

[78] Y. Zhang, J. Xiao, Q. Lv, L. Wang, X. Dong, M. Asif, J. Ren, W. He, Y. Sun, F. Xiao, S. Wang, In situ electrochemical sensing and real-time monitoring live cells based on freestanding nanohybrid paper electrode assembled from 3D functionalized graphene framework, ACS Appl. Mater. Interfaces. 9 (2017) 38201-38210. https://doi.org/10.1021/acsami.7b08781

[79] S. Priyadarsini, S. Mohanty, S. Mukherjee, S. Basu, M. Mishra, Graphene and graphene oxide as nanomaterials for medicine and biology application, J. Nanostructure Chem. 8 (2018) 123-137. https://doi.org/10.1007/s40097-018-0265-6

[80] M. Ehsani, Y. Gao, S. Longo, K. Ebrahimi, Modern electric, hybrid electric, and fuel cell vehicles, CRC Press, Boca Raton. (2018).

[81] E. Kjeang, N. Djilali, D. Sinton, Microfluidic fuel cells: a review, J. Power Sources. 186 (2009) 353-369. https://doi.org/10.1016/j.jpowsour.2008.10.011

[82] F. Qian, Z. He, M.P. Thelen, Y. Li, A microfluidic microbial fuel cell fabricated by soft lithography, Bioresour. Technol. 102 (2011) 5836-5840. https://doi.org/10.1016/j.biortech.2011.02.095

[83] M.D. Khan, N. Khan, S. Sultana, R. Joshi, S. Ahmed, E. Yu, K. Scott, A. Ahmad, M.Z. Khan, Bioelectrochemical conversion of waste to energy using microbial fuel cell technology, Process Biochem. 57 (2017) 141-158. https://doi.org/10.1016/j.procbio.2017.04.001

[84] S.V. Mohan, G. Velvizhi, J.A. Modestra, S. Srikanth, Microbial fuel cell: critical factors regulating bio-catalyzed electrochemical process and recent advancements, Renew. Sust. Energ. Rev. 40 (2014) 779-797. https://doi.org/10.1016/j.rser.2014.07.109

[85] R. Kumar, L. Singh, A.W. Zularisam, Exoelectrogens: recent advances in molecular drivers involved in extracellular electron transfer and strategies used to improve it for microbial fuel cell applications, Renew. Sust. Energ. Rev. 56 (2016) 1322-1336. https://doi.org/10.1016/j.rser.2015.12.029

[86] G. Pasternak, J. Greenman, I. Ieropoulos, Comprehensive study on ceramic membranes for low-cost microbial fuel cells, ChemSusChem. 9 (2016) 88-96. https://doi.org/10.1002/cssc.201501320

[87] V. Yousefi, D. M. Kalhori, A. Samimi, Ceramic-based microbial fuel cells (MFCs): a review, Int. J. Hydrogen Energ. 42 (2017) 1672-1690. https://doi.org/10.1016/j.ijhydene.2016.06.054

[88] J. Winfield, L.D. Chambers, J. Rossiter, I. Ieropoulos, Comparing the short and long term stability of biodegradable, ceramic and cation exchange membranes in microbial fuel cells, Bioresour. Technol. 148 (2013) 480-486. https://doi.org/10.1016/j.biortech.2013.08.163

[89] E. Antolini, Composite materials for polymer electrolyte membrane microbial fuel cells, Biosens. Bioelectron. 69 (2015) 54-70. https://doi.org/10.1016/j.bios.2015.02.013

[90] S. Das, K. Dutta, D. Rana, Polymer electrolyte membranes for microbial fuel cells: a review, Polym. Rev. (2018) 1-20. https://doi.org/10.1080/15583724.2017.1418377

[91] K. Dutta, Polymer-inorganic nanocomposites for polymer electrolyte membrane fuel cells, Polymer-Engineered Nanostructures for Advanced Energy Applications, Springer, Cham. (2017) pp. 577-606.

[92] Y.C. Yong, X.C. Dong, M.B. Chan-Park, H. Song, P. Chen, Macroporous and monolithic anode based on polyaniline hybridized three-dimensional graphene for high-performance microbial fuel cells, ACS Nano. 6 (2012) 2394-2400. https://doi.org/10.1021/nn204656d

[93] Y. Zhang, G. Mo, X. Li, W. Zhang, J. Zhang, J. Ye, X. Huang, C. Yu, A graphene modified anode to improve the performance of microbial fuel cells, J. Power Sources. 196 (2011) 5402-5407. https://doi.org/10.1016/j.jpowsour.2011.02.067

[94] S. Khilari, S. Pandit, M.M. Ghangrekar, D. Pradhan, D. Das, Graphene oxide-impregnated PVA–STA composite polymer electrolyte membrane separator for power generation in a single-chambered microbial fuel cell, Ind. Eng. Chem. Res. 52 (2013) 11597-11606. https://doi.org/10.1021/ie4016045

[95] C.E. Zhao, J. Wu, S. Kjelleberg, J.S.C. Loo, Q. Zhang, Employing a flexible and low-cost polypyrrole nanotube membrane as an anode to enhance current generation in microbial fuel cells, Small. 11 (2015) 3440-3443. https://doi.org/10.1002/smll.201403328

[96] Y. Zou, J. Pisciotta, I.V. Baskakov, Nanostructured polypyrrole-coated anode for sun-powered microbial fuel cells, Bioelectrochemistry. 79 (2010) 50-56. https://doi.org/10.1016/j.bioelechem.2009.11.001

[97] Y. Zou, C. Xiang, L. Yang, L.X. Sun, F. Xu, Z. Cao, A mediatorless microbial fuel cell using polypyrrole coated carbon nanotubes composite as anode material, Int. J. Hydrogen Energ. 3 3 (2 0 0 8) 4 8 5 6 – 4 8 6 2.

[98] A.N. Lai, L.S. Wang, C.X. Lin, Y.Z. Zhuo, Q.G. Zhang, A.M. Zhu, Q.L. Liu, Phenolphthalein-based poly(arylene ether sulfone nitrile) s multiblock copolymers as anion exchange membranes for alkaline fuel cells, ACS Appl. Mater. Interfaces. 7 (2015) 8284-8292. https://doi.org/10.1021/acsami.5b01475

[99] E.N. Hu, C.X. Lin, F.H. Liu, X.Q. Wang, Q.G. Zhang, A.M. Zhu, Q.L. Liu, Poly(arylene ether nitrile) anion exchange membranes with dense flexible ionic side chain for fuel cells, J. Membr. Sci Technol. 550 (2018) 254-265. https://doi.org/10.1016/j.memsci.2018.01.010

[100] S. Angioni, L. Millia, G. Bruni, D. Ravelli, P. Mustarelli, E. Quartarone, Novel composite polybenzimidazole-based proton exchange membranes as efficient and sustainable separators for microbial fuel cells, J. Power Sources. 348 (2017) 57-65. https://doi.org/10.1016/j.jpowsour.2017.02.084

[101] A. Kalathil, A. Raghavan, B. Kandasubramanian, Polymer fuel cell based on polybenzimidazole membrane: A Review, Polym. Plast. Technol. Eng. (2018) 1-33. https://doi.org/10.1080/03602559.2018.1482919

Microbial Fuel Cells: Materials and Applications   Materials Research Forum LLC
Materials Research Foundations **46** (2019) 125-150   doi: http://dx.doi.org/10.21741/9781644900116-6

Chapter 6

Carbon Nanotube Based Anodes and Cathodes for Microbial Fuel Cells

Naveen Patel[1*], Dhananjai Rai[2], Deepak Chauhan[2], Shraddha Shahane[1], Umesh Mishra[1], and Biswanath Bhunia[3]

[1]Department of Civil Engineering, National Institute of Technology, Agartala, Agartala-799046, Tripura, India

[2]Department of Civil Engineering, Bundelkhand Institute of Engineering and Technology, Jhansi, Jhansi-284128, Uttar Pradesh, India

[3]Department of Bio Engineering, National Institute of Technology Agartala, Agartala-799046, Tripura, India

naveenrbverma30@gmail.com

Abstract

Microbial fuel cell (MFC) is one of the natural cordial efficient power vitality sources which have the capability to convert chemical energy into electrical energy from wastewater and microorganisms as a biocatalyst. However, the low power production and the high cost of electrodes have limited application of MFCs. One of the important factors which affect the overall performance of MFC is an electrode. The carbon nanotube has become a potent electrode material owing to its exceptional features. This book chapter provides an overview of electrode materials based on carbon nanotube for MFC operations, which will be the promising candidates for better MFC operations and other bio-electrochemical systems. Hence, these electrodes will ultimately help in achieving sustainable water/wastewater treatment and bioenergy production.

Keywords

Biocatalyst, Electrode, Carbon Nanotube, Bio-Electrochemical Systems and Bioenergy

List of Abbreviation

COD   Chemical oxygen demand

CE   Coulombic efficiency

CHIT   Chitosan

CNTs   Carbon nanotubes

Microbial Fuel Cells: Materials and Applications   Materials Research Forum LLC
Materials Research Foundations **46** (2019) 125-150   doi: http://dx.doi.org/10.21741/9781644900116-6

CVD	Chemical vapour deposition
DMFC	Direct methanol fuel cell
HNQ	2-hydroxy-1,4-naphthoquinone
LbL	Layer-by-layer
MFC	Microbial fuel cell
MWCNTs	Multi-walled CNTs
NPs	Nanoparticles
NR	Neutral red
NRES	Non-renewable energy sources
OCV	Open circuit voltage
ORR	Oxygen reduction reaction
PANI	Polyaniline
Pc	Phthalocyanine
PEM	Proton exchange membrane
PEI	Polyethyleneimine
PEMFC	Proton exchange membrane fuel cell
PPy	Polypyrrole
Pt	Platinum
SSLbL	Spin-spray layer-by-layer
SWCNTs	Single-walled CNTs
TMPP	Tetramethoxyphenylporphyrin

Contents

1.   Introduction

In modern days energy issues have earned a lot of attraction all around the globe. For the world to thrive, perfect, proficient and reasonable energy administrations must be accessible. Non-renewable energy sources (NRES) are the major energy producing resources, near about 87% of the total energy generation around the world originates from NRES, which are currently being depleted, which may lead a worldwide energy emergency in the near future [1]. In addition to this, fossil fuel combustion emits greenhouse gases and carbon dioxide which may lead to the rise of the sea water level and global climate change, thus threatening food security [2]. The increase in the price of fossil fuel due to high demand, energy scarcity and global climate change problems have started the question about the need for development of other sustainable energy sources. Nuclear power delivers a carbon-free method of energy production, but because of non-availability of proper disposal techniques of radioactive wastes and leakages, it also becomes a serious threat to the society [3]. Energy sources such as hydropower, solar energy, wind power and biomass are carbon-neutral renewable energy sources which are continually being researched [1]. From the different energy alternatives, one of the proposed future energy sources is biomass, as it is carbon neutral. However, the combustion process, which creates pollution, is currently being utilised for biomass energy development [1]. Therefore, new methods without involving the combustion process are required to be developed to capture the energy from biomass and to provide sustainable energy for a global society without creating any pollution problems. Microorganisms can be used for channelizing the electrons and energy, which leads to the production of methane, hydrogen and electricity [4]. Electricity, among these energy forms, is considered to be more attractive because of its ability to be utilised directly without involving any conversion process. To the present date, wastewater is considered to be one of the most primarily available sources of biomass. In recent years, a new tendency has been developed where researchers have started considering wastewater not as a waste only which needs to be discarded but it can also be harvested as a source of energy [5]. Therefore, to find a technology that allows effective wastewater treatment to generate electricity directly from wastewater has become very important.

1.2   Microbial Fuel Cell Technology

The microbial fuel cell (MFC) is a widely used technology as it is eco-friendly and applicable to wastewater treatment with simultaneous recovery of electrical energy. MFC uses bacteria, which have the ability to convert organic matters into energy, carbon dioxide and water directly. The bacterial growth and their metabolism in MFCs can be promoted and maintained by the energy generated by these. However, MFCs allow a portion of this energy to be diverted and harvested which can be further utilized for developing a biofilm on the anode. Accordingly, MFC could be an ideal innovation for wastewater treatment, sustainable power source generation and convenient power supply system [6, 7]. Due to potential advantages, MFCs have gathered an impressive consideration by the specialists in current years. Many factors including electrode material [8], catalyst [9], microbial inoculum [10], mediator [11], reactor design [10] electrode spacing [10] resistance in both external and internal [12], proton exchange membrane (PEM) [13] and substrate [14] affect the performance of MFCs. As being an important part of MFCs, both anode and cathode, perform integral roles in the generation of electricity and increasing the removal efficiency of organic matter. This chapter deals with the issues related to anode and cathode electrodes.

2.   Anode

Being one of the most indispensable parts of MFCs, anode materials help in amplifying the performance of MFCs by improving the electron transfer, anode resistance, bacterial attachment, and oxidation of substrate [9, 15]. Basically, enhanced conductivity, improved specific surface area, elevated porosity, robust biocompatibility, chemical stability, fabrication ease, congruous mechanical strength and cost-effectiveness are the

Microbial Fuel Cells: Materials and Applications   Materials Research Forum LLC
Materials Research Foundations **46** (2019) 125-150   doi: http://dx.doi.org/10.21741/9781644900116-6

primary requirements of anode materials [9]. The non-carbon- and carbon- anode materials are discussed below.

2.1   Non-carbon Anode Materials

Till date, limited investigations based on non-carbon anode materials for MFCs have been reported [16-19]. In 2008, Dumas and colleagues utilized stainless steel as suitable material for the anode with *Geobacter sulfurreducens* for sediment MFCs. Shockingly, the outcomes showed the low P_{max} at 4 mW/m^2, and this was because of the way that the smooth surface of treated steel couldn't encourage bacterial connection. Richter et al. demonstrated the likelihood of gold anode with *Geobacter sulfurreducens* [19]. These MFCs gave a current, similar to that of the control graphite MFCs i.e. 0.4 - 0.7 mA. Tragically, no further information on the value of open circuit voltage (OCV), P_{max} and internal resistance was provided. An earlier work by Crittenden et al. has demonstrated the application of gold anode covered with a self-gathered monolayer with *Shewanella putrefaciens* [16]. MFCs current generation is correlated with the molecular chain length of monolayer and head-group of the monolayer molecules. However, the practical applications of gold based anodes are limited to small-scale due to higher prices. A comparative study, in terms of application as an anode, was performed between titanium and graphite. And no current generation was obtained when titanium was utilised as anode [20, 18].

2.2   Carbon Anode Materials

Generally, carbon is considered as most adaptable anode material, which incorporates graphite plate, graphite pole, graphite granule, smooth carbon, reticulated vitreous carbon (RVC), and fibrous materials as felt, fabric, paper, fiber and foam. Table-1 records various carbon-based materials along with their unique properties, which enable them to be utilised as anode materials. Among these carbon materials, carbon fabric and carbon paper have been more frequently utilized anode materials in MFC because of their unfaltering quality in microbial cultures and high specific surface area along with high conductivity [7, 21]. Therefore, it became very important to develop new types of materials or to modify these carbon materials for improving the anode performance. Surface-based modification methods for the anode are discussed in the following section.

Table.1 *Comparison of some carbon anode materials in MFCs.*

Anode material	Advantage	Disadvantage	Reference
Graphite rod	Chemical stability high conductivity and low cost	Low surface area and porosity	[13]
Carbon felt	Large aperture	High resistance and low specific surface area	[22]
RVC	High conductivity, plasticity and porosity	Brittle	[23]
Carbon cloth	High porosity and Flexibility	Expensive	[24]
Graphite fiber brush	High specific surface area and conductivity	Clogging	[25]
Carbon granule	High specific surface area	Low porosity	[26]

2.3   Surface Treatment and Coating for Anode Materials

Anode material surface characteristics are one of the main factors that affect the attachment of bacteria and transfer of electron. Two different modification methods have been studied in order to enhance the performance of the anode.

a) Surface Treatment

Cheng and Logan have used ammonia gas and conductive electrolyte to treat carbon cloth [27] and observed an increase in power yield by 48% on using the treated carbon cloth in comparison to that of plain/untreated carbon cloth. The power output improvement was due to the increase in the surface charge of the anode from 0.38 to 3.99 meq/m^2 which further lead to expand the positive surface charge resulting in the development of a biofilm. However, this method requires particular equipment. Wang et al. proposed a more suitable strategy which included warming of carbon in a muffle furnace, resulting in the expansion of 3% in the value of P$_{max}$. This could be due to the development of an active surface area because of the warming process. In 2010, Feng et al. proposed carbon brush treatment by utilizing sulfuric acid along with heat, that caused 25% expansion in the P$_{max}$ value because of the expansion in specific surface area and a positive charge on the anode [28]. Interestingly, the same research group proposed that carboxyl groups can be functionalized on the graphite felt anode by electrochemical oxidation at a consistent current density of 30 mA/cm^2 [29]. Such approach resulted in the supply of higher current up to 40% than that of an untreated anode because of the consideration that the carboxyl groups on the graphite felt have the capability to furnish strong hydrogen bonding with peptide in bacterial cytochromes, which further result in better electron transfer.

b) Surface Coating

Recently, various mediators, metals and their composites have represented as materials for surface coating [30, 31]. A number of mediators consisting of neutral red (NR), 2-hydroxy-1,4-naphthoquinone (HNQ) and anthraquinone-1,6-disulfonic acid, have been immobilized on the anode. The results obtained from immobilization of graphite with mediators demonstrated the fundamental increment in P_{max} value because of the improvement in electron exchange [32, 33]. However, while using these immobilized mediators in long-term MFC operations, easy degradation is the main problem.

Metal and metal oxide-coated anodes which include Au, Pd and Fe_3O_4 nanoparticles, have also been reported. Recently, Fan et al. proposed gold nanoparticles covered graphite anodes which helped in enhancing the value of current density by 20 times than that of the plain graphite, compared to control graphite the current density was increased 1.5-2.5 times when graphite coated with Pd nanoparticles were used [30]. In another investigation with iron composites, Kim et al. detailed the importance of coated ferric oxide on the anodes, which helped them in improving the properties of iron-reducing bacteria. The outcomes were higher P_{max} and improved coulombic efficiency (CE) than the plain anode [34]. In outline, specific equipment, high temperature, multiple steps and long treatment time are the basic requirements for most of the anode modification methods. Therefore, further researches are required so that simple and more effective modification method can be developed.

3. Cathode

The majority of the materials referred under segment 2.3 for the anode, can possibly be used as cathode materials. At present, carbon fabric and carbon paper are most gigantically used as cathode materials. Oxygen, in perspective of its high oxidation potential, is normally used as an electron acceptor [35]. In order to improve the kinetics of oxygen reduction at the electrode surface along with a decrease in the activation energy barrier catalysts are required [36]. For the most part, the catalysts required in MFCs need to have; high catalytic activity towards the oxidation at a temperature and pH range of 10-40°C and 5-7 respectively, chemical and electrochemical solidness, great biocompatibility, and financially savvy [37]. Platinum (Pt) is the widely used metal catalyst for oxygen reduction reaction (ORR) in MFCs due to its excellent catalytic ability.

4.   Problem Statements

4.1   Need of CNT-based Anodes

Among the different components that influence MFCs performance, the anode is viewed as the basic parameter because of its ability to manage the organic removal efficiency as well as power generation [5, 8]. High electrical conductivity, improved specific surface area, adequate surface roughness, higher porosity, chemical stability, ease of designing, suitable mechanical quality, strong biocompatibility and cost-ampleness are the critical properties which are required for a material to be used as anode for the MFCs [8, 38, 39]. It has been reported that the colony forming capacity of microorganisms, the capability of electron transfer and anode surface response have been influenced by different anode materials [8, 39]. Carbon cloth and carbon paper in view of their high conductivity and solid chemical stability in microbial culture, have been regularly used MFC anode materials [40]. However, the shortcoming associated with these materials include the limited catalytic activity, inferior porosity and low specific surface area [26]. Additionally, carbon paper due to its delicate property cannot be used for long-term MFC operations [41]. In this way, it is important to grow new sorts of anode materials for MFCs and in this sense, carbon nanotubes (CNTs) have demonstrated as promising contenders due to their novel structural, electrical, physical and chemical properties [15, 42].

4.2   Cathodes based on CNTs

At the cathode of a MFC, one of the most reasonable and sensible electron acceptors is oxygen; however, its moderate oxygen reduction reaction (ORR) rate prompts high decrease overpotential, which acts as one of the most generally perceived compelling factors in the MFCs performance. A catalyst is generally required to improve the kinetics of ORR and reduction in the activation energy barrier [36]. The steadiness in the catalyst performance is highly influenced by the assistance of catalyst support because of the diverse association between the catalyst particles and supports [43]. High conductivity, specific surface area, resistance against corrosion and capability to create anchoring sites with metal catalyst are considered as specific requirements for catalyst support [43, 44]. CNTs due to their wonderful electrical, physical and chemical properties are considered to be promising catalyst support and cathode material.

5.   Carbon Nanotubes (CNTs)

Carbon nanotubes (CNTs) contain graphene sheet layers, with expansiveness in nanometres and length in microns that are rolled-up into the shape of cylinders that are

Microbial Fuel Cells: Materials and Applications   Materials Research Forum LLC
Materials Research Foundations **46** (2019) 125-150   doi: http://dx.doi.org/10.21741/9781644900116-6

often closed at both ends as shown in Fig. 1. These rolls are laid out in such a way so that CNTs obtained are extensive in bundles along with their walls formed by hexagonal carbon rings. The closures/ terminations of CNTs are domed structured that is shaped of six-membered rings and capped by a five-membered ring. Variety of techniques can be used for the synthesis of CNTs. In general, arc discharge, chemical vapour deposition (CVD), and laser vaporization (laser ablation) are probably the most well-known techniques [45, 46].

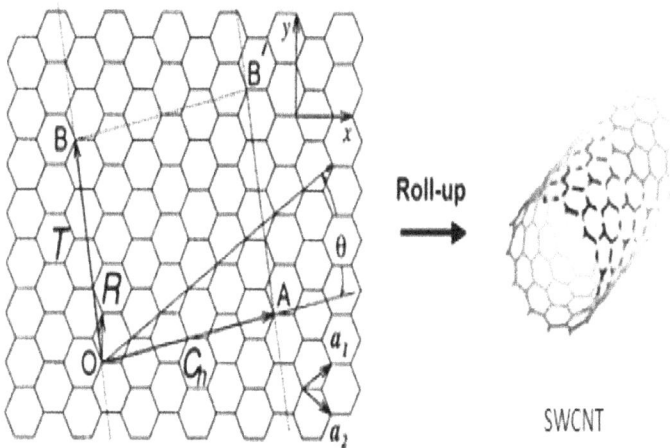

Fig. 1- Schematic representation of the roll-up of a graphene sheet to form a single-walled carbon nanotube structure [47]

Generally, CNTs have two key classifications [48]. a) Single-walled CNTs (SWCNTs); which typically contain single cylindrical graphene layer in the wall of the tube as shown in Fig. 2, with diameters in the range of 0.4 to 2 nm and length up to 1.5 cm.

b) Multi-walled CNTs (MWCNTs) are those which contain several layers of graphene telescoped one over another with a spacing of approximately 0.34 nm in the middle of them, while the diameters range is from 10 to 200 nm, and lengths up to hundreds of microns (Fig. 2).

To date, CNTs in view of their exceptional physical and chemical properties are getting tremendous ubiquity and these properties incorporate high specific surface area, porosity, conductivity and tensile strength, in addition of biocompatibility and possibility to create anchoring sites with catalytic nanoparticles etc. [44, 49]. Still, more researches are required to be carried out by using CNTs as electrode materials in proton exchange membrane fuel cell (PEMFC), direct methanol fuel cell (DMFC), and MFC [39, 50, 51].

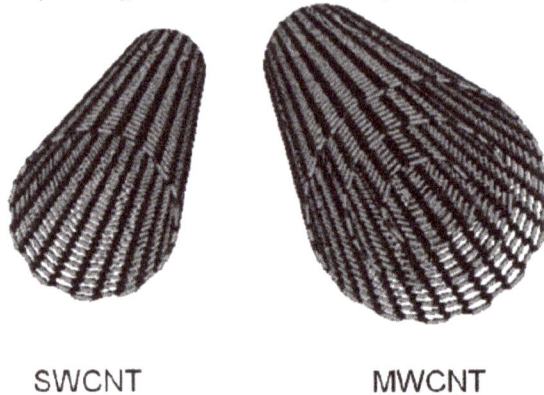

Fig. 2-Molecular structures of SWCNTs and MWCNTs [52].

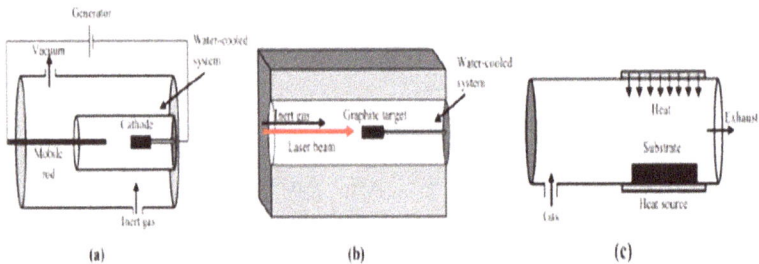

Fig. 3- Process illustration for the formation of by pure CNTs [53]

5.1. Synthesis of pristine CNTs

With the help of three techniques, to be specific, arc-discharge [53], laser ablation [54] and chemical vapour deposition (CVD) diverse sorts of individual CNTs have been delivered [55]. CNTs have been created with the help of arc-discharge and laser ablation by vaporization of a solid carbon source and condensation of inert gases like argon or helium (Fig. 3). In laser ablation, the energy of high temperature used for evaporation of carbon sources is provided by laser energy whereas, in case of arc discharge, the energy of a high-temperature plasma is discharged between two graphite electrodes [53, 54]. Laser ablation and arc discharge have genuine troubles for being used as cost-effective CNT producing procedures because of the high temperature (3000 °C) necessity for the reaction environment. Hence, limiting the number of CNTs which can be synthesized using these two processes when compared with CV [55].

5.2   Functionalization of CNTs

CNTs are set up of changing diameter, length, structure, and more importantly sometimes contain contaminations of metal or carbonaceous deposits that impact their exceptional properties. Thus, in order to eliminate these impurities, different routes of CNT functionalization are required in order to the synthesis of other CNT-based hybrid materials reasonable for a coveted application. Covalent attachment of functional groups with various groups like polymers, biomolecules, and metal nanoparticles through a chemical reaction, or adsorption or wrapping the CNTs helps in functionalizing the CNTs [56]. By using covalent linkage, or direct methodology like creating or depositing these metal nanoparticles (NPs) onto the CNT structure, these NPs get related easily with the CNT structure [56, 57]. Vander Waals collaboration is critical in the direct procedure, through which metal Ps get associated with CNTs by means of a reduction process in the presence of CNTs where noble metals are used as the precursor. Heat, light, and reducing agents are used in this process [57]. Metal NPs collaboration with CNTs by means of covalent linkage can be accomplished through functional groups on NPs known as linkers [58, 59]. These linkers may incite the advancement of covalent bonds with the functional groups presentation on the CNT surface, or bond with frail intermolecular cooperation particularly stacking, hydrophobic, and electrostatic associations. CNTs may likewise be functionalized with polymer utilitarian gatherings that are physically adsorbed onto the CNT surface, or covalently connected to its structure. The regularly used strategy of polymer functionalization of CNTs is the solution casting techniques, where CNTs and polymers are blended in a solvent, trailed by evaporation of the solvent to acquire the polymer/CNT composite [59, 60]. However, these customary systems and low-tech strategies of casting or dip-coating for the fabrication of conductive cross-linked polymer lattice, show low command over the completed composite and material waste. New procedures include the classification of layer-by-layer (LbL) methods that have been comprehensively used for the synthesis of highly functional electrodes for batteries and fuel cells [61-63]. Many assortments of LbL have been proposed among which spin-spray layer-by-layer (SSLbL) presented by Merrill et al. has been remarkable [64]. Their union system was later upgraded by Gittleson et al. as computerized and highly fast assembly of polyelectrolyte/CNT films with the possibility of lessening the process durations to even sub-seconds. Various other features of this system incorporate material waste reduction, homogenization of thin film conductance, and nanolevel control over the film advancement [65].

5.3 CNT-based Electrodes in MFCs

For a real MFC system, just a couple of attempts have been accounted on CNT-based electrodes, Table 2 shows the applications for anode whereas Table 3 discusses cathode applications.

Table.2 *Summary of the various studies performed on CNT-based anodes in MFC applications.*

Anode Material	Preparation Method	Maximum power density (mW/m^2)	Reference
Ni foam with PANI/CNTs	Chemical polymerization	42	[15]
Carbon paper with Sn-Pt /MWCNT-COOH	Not reported	1994	[42]
Carbon cloth with MWCNTs	Dipping and drying	65	[51]
Carbon paper with MWCNT-COOH	Dipping	290	[68]
Polyester fabric with SWCNTs	Dipping and drying	1098	[69]
MWCNT/Silicon	Vertically aligned/PECVD	19.36	[71]
CNT+Mo$_2$C +PTFE/ Carbon felt	Pressure coating	1050	[72]
CNT/Carbon paper	3D conductive network/Dispersion	1760	[73]
MWCNT+ Mo$_2$C +Na fion/Plain graphite	Pressure coating	109.1	[74]
MWCNT/SSM	NA/CVD	450	[75]
MWCNT+PANI/Graphite felt	Electro deposition	257	[76]

Table.3 *List of various studies performed on CNT-based anodes in MFC applications.*

Cathode material	Preparation method	Maximum power density (mW/m^2)	Reference
Carbon papers with Pt/MWCNTs and Sn-Pt/MWCNTs	Not reported	1342 1430	[42]
Carbon cloth with Pt/MWCNTs	Dipping and drying	65	[51]
Carbon cloth with CoTMPP-FePc/ MWCNT-COOH	Paste painting	751	[80]
Carbon cloth with β-MnO2/CNTs	Spraying	98	[91]
Carbon paper with CNTs	Not reported	8	[81]
CNT+Pt/Carbon paper	Dip coating	169.7	[92]
Mn+PPy+CNT+Nafion/GCE	Drop casting	213	[93]
MWCNT/Graphite felt	Electrophoretic deposition	214.7	[94]
CNT+PPy/Carbon paper	NA	113.5	[95]
N–CNT/Carbon cloth	CVD	135	[96]

5.3.1 CNT-based Anodes in MFCs

It has been accounted through literature that CNTs have a cell harmfulness that may prompt expansion restraint and cell damage [66]. Subsequently, a few analysts have changed the configuration of CNT-based anodes with the help of biocompatible polymers that will help in decreasing the cell toxicity [15, 39]. They prepared anode materials based on CNT/conductive polymer nanocomposites coated on nickel foam and carbon paper. It was concluded that the additional CNTs along with the conductive polymers have enhanced the power output as compared with the polymer only. These conductive polymers include; namely polyaniline (PANI) and polypyrrole (PPy). However, low conductivity and poor electron transfer properties of polymers limit their application in MFCs. Most recently, Higgins et al. masterminded carbon felt impregnated with chitosan (CHIT)/MWCNT platforms, which brought out open circuit voltage (OCV) of 0.600 V and the P_{max} of 4.75 W/m^3 in contrast to that of plain carbon felt (3.5 W/m^3). However, the preparation strategy for CHIT/CNT nanocomposite was confounded and tedious [38].

A Sn-Pt/MWCNT-COOH coated carbon paper along with a novel electron mediator without using any conductive polymers was developed by Sharma et al. Further, using blend arrangement of Nafion solution and MWCNT-COOH suspension for coating carbon paper had resulted in the development of these CNT-based anodes. The noteworthy outcomes were obtained on using such CNT-based anodes in MFCs which have produced six times higher P_{max} than that of the control graphite anode [42]. However, certain limitations of Nafion i.e. it causes an increase in resistivity and a decrease in catalytic activity have been reported by some researchers [20, 67].

By looking at the performance of MFCs on the application of MWCNT-coated carbon cloth and plain carbon cloth as an anode, Tsai et al. presumed that the additional CNTs in comparison to that of carbon cloth anodes lead to significant upgradation in the values of P_{max}, CE and COD removal efficiency [51]. Unfortunately, the thin films of CNTs on carbon cloth were not stable at a higher flow rate and for long-term operation. In another attempt, Sun et al., have examined a self-assembled CNT-based anode which was developed by using novel layer-by-layer (LBL) method [68]. These kinds of anodes were prepared by assembling negatively charged MWCNT-COOH and positively charged polyethyleneimine (PEI) on carbon paper, which further resulted in increasing the value of P_{max} by 20% than that of obtained by utilizing plain carbon paper. However, the value of OCV was adversely influenced on using these kinds of anodes.

More importantly, in recent years with the development of three-dimensional CNT-textile anodes by a researcher group using the dipping and drying method had made these CNT based electrodes more famous and useful [69]. Polyester fabric was dipped into a blend solution of SWCNT suspension and SDBS solution, trailed by drying at 120°C. This CNT-textile anode reactor demonstrated the P_{max} 68% higher than that acquired from the carbon cloth. However, enormous experimentations are required to carry out regarding the catalytic activity of CNT-textile anode. Liang et al. had explored the impact of the addition of MWCNT-COOH powders in anode chamber on MFC performance [70]. Results demonstrated that both the start-up time for biofilm development and the internal resistance during 40 days of operation were decreased. However, no noteworthy increment of the biomass in anodic biofilm was recognized from the examinations on various doses of CNTs. Subsequently, more in-depth investigations on the measurement of the electrochemical activity of the biomass per unit mass are quite important.

5.3.2 CNT-based Cathodes in MFCs

Basically, catalyst supports for MFCs are selected on the basis of their conductivity, framework, specific surface area, and corrosion resistance properties [44, 49]. As of late, diverse nanostructures developed from carbon materials with graphitic structure have been likewise utilized for catalyst support applications [77, 78]. Carbon black, for example, Vulcan XC-72 has been broadly utilized in a fuel cell as a catalyst support, because of its high conductivity, adequate surface area, easy accessibility and minimal cost [78, 79]. Recently, CNTs have been of great interest as a cathode material in fuel cell applications because of their exceptional electrical, physical and substance properties [43, 50]. However, in spite of being the recent area of interest, still, their limited investigations have been reported on the CNT-based cathode for MFCs [42, 80, 81].

Some researchers have compared performances of catalyst supports such as CNTs and carbon black utilized in fuel cell applications [82, 83]. In 2006, Xin exhibited that MWCNTs showed 30% more resistance towards corrosion as compared to Vulcan XC-72, achieving solidness of the PEMFC [84]. In another study, Matsumoto et al. reported that a better PEMFC performance can be obtained by using Pt/CNTs as compared to commercial Pt/carbon black with reducing 60% Pt loading per cathode area [83]. Sharma et al. have developed a new configuration of the catalyst by depositing Pt and Pt alloy catalysts on MWCNTs and further coating them on carbon paper. Cathode reactor of Sn-Pt/MWCNT has provided higher P_{max} as compared to other reactors of Ru-Pt/MWCNT and Pt/MWCNT [42]. However, in the same study, application of nafion as a binder has shown the ability to isolate MWCNTs from the catalyst layers in the MWCNT suspension [85]. Furthermore, nafion could cause an expansion in resistivity and an abatement in catalytic activity [67].

In 2008, a new category of anode materials which consists of MWCNT-coated carbon cloth was prepared by a group of researchers. Accordingly, Pt catalysts were coated straightforwardly onto the MWCNT-coated carbon cloth cathode. Shockingly, no major difference in the value of P_{max} and COD removal efficiency was obtained when MWCNT-coated carbon cloth reactors with or without Pt catalysts were operated. This might be because of weak collaboration between the Pt and MWCNT-coated carbon cloth, which have the potential to cause the disengagement of Pt catalysts from the MWCNTs [51]. In a recent study, SWCNT and MWCNT modified cathodes with/without Pt catalyst were fabricated by Wang et al. using some new methods [86]. Further, these cathodes were developed and produced by filtration of the CNT suspension by employing PTFE membrane. Microwave method operated at 250W and 140°C for the 90s was used for coating of Pt catalysts on the surface of CNTs. As indicated by their P_{max} and OCV outcomes, MFCs could be positioned as follows Pt/SWCNT > Pt/MWCNT > SWCNT cathode. Beside this, on the application of Pt/SWCNT based cathode reactor, about three times lower internal resistance was obtained as that acquired from Pt/MWCNT reactor. Indeed, even the performance of SWCNT cathode reactors in contrast to those from carbon cloth indicates incredible MFC response. Chemical deposition and microbial adherence on the cathode surfaces have caused a steady decline in SWCNT performance, as they knock down the charge transfer and catalytic activity of the reactor. Interestingly, it is known that tetramethoxyphenylporphyrin (TMPP) and phthalocyanine (Pc) with Fe or Co acts as fine ORR catalysts and can be also explored as alternatives to Pt [87-89]. Deng et al. were the first to use CoTMPP and FePc for deposition on MWCNT-COOH which can be further utilized as MFC cathodes. These types of developed catalysts showed higher catalytic activity than that of CoTMPP-FePc/graphite [80]. Results on the application of these cathodes for MFCs application had demonstrated that the value P_{max} obtained was 1.5 times higher than that of obtained on using commercial Pt-carbon cloth cathodes. But still, the performance of CoTMPP-FePc/MWCNT-COOH on carbon cloth could be further improved by employing air splashing in place of paste painting. Additionally, the deposition method of CoTMPP-FePc on MWCNT-COOH was found to be convoluted and tedious. On the further extension of the study, Yuan et al. have used a catalyst of FePc as a support on MWCNT-NH_2 and MWCNT-COOH. MFCs could be ranked as follows in terms of P_{max} value: FePc/MWCNT-NH_2 > FePc/MWCNT-COOH > FePc/MWCNT [90]. In one of the alternative studies regarding catalyst by Lu et al. and Zhang et al. shown a new possibility about utilizing the MnO_2/CNT-coated carbon support as the cathode of MFCs [81, 91]. Lu et al. compared three types of MnO_2 namely α-MnO_2, β-MnO_2, γ-MnO_2/CNT spray-coated carbon cloths on the basis of MFC performance [91]. Based on their results, the highest P_{max}, OCV and COD removal efficiency was obtained by β-MnO_2/CNT cathode reactors among other MnO_2/CNT reactors. Unfortunately, the P_{max} value obtained by the β-MnO_2/CNT reactor was 1.56 times lesser than that obtained by commercial-Pt carbon cloth. With likewise catalyst, Zhang et al. also affirmed that CNTs were perfect catalyst support material for MnO_2 in which MnO_2/CNT acted as a good substitute for the Pt/carbon black in MFC applications [81]. As such, further studies are required to be carried out so that highly improved, efficient, cost-effective and easily prepared CNT-modified electrodes can be developed.

Conclusion

Outlining of the electrode in light of the savvy innovation is the best test in assembling of MFCs. Surface zone, morphology, biocompatibility, conductivity and stability are the most imperative properties on which MFCs performance depends upon. CNTs and CNT composites because of their unique properties have been explored to develop anodes and cathodes, and several electrode modification methods have been developed so that power generation can be improved. To date, various changes are in progress, for example, surface treatment, coating with nanomaterials and composites. The power generation and electrode cost have not reached the level for commercial use. Additionally, investigation on more successful electrode materials and optimization of the configuration are expected to address these challenges.

References

[1]   E.I. Administration, Annual energy review 2011, U.S. Energy Information, Administration, Office of Energy Statistics, U.S. Department of Energy, DOE/EIA-0384 (2011).

[2]   W.G.I.P.O.C. Change, I.P.O.C. Change, Climate change 2007: contribution to the fourth assessment report of the Intergovernmental Panel on Climate Change. 2. Impacts, adaptation and vulnerability: contribution of Working Group II to the fourth assessment report of the Intergovernmental Panel on Climate Change, Cambridge University Press 2007.

[3]   M. Greenberg, Energy sources, public policy, and public preferences: analysis of US national and site-specific data, Energy Policy. 37 (2009) 3242-3249. https://doi.org/10.1016/j.enpol.2009.04.020

[4]   H.S. Lee, P. Parameswaran, A.K. Marcus, C.I. Torres, B.E. Rittmann, Evaluation of energy-conversion efficiencies in microbial fuel cells (MFCs) utilizing fermentable and non-fermentable substrates, Water Res. 42 (2008) 1501-1510. https://doi.org/10.1016/j.watres.2007.10.036

[5]   B. Logan, S. Cheng, V. Watson, G. Estadt, Graphite fiber brush anodes for increased power production in air-cathode microbial fuel cells, Environ. Sci. Technol. 41 (2007) 3341-3346. https://doi.org/10.1021/es062644y

[6]   H. Liu, R. Ramnarayanan, B.E. Logan, Production of electricity during wastewater treatment using a single chamber microbial fuel cell, Environ. Sci. Technol. 38 (2004) 2281-2285. https://doi.org/10.1021/es034923g

[7]   B.E. Logan, B. Hamelers, R. Rozendal, U. Schröder, J. Keller, S. Freguia, P. Aelterman, W. Verstraete, K. Rabaey, Microbial fuel cells: methodology and technology, Environ. Sci. Technol. 40 (2006) 5181-5192. https://doi.org/10.1021/es0605016

[8]   J. Wei, P. Liang, X. Huang, Recent progress in electrodes for microbial fuel cells, Bioresour Technol. 102 (2011) 9335-9344. https://doi.org/10.1016/j.biortech.2011.07.019

[9]   M. Zhou, M. Chi, J. Luo, H. He, T. Jin, An overview of electrode materials in microbial fuel cells, J. Power Sources. 196 (2011) 4427-4435. https://doi.org/10.1016/j.jpowsour.2011.01.012

[10]   A. Rinaldi, B. Mecheri, V. Garavaglia, S. Licoccia, P.D. Nardo, E. Traversa, Engineering materials and biology to boost performance of microbial fuel cells: a critical review, Environ. Sci. Technol. 1 (2008) 417-429.

[11]   N. Kim, Y. Choi, S. Jung, S. Kim, Effect of initial carbon sources on the performance of microbial fuel cells containing Proteus vulgaris, Biotechnol. Bioeng. 70 (2000) 109-114. https://doi.org/10.1002/1097-0290(20001005)70:1<109::AID-BIT11>3.0.CO;2-M

[12]   P. Aelterman, M. Versichele, M. Marzorati, N. Boon, W. Verstraete, Loading rate and external resistance control the electricity generation of microbial fuel cells with different three-dimensional anodes, Bioresour. Technol. 99 (2008) 8895-8902. https://doi.org/10.1016/j.biortech.2008.04.061

[13]   H. Liu, S. Cheng, B.E. Logan, Power generation in fed-batch microbial fuel cells as a function of ionic strength, temperature, and reactor configuration, Environ. Sci. Technol. 39 (2005) 5488-5493. https://doi.org/10.1021/es050316c

[14]   B. Min, B.E. Logan, Continuous electricity generation from domestic wastewater and organic substrates in a flat plate microbial fuel cell, Environ. Sci. Technol. 38 (2004) 5809-5814. https://doi.org/10.1021/es0491026

[15]   Y. Qiao, C.M. Li, S.J. Bao, Q.L. Bao, Carbon nanotube/polyaniline composite as anode material for microbial fuel cells, J. Power Sources. 170 (2007) 79-84. https://doi.org/10.1016/j.jpowsour.2007.03.048

[16]   S.R. Crittenden, C.J. Sund, J.J. Sumner, Mediating electron transfer from bacteria to a gold electrode via a self-assembled monolayer, Langmuir. 22 (2006) 9473-9476. https://doi.org/10.1021/la061869j

[17]   C. Dumas, R. Basseguy, A. Bergel, Electrochemical activity of Geobacter sulfurreducens biofilms on stainless steel anodes, Electrochim. Acta. 53 (2008) 5235-5241. https://doi.org/10.1016/j.electacta.2008.02.056

[18]   A.T. Heijne, H.V. Hamelers, M. Saakes, C.J. Buisman, Performance of non-porous graphite and titanium-based anodes in microbial fuel cells, Electrochim. Acta. 53 (2008) 5697-5703. https://doi.org/10.1016/j.electacta.2008.03.032

[19] H. Richter, K. McCarthy, K.P. Nevin, J.P. Johnson, V.M. Rotello, D.R. Lovley, Electricity generation by Geobacter sulfurreducens attached to gold electrodes, Langmuir. 24 (2008) 4376-4379. https://doi.org/10.1021/la703469y

[20] L.C. Nagle, J.F. Rohan, Aligned carbon nanotube–Pt composite fuel cell catalyst by template electrodeposition, J. Power Sources. 185 (2008) 411-418. https://doi.org/10.1016/j.jpowsour.2008.06.067

[21] C. Du, J. Yeh, N. Pan, Carbon nanotube thin films with ordered structures, J. Mater. Chem. 15 (2005) 548-550. https://doi.org/10.1039/b414682d

[22] H.J. Kim, H.S. Park, M.S. Hyun, I.S. Chang, M. Kim, B.H. Kim, A mediator-less microbial fuel cell using a metal reducing bacterium, Shewanella putrefaciens, Enzyme Microb. Technol. 30 (2002) 145-152. https://doi.org/10.1016/S0141-0229(01)00478-1

[23] Z. He, S.D. Minteer, L.T. Angenent, Electricity generation from artificial wastewater using an upflow microbial fuel cell, Environ. Sci. Technol. 39 (2005) 5262-5267. https://doi.org/10.1021/es0502876

[24] S. Ishii, K. Watanabe, S. Yabuki, B.E. Logan, Y. Sekiguchi, Comparison of electrode reduction activities of Geobacter sulfurreducens and an enriched consortium in an air-cathode microbial fuel cell, Appl. Environ. Microbiol. 74 (2008) 7348-7355. https://doi.org/10.1128/AEM.01639-08

[25] Y. Ahn, B.E. Logan, Effectiveness of domestic wastewater treatment using microbial fuel cells at ambient and mesophilic temperatures, Bioresour. Technol. 101 (2010) 469-475. https://doi.org/10.1016/j.biortech.2009.07.039

[26] K. Rabaey, W. Verstraete, Microbial fuel cells: novel biotechnology for energy generation, Trends Biotechno. 23 (2005) 291-298. https://doi.org/10.1016/j.tibtech.2005.04.008

[27] S. Cheng, B.E. Logan, Sustainable and efficient biohydrogen production via electrohydrogenesis, Proc. Natl. Acad. Sci. 104 (2007) 18871-18873. https://doi.org/10.1073/pnas.0706379104

[28] X. Wang, S. Cheng, Y. Feng, M.D. Merrill, T. Saito, B.E. Logan, Use of carbon mesh anodes and the effect of different pretreatment methods on power production in microbial fuel cells, Environ. Sci. Technol. 43 (2009) 6870-6874. https://doi.org/10.1021/es900997w

[29] X. Tang, K. Guo, H. Li, Z. Du, J. Tian, Electrochemical treatment of graphite to enhance electron transfer from bacteria to electrodes, Bioresour. Technol. 102 (2011) 3558-3560. https://doi.org/10.1016/j.biortech.2010.09.022

[30] Y. Fan, S. Xu, R. Schaller, J. Jiao, F. Chaplen, H. Liu, Nanoparticle decorated anodes for enhanced current generation in microbial electrochemical cells, Biosens. Bioelectron. 26 (2011) 1908-1912. https://doi.org/10.1016/j.bios.2010.05.006

[31] K. Scott, G. Rimbu, K. Katuri, K. Prasad, I. Head, Application of modified carbon anodes in microbial fuel cells, Process Saf. Environ. Prot. 85 (2007) 481-488. https://doi.org/10.1205/psep07018

[32] D.A. Lowy, L.M. Tender, J.G. Zeikus, D.H. Park, D.R. Lovley, Harvesting energy from the marine sediment–water interface II: Kinetic activity of anode materials, Biosens. Bioelectron. 21 (2006) 2058-2063. https://doi.org/10.1016/j.bios.2006.01.033

[33] K. Wang, Y. Liu, S. Chen, Improved microbial electrocatalysis with neutral red immobilized electrode, J. Power Sources. 196 (2011) 164-168. https://doi.org/10.1016/j.jpowsour.2010.06.056

[34] J.R. Kim, B. Min, B.E. Logan, Evaluation of procedures to acclimate a microbial fuel cell for electricity production, Appl. Microbiol. Biotechnol. 68 (2005) 23-30. https://doi.org/10.1007/s00253-004-1845-6

[35] K. Watanabe, Recent developments in microbial fuel cell technologies for sustainable bioenergy, J. Biosci. Bioeng. 106 (2008) 528-536. https://doi.org/10.1263/jbb.106.528

[36] H. Yazdi, S.M. Carver, A.D. Christy, O.H. Tuovinen, Cathodic limitations in microbial fuel cells: an overview, J. Power Sources. 180 (2008) 683-694. https://doi.org/10.1016/j.jpowsour.2008.02.074

[37] M. Rosenbaum, F. Zhao, M. Quaas, H. Wulff, U. Schröder, F. Scholz, Evaluation of catalytic properties of tungsten carbide for the anode of microbial fuel cells, Appl. Catal. B 74 (2007) 261-269. https://doi.org/10.1016/j.apcatb.2007.02.013

[38] S.R. Higgins, D. Foerster, A. Cheung, C. Lau, O. Bretschger, S.D. Minteer, K. Nealson, P. Atanassov, M.J. Cooney, Fabrication of macroporous chitosan scaffolds doped with carbon nanotubes and their characterization in microbial fuel cell operation, Enzyme Microb. Technol. 48 (2011) 458-465. https://doi.org/10.1016/j.enzmictec.2011.02.006

[39] Y. Zou, C. Xiang, L. Yang, L.X. Sun, F. Xu, Z. Cao, A mediatorless microbial fuel cell using polypyrrole coated carbon nanotubes composite as anode material, Int. J, Hydrogen Energ. 33 (2008) 4856-4862. https://doi.org/10.1016/j.ijhydene.2008.06.061

[40] B.E. Logan, J.M. Regan, Electricity-producing bacterial communities in microbial fuel cells, Trends Microbiol. 14 (2006) 512-518. https://doi.org/10.1016/j.tim.2006.10.003

[41] J.R. Kim, S.H. Jung, J.M. Regan, B.E. Logan, Electricity generation and microbial community analysis of alcohol powered microbial fuel cells, Bioresour. Technol. 98 (2007) 2568-2577. https://doi.org/10.1016/j.biortech.2006.09.036

[42] T. Sharma, A.L.M. Reddy, T. Chandra, S. Ramaprabhu, Development of carbon nanotubes and nanofluids based microbial fuel cell, Int. J. Hydrogen Energ. 33 (2008) 6749-6754. https://doi.org/10.1016/j.ijhydene.2008.05.112

[43] J. Shen, Y. Hu, C. Li, C. Qin, M. Ye, Pt–Co supported on single-walled carbon nanotubes as an anode catalyst for direct methanol fuel cells, Electrochim. Acta. 53 (2008) 7276-7280. https://doi.org/10.1016/j.electacta.2008.04.019

[44] S. Timur, U. Anik, D. Odaci, L. Gorton, Development of a microbial biosensor based on carbon nanotube (CNT) modified electrodes, Electrochem. Commun. 9 (2007) 1810-1815. https://doi.org/10.1016/j.elecom.2007.04.012

[45] H. Dai, Carbon nanotubes: synthesis, integration, and properties, Acc. Chem. Res. 35 (2002) 1035-1044. https://doi.org/10.1021/ar0101640

[46] M. Trojanowicz, Analytical applications of carbon nanotubes: a review, Trends Analyt. Chem. 25 (2006) 480-489. https://doi.org/10.1016/j.trac.2005.11.008

[47] T.W. Odom, J.L. Huang, P. Kim, C.M. Lieber, Structure and electronic properties of carbon nanotubes, J. Phys. Chem. B 104 (2000) 2794-2809. https://doi.org/10.1021/jp993592k

[48] M. Trojanowicz, Analytical applications of carbon nanotubes: a review, Trends Analyt. Chem. 25 (2006) 480-489. https://doi.org/10.1016/j.trac.2005.11.008

[49] H. Tang, J. Chen, Z. Huang, D. Wang, Z. Ren, L. Nie, Y. Kuang, S. Yao, High dispersion and electrocatalytic properties of platinum on well-aligned carbon nanotube arrays, Carbon. 42 (2004) 191-197. https://doi.org/10.1016/j.carbon.2003.10.023

[50] A. Kannan, P. Kanagala, V. Veedu, Development of carbon nanotubes based gas diffusion layers by in situ chemical vapor deposition process for proton exchange

membrane fuel cells, J. Power Sources. 192 (2009) 297-303.
https://doi.org/10.1016/j.jpowsour.2009.03.022

[51]   H.Y. Tsai, C.C. Wu, C.Y. Lee, E.P. Shih, Microbial fuel cell performance of multiwall carbon nanotubes on carbon cloth as electrodes, J. Power Sources. 194 (2009) 199-205. https://doi.org/10.1016/j.jpowsour.2009.05.018

[52]   G. Pastorin, K. Kostarelos, M. Prato, A. Bianco, functionalized carbon nanotubes: towards the delivery of therapeutic molecules, J. Biomed. Nanotech. 1 (2005) 1-10. https://doi.org/10.1166/jbn.2005.017

[53]   C. Journet, P. Bernier, Production of carbon nanotubes, Appl. Phys. A Mater. Sci. Process. 67 (1998) 1-9. https://doi.org/10.1007/s003390050731

[54]   A. Thess, R. Lee, P. Nikolaev, H. Dai, P. Petit, J. Robert, C. Xu, Y.H. Lee, S.G. Kim, A.G. Rinzler, Crystalline ropes of metallic carbon nanotubes, Science. 273 (1996) 483-487. https://doi.org/10.1126/science.273.5274.483

[55]   A.M. Cassell, J.A. Raymakers, J. Kong, H. Dai, Large scale CVD synthesis of single-walled carbon nanotubes, J. Phys. Chem. 103 (1999) 6484-6492. https://doi.org/10.1021/jp990957s

[56]   N. Karousis, N. Tagmatarchis, D. Tasis, Current progress on the chemical modification of carbon nanotubes, Chem. Rev. 110 (2010) 5366-5397. https://doi.org/10.1021/cr100018g

[57]   V. Georgakilas, D. Gournis, V. Tzitzios, L. Pasquato, D.M. Guldi, M. Prato, Decorating carbon nanotubes with metal or semiconductor nanoparticles, J. Mater. Chem. 17 (2007) 2679-2694. https://doi.org/10.1039/b700857k

[58]   D.P. Dubal, G.S. Gund, C.D. Lokhande, R. Holze, Decoration of spongelike Ni (OH)2 nanoparticles onto MWCNTs using an easily manipulated chemical protocol for supercapacitors, ACS Appl. Mater. Interf. 5 (2013) 2446-2454. https://doi.org/10.1021/am3026486

[59]   S.W. Lee, N. Yabuuchi, B.M. Gallant, S. Chen, B.S. Kim, P.T. Hammond, Y.S. Horn, High-power lithium batteries from functionalized carbon-nanotube electrodes, Nat. Nanotechnol. 5 (2010) 531. https://doi.org/10.1038/nnano.2010.116

[60]   M.T. Byrne, Y.K. Gun'ko, Recent advances in research on carbon nanotube–polymer composites, Adv. Mater. 22 (2010) 1672-1688. https://doi.org/10.1002/adma.200901545

[61] M.N. Hyder, S.W. Lee, F.Ç. Cebeci, D.J. Schmidt, Y.S. Horn, P.T. Hammond, Layer-by-layer assembled polyaniline nanofiber/multiwall carbon nanotube thin film electrodes for high-power and high-energy storage applications, ACS Nano. 5 (2011) 8552-8561. https://doi.org/10.1021/nn2029617

[62] S.W. Lee, B.S. Kim, S. Chen, Y.S. Horn, P.T. Hammond, Layer-by-layer assembly of all carbon nanotube ultrathin films for electrochemical applications, J. Am. Chem. Soc. 131 (2008) 671-679. https://doi.org/10.1021/ja807059k

[63] H. Paloniemi, M. Lukkarinen, T. Ääritalo, S. Areva, J. Leiro, M. Heinonen, K. Haapakka, J. Lukkari, Layer-by-layer electrostatic self-assembly of single-wall carbon nanotube polyelectrolytes, Langmuir. 22 (2006) 74-83. https://doi.org/10.1021/la051736i

[64] M. Merrill, C. Sun, Fast, simple and efficient assembly of nanolayered materials and devices, Nanotechnology. 20 (2009) 075606. https://doi.org/10.1088/0957-4484/20/7/075606

[65] F.S. Gittleson, D.J. Kohn, X. Li, A.D. Taylor, Improving the assembly speed, quality, and tunability of thin conductive multilayers, ACS Nano. 6 (2012) 3703-3711. https://doi.org/10.1021/nn204384f

[66] A. Magrez, S. Kasas, V. Salicio, N. Pasquier, J.W. Seo, M. Celio, S. Catsicas, B. Schwaller, L. Forró, Cellular toxicity of carbon-based nanomaterials, Nano Lett. 6 (2006) 1121-1125. https://doi.org/10.1021/nl060162e

[67] J. Chen, W. Li, D. Wang, S. Yang, J. Wen, Z. Ren, Electrochemical characterization of carbon nanotubes as electrode in electrochemical double-layer capacitors, Carbon. 40 (2002) 1193-1197. https://doi.org/10.1016/S0008-6223(01)00266-4

[68] J.J. Sun, H.Z. Zhao, Q.Z. Yang, J. Song, A. Xue, A novel layer-by-layer self-assembled carbon nanotube-based anode: Preparation, characterization, and application in microbial fuel cell, Electrochim. Acta. 55 (2010) 3041-3047. https://doi.org/10.1016/j.electacta.2009.12.103

[69] X. Xie, L. Hu, M. Pasta, G.F. Wells, D. Kong, C.S. Criddle, Y. Cui, Three-dimensional carbon nanotube–textile anode for high-performance microbial fuel cells, Nano Lett. 11 (2010) 291-296. https://doi.org/10.1021/nl103905t

[70] P. Liang, H. Wang, X. Xia, X. Huang, Y. Mo, X. Cao, M.J.B. Fan, Carbon nanotube powders as electrode modifier to enhance the activity of anodic biofilm

in microbial fuel cells, Biosens. Bioelectron. 26 (2011) 3000-3004.
https://doi.org/10.1016/j.bios.2010.12.002

[71] J.E. Mink, M.M. Hussain, Sustainable design of high-performance microsized
microbial fuel cell with carbon nanotube anode and air cathode, ACS Nano. 7
(2013) 6921-6927. https://doi.org/10.1021/nn402103q

[72] Y. Wang, B. Li, D. Cui, X. Xiang, W. Li, Nano-molybdenum carbide/carbon
nanotubes composite as bifunctional anode catalyst for high-performance
Escherichia coli-based microbial fuel cell, Biosens. Bioelectron. 51 (2014) 349-
355. https://doi.org/10.1016/j.bios.2013.07.069

[73] S. Matsumoto, K. Yamanaka, H. Ogikubo, H. Akasaka, N. Ohtake, Carbon
nanotube dispersed conductive network for microbial fuel cells, Appl. Phys. Lett.
105 (2014) 083904. https://doi.org/10.1063/1.4894259

[74] Y. Fu, J. Yu, Y. Zhang, Y. Meng, Graphite coated with manganese
oxide/multiwall carbon nanotubes composites as anodes in marine benthic
microbial fuel cells, Appl. Surf. Sci. 317 (2014) 84-89.
https://doi.org/10.1016/j.apsusc.2014.08.044

[75] C. Erbay, G. Yang, P. de Figueiredo, R. Sadr, C. Yu, A. Han, Three-dimensional
porous carbon nanotube sponges for high-performance anodes of microbial fuel
cells, J. Power Sources. 298 (2015) 177-183.
https://doi.org/10.1016/j.jpowsour.2015.08.021

[76] H.F. Cui, L. Du, P.B. Guo, B. Zhu, J.H. Luong, Controlled modification of carbon
nanotubes and polyaniline on macroporous graphite felt for high-performance
microbial fuel cell anode, J. Power Sources. 283 (2015) 46-53.
https://doi.org/10.1016/j.jpowsour.2015.02.088

[77] E. Yoo, T. Okada, T. Kizuka, J. Nakamura, Effect of carbon substrate materials as
a Pt–Ru catalyst support on the performance of direct methanol fuel cells, J. Power
Sources. 180 (2008) 221-226. https://doi.org/10.1016/j.jpowsour.2008.01.065

[78] T. Ralph, M. Hogarth, Catalysis for low temperature fuel cells, Platin. Met. Rev.
46 (2002) 117-135.

[79] U. Paulus, A. Wokaun, G. Scherer, T. Schmidt, V. Stamenkovic, V. Radmilovic,
N. Markovic, P. Ross, Oxygen reduction on carbon-supported Pt− Ni and Pt− Co
alloy catalysts, J. Phys. Chem. B 106 (2002) 4181-4191.
https://doi.org/10.1021/jp013442l

[80]   L. Deng, M. Zhou, C. Liu, L. Liu, C. Liu, S. Dong, Development of high performance of Co/Fe/N/CNT nanocatalyst for oxygen reduction in microbial fuel cells, Talanta. 81 (2010) 444-448. https://doi.org/10.1016/j.talanta.2009.12.022

[81]   Y. Zhang, Y. Hu, S. Li, J. Sun, B. Hou, Manganese dioxide-coated carbon nanotubes as an improved cathodic catalyst for oxygen reduction in a microbial fuel cell, J. Power Sources. 196 (2011) 9284-9289. https://doi.org/10.1016/j.jpowsour.2011.07.069

[82]   W. Li, X. Wang, Z. Chen, M. Waje, Y. Yan, Pt– Ru supported on double-walled carbon nanotubes as high-performance anode catalysts for direct methanol fuel cells, J. Phys. Chem. B 110 (2006) 15353-15358. https://doi.org/10.1021/jp0623443

[83]   T. Matsumoto, T. Komatsu, H. Nakano, K. Arai, Y. Nagashima, E. Yoo, T. Yamazaki, M. Kijima, H. Shimizu, Y. Takasawa, Efficient usage of highly dispersed Pt on carbon nanotubes for electrode catalysts of polymer electrolyte fuel cells, Catal. Today. 90 (2004) 277-281. https://doi.org/10.1016/j.cattod.2004.04.038

[84]   Y. Yan, X. Wang, W. Li, Z. Chen, M.M.J.E.T. Waje, Characterization of support corrosion in PEM fuel cell: improved durability of carbon nanotube based electrode, ECS Trans. 1 (2006) 33-40.

[85]   W. Zhu, J. Zheng, R. Liang, B. Wang, C. Zhang, G. Au, E. Plichta, Durability study on SWNT/nanofiber buckypaper catalyst support for PEMFCs, J. Electrochem. Soc. 156 (2009) B1099-B1105. https://doi.org/10.1149/1.3160572

[86]   H. Wang, Z. Wu, A. Plaseied, P. Jenkins, L. Simpson, C. Engtrakul, Z. Ren, Carbon nanotube modified air-cathodes for electricity production in microbial fuel cells, J. Power Sources. 196 (2011) 7465-7469. https://doi.org/10.1016/j.jpowsour.2011.05.005

[87]   S. Cheng, H. Liu, B.E. Logan, Power densities using different cathode catalysts (Pt and CoTMPP) and polymer binders (Nafion and PTFE) in single chamber microbial fuel cells, Environ. Sci. Technol. 40 (2006) 364-369. https://doi.org/10.1021/es0512071

[88]   E. HaoYu, S. Cheng, K. Scott, B. Logan, Microbial fuel cell performance with non-Pt cathode catalysts, J. Power Sources. 171 (2007) 275-281. https://doi.org/10.1016/j.jpowsour.2007.07.010

[89] F. Zhao, F. Harnisch, U. Schröder, F. Scholz, P. Bogdanoff, I. Herrmann, Application of pyrolysed iron(II) phthalocyanine and CoTMPP based oxygen reduction catalysts as cathode materials in microbial fuel cells, Electrochem. Commun. 7 (2005) 1405-1410. https://doi.org/10.1016/j.elecom.2005.09.032

[90] Y. Yuan, B. Zhao, Y. Jeon, S. Zhong, S. Zhou, S. Kim, Iron phthalocyanine supported on amino-functionalized multi-walled carbon nanotube as an alternative cathodic oxygen catalyst in microbial fuel cells, Bioresour. Technol. 102 (2011) 5849-5854. https://doi.org/10.1016/j.biortech.2011.02.115

[91] M. Lu, S. Kharkwal, H.Y. Ng, S.F.Y. Li, Carbon nanotube supported MnO2 catalysts for oxygen reduction reaction and their applications in microbial fuel cells, Biosens. Bioelectron. 26 (2011) 4728-4732. https://doi.org/10.1016/j.bios.2011.05.036

[92] M. Ghasemi, M. Ismail, S.K. Kamarudin, K. Saeedfar, W.R.W. Daud, S.H. Hassan, L.Y. Heng, J. Alam, S.-E. Oh, Carbon nanotube as an alternative cathode support and catalyst for microbial fuel cells, Appl. Energ. 102 (2013) 1050-1056. https://doi.org/10.1016/j.apenergy.2012.06.003

[93] M. Lu, L. Guo, S. Kharkwal, H.Y. Ng, S.F.Y. Li, Manganese–polypyrrole–carbon nanotube, a new oxygen reduction catalyst for air-cathode microbial fuel cells, J. Power Sources. 221 (2013) 381-386. https://doi.org/10.1016/j.jpowsour.2012.08.034

[94] D. Zhu, D. B. Wang, T. S. Song, T. Guo, P. Ouyang, P. Wei, J. Xie, Effect of carbon nanotube modified cathode by electrophoretic deposition method on the performance of sediment microbial fuel cells, Biotechnol. Lett. 37 (2015) 101-107. https://doi.org/10.1007/s10529-014-1671-6

[95] M. Ghasemi, W.R.W. Daud, S.H. Hassan, T. Jafary, M. Rahimnejad, A. Ahmad, M.H.J.I.J.O.H.E. Yazdi, Carbon nanotube/polypyrrole nanocomposite as a novel cathode catalyst and proper alternative for Pt in microbial fuel cell, Int. J. Hydrogen Energ. 41 (2016) 4872-4878. https://doi.org/10.1016/j.ijhydene.2015.09.011

[96] Y.R. He, F. Du, Y.X. Huang, L.M. Dai, W.W. Li, H.Q. Yu, Preparation of microvillus-like nitrogen-doped carbon nanotubes as the cathode of a microbial fuel cell, J. Mater. Chem. A Mater 4 (2016) 1632-1636. https://doi.org/10.1039/C5TA06673E

Microbial Fuel Cells: Materials and Applications
Materials Research Foundations **46** (2019) 151-176

Materials Research Forum LLC
doi: http://dx.doi.org/10.21741/9781644900116-7

Chapter 7

Use of Carbon-Nanotube Based Materials in Microbial Fuel Cells

Hakan Burhan[1], Gazi Yilmaz[1], Ahmet Zeytun[1], Harbi Calimli[1,2], Fatih Sen[1*]

[1]Sen Research Group, Department of Biochemistry, Dumlupınar , University, 43100 Kütahya, Turkey

[2]Tuzluca Vocational High School, Igdir University, Igdir, Turkey

fatihsen1980@gmail.com

Abstract

Today, carbon nanotubes are extensively used for energy suppy in electrical devices due to their specific mechanical, thermal, and electrical properties. Additionally, carbon nanotubes are widely used in microbial fuel cells which provide zero carbon emission in energy supply. Carbon nanotube-based materials have good electrochemical properties for the microbial fuel cells. Also, the use of carbon nanotube-based composites in the design of microbial fuel cells is very important in a large area of application from medical devices to scaled power generation. For this purpose, in this chapter, microbial fuel cells formed by carbon nanotube-based composites were evaluated by using microbial fuel cells and the effects of microbial fuel cells.

Keywords

Carbon-Based Materials, Electrochemistry, Fuel Cells, Nanomaterial, Nanotechnology

Contents

1.   Introduction

Since Iijima and his colleagues [1] discovered the carbon nanotubes in 1991, they have attracted considerable interest due to their unique characteristics. The general structure of the carbon nanotube is such that it has a narrow, long, and cylindrical carbon atom at each end [2–3]. These tubular structures are often multi-walled (MWCNT) and single-tube or single-walled (SWCNT) and have a very high degree of electrical conductivity [4], thermal conductivity [5], high tensile strength [6], and elastic modulus. Materials treated with carbon nanotubes gain unique properties and therefore, are highly preferred as the basic building blocks. Carbon nanotube-based materials have unique properties such as magnetic, adsorption, large surface area, and biocompatibility. Thus, carbon nanotube-based materials are extensively used in the design of materials. These structures can be used in many areas including tissue engineering [7], energy collection materials [8–9], biosensors [10–14], biological distribution systems [15–17], and microelectronics [18–20]. Among these applications, microbial fuel cells are also very important and they use microorganisms in their platform. Microorganisms used in microbial fuel cell produce electricity due to their capability of exocellular electron transfer, so they are known as exoelectrogen [21]. To generate electricity in microbial fuel cells, some organic substances, such as lactase, glucose, and acetate are oxidized with the help of an electrocatalyst. Reduction of the organic substance (e-transmitter) to the carbon dioxide occurs by metabolic events and bacterial respiration. Upon the electrons transferred on the surface of an anode, they pass to an external surface to generate electricity. Finally, in the electrolyte, the electrons pass through a membrane and come together with protons that pass from the anode to the cathode, and a cathode is collected at the tip, whereby water from the oxygen (e-receiver) is formed. Due to the possibility of using different bacterial sources in energy recovery applications, microbial fuel cells can be used in small power sources [22–25], biomass [26–30], hydrogen production [31–35], biosensor studies [36–40], and chemical synthesis systems [41–42].

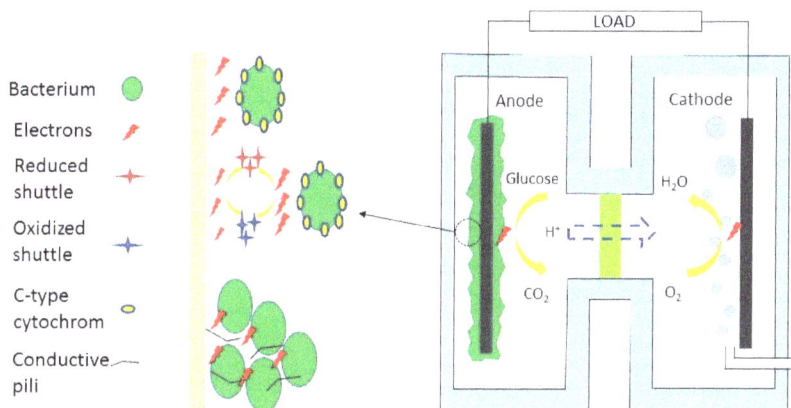

Fig. 1. (Right) In the general working principle of microbial fuel cells; It is catalyzed by the biofilm (bacterial community) that produce protons and carbon dioxide from the oxidations of organic molecules; as a result, the electrons are passed to the anode. It is shown in the figure that it is substantially reduced by the use of metal catalysts in conventional microbial fuel cells to generate oxygen water in the cathode. In general, a membrane (green) can be used to increase the filtration of the protons between the cathode and the anode parts, but in laminar flow microbial fuel cells, the membranes are removed because the cations move at high speed and relative flow rates cause low convector. In a multilayered biofilm, the electrons for the outer layers are carried by their conducting batteries, mainly known as nanowires. At the same time, electrons are transferred to the surface of anode thank to contact directly or electron shuttles (on the left side).

Among the materials that can be used for microbial fuel cells, carbon nanotube-based materials are very important. Especially, 10 years ago, carbon nanotube-based composites and carbon nanotubes were used as components of microbial fuel cells. The resulting nano-engineering materials, which help the oxygen reduction reaction (ORR) in the cathode and the bacterial population in the anode, are aimed to develop the microbial fuel cells in these two ways. We should note that in this classification, only the biocathodes and bacterial communities act as electron acceptors [43–45]. Besides, Antolini et al. [46] have investigated microbial fuel cells composite applications containing polymer/carbon, carbon/metal, polymer/metal composites, while another article deals with a broader approach to electrode materials [47]. While emphasizing the importance of their work, the study included carbon nanotubes in microbial fuel cells and stated that they would benefit both nanomaterial researchers and the microbial fuel cell research community due to the high demand for new carbon nanotube-based chemicals to produce bioenergy. In this chapter, we start by monitoring the carbon nanotubes that are initially functional and continue with the methods that must be applied for the preparation of carbon nanotube-based microbial fuel cell electrodes and then, the synthesis process of the carbon nanotubes. In the next step, we evaluated the impact of nanomaterials on microbial fuel cell and finally, we discussed the problems and possibilities of microbial fuel cell's scaling and commercialization with the help of carbon nanotubes.

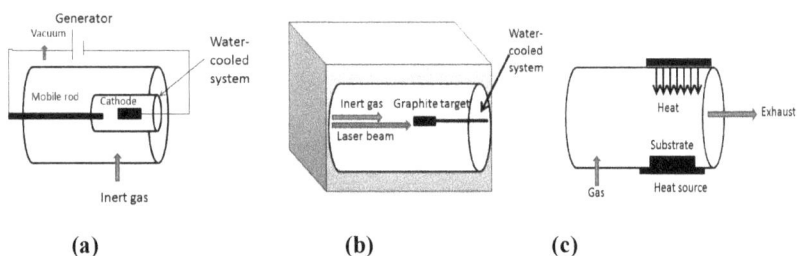

(a)   (b)   (c)

Fig. 2. In the arc discharge shown in Figure 2, the energy present in a plasma having high temperature released from the electrodes as a result of evaporation in compressed of carbon nanotubes. (b) Laser energy is produced by Laser ablation having high temperature and convert a carbon source upon the carbon vapor occurs in a helium and argon stream. (c) In chemical vapor deposition, the gaseous hydrocarbon is crushed into hydrogen and carbon and deposited on the catalyst bed as carbon nanotubes and introduced into the furnace.

2   Design of carbon nanotube-based electrodes

2.1   Fabrication of pristine carbon nanotubes

The individual carbon nanotubes can be produced in high volumes using these three methods, laser ablation [48], arc–discharge [49] and chemical vapour deposition (CVD) [50]. Laser ablation and arc discharge produce high-quality carbon nanotubes by compressing an inert gas, such as helium or argon and by drying a solid carbon source

(Figure 2). The high temperature in laser ablation is achieved by laser energy [48–49]; hence, the drying of the carbon source is supplied, and the energy present on plasma is discharged between the anode and the cathode. In the studies which aim to produce cost-effective carbon nanotubes, the efficiency is low due to high temperatures. Therefore, arc discharge and laser ablation are disruptive. Furthermore, as shown in Fig. 2, high-quality carbon nanotubes in chemical vapour deposition can be produced in the result of the reaction occurred among catalyst and hydrocarbon feedstock [51].

2.2   Functionalization of carbon nanotubes

Carbon nanotubes differ in length, structure, diameter, etc. The carbonaceous by-products and metal impurities are generally obtained with their impurities. Therefore, some functionalization pathways are needed to eliminate impurities and to offer a way to synthesize carbon nanotubes containing hybrid materials to design some applications. Carbon nanotubes are functionalized by covalent bonding formed from a chemical reaction or by adsorption or coagulation of carbon nanotubes with different groups, such as polymers, metal nanoparticles [52–61], and biomolecules metal nanoparticles [62]. The preformed nanoparticles (NPs) can be enlarged in addition to the carbon nanotubes structure which is the direct processor that can be linked to the carbon nanotubes forms by covalent and weaker bonds as the indirect approach [62–63]. In the procedure, a reduction process is carried out using carbon nanotubes where containing noble metals as precursors, and metal nanoparticles are adhered by Van der Waals interaction. This process is supported by light, reducing agents, and heat [63–64]. The carbon nanotubes interact with metal nanoparticles using the covalent connection path. These binders provide the formation of covalent bonds with weak intercellular bonds, such as hydrophobic and electrostatic interactions, and functional groups on the carbon nanotubes surface. Carbon nanotubes can also be functionalized by covalently bonded polymer functional groups or by adsorption on the surface of the carbon nanotubes. Carbon nanotubes and polymers will be mixed in a solvent to obtain polymers/carbon nanotube composite, and solvent-based solution casting techniques, which are very common, are used for the polymerization process of the carbon nanotubes [65–66]. Furthermore, other than these conventional techniques, the manufacture of conductive crosslinked polymer materials, low-tech dip coating or casting methods show low control over surface-coated on the surface of the composite material. Current studies commonly utilize layer-layer (LBL) assembly techniques in the producing of electrodes having functional groups for fuel cells and batteries [67–69]. These recommendations were then reported by Gittleson et al. [70], and it was possible to install polyelectrolyte/carbon nanotube films automatically and very quickly, as in 13 seconds, thanks to reducing the cycle time to less than seconds. Other properties for nano level control are film growth, material waste cost, and thin film conductivity.

2.3   The use of carbon nanotube-based materials

Carbon nanotubes have been produced for aerogels, foams, wafers, arrays, sheets, yarns, and sponges with macroscopic structure. Each of these structures offers substantially different properties, functionality, and physical parameters from the functionalized or crafted carbon nanotubes. Carbon nanotubes have a broad usage area, such as energy conversion and storage systems in devices. Additionally, they are used in the forming 2D thin films decorated on a conductive surface and a 3D monolith [71].

2.3.1   Carbon nanotube-based thin films

In a chemical vapour deposition-baked substrate, such as silicon oxide, the carbon nanotube thin films can be modified or modified directly with functional groups usually at a temperature of 500 to 1200 °C. Catalyst nanoparticles, thanks to the help of chemical vapour deposition parameters and the surface properties of the substrate, make the surface of materials functionalized due to having a network that allows the properties of materials to be controlled by a network regarding density, porosity, and alignment. To summarize, previously prepared carbon nanotubes can be collected or coated on a desired substrate at the ambient condition to produce a thin film to use as electrodes [72].

This process is usually conducted using an organic solvent containing a surfactant and comprises a multi-step wet-chemical process involving acid treatment of intact carbon nanotubes. This method can be classified regarding commercial values where growths on the surface of carbon nanotubes are taken as a scale, but if left in the stored thin film, the electrical conductivity of carbon nanotube networks is adversely affected by the use of harsh chemicals and surfactants. Many autoclavable and chemically resistant polymers are profoundly retarded for laboratory and production of microbial fuel cells [73–74]. However, some of the carbon nanotube film production techniques mentioned above cannot be used directly. For this reason, these considerations require the shaping of the carbon nanotube web on the polymer substrate according to the requirements of the microbial fuel cell containing carbon nanotube thin film printing methods on a receiving substrate (Figure 3). From a general point of view, carbon nanotube thin film transfer, including wet and dry processes, is divided into two sections. These techniques are supported by factors such as pressure, heat, laser, etc., to remove impurities and improve the process activity of the desired model. Negligible substrate or substrate bearing carbon nanotubes, and the dry transfer in the carbon nanotubes are generally controlled by the energy produced from the substrates present in the experiments [75–87]. PDMS is an

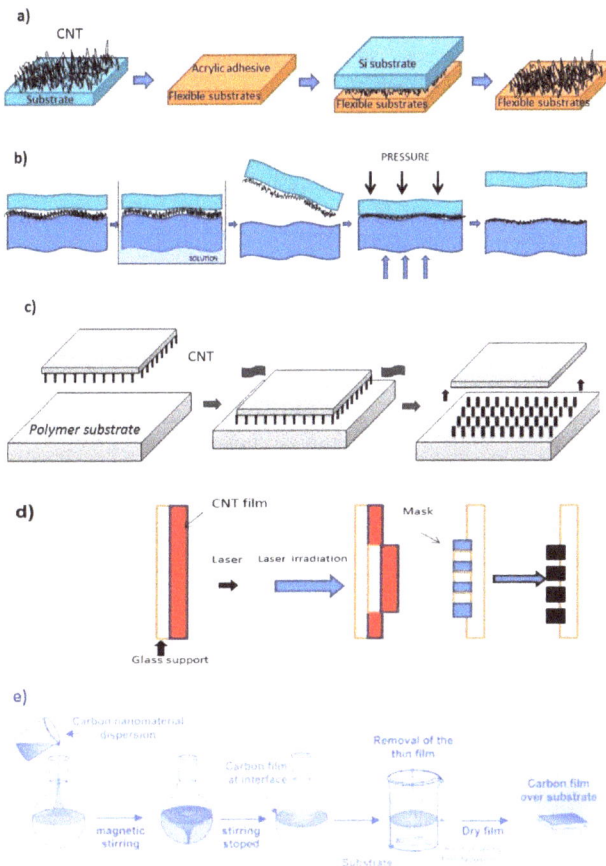

Figure 3(a). Lin and his colleagues used an acrylic substrate to provide adhesive transfer of carbon nanotubes from the surface of the silicon substrate. (b)Transfer pressure; It exposes carbon nanotube film by spraying and submerging on a glass surface with PDMS stamp. After this step, the PMLS is peeled, and the temperature and pressure are applied to transfer it in a substep. (c) A polymer containing carbon nanotube as a carrier is loaded upon the microwave -assisted transfer applies for the enlargement of silicon. The resulting pattern comprising carbon nanotubes is heated by microwave (d). A laser impact energy is conducted for the evaporation of some impurities for carbon nanotube present on the first substrate. (e) The liquid-liquid interface transfer is performed by resting the carbon nanotube with an aqueous solution and an organic solution, and then a film has appeared at the liquid interface. To adhere to the thin film to the surface, a substrate present in the mixture is drawn from bottom to top.

appropriate material to apply for transferring of film from the initial and the desired substrate. The PDMS undergoes soft lithographic processing to transfer a carbon nanotube network pattern specified by the configuration of the microbial fuel cells compartments. Also, the potential for the energy of microbial fuel cells is lower than the polymers used in its design and manufacture. The more precise and difficult part of the fabrication of the thin film is a wet transfer using many commercially available carbon nanotube solutions. This process is simultaneously applied with thermal heating to eliminate surfactants in the solution [71].

2.3.2 Carbon nanotubes–based monolith structures

Current developments have shown that 3D monolithic structures have a high conductivity for freestanding electrodes in terms of exposing the surface of the carbon nanotubes to the adsorption of chemical groups. While the variation of 3D materials was improved using carbon nanotubes, these materials can be separated into three sections, which are the synthesis of foams [87], aerogels [88], and sponges [89]. They are identified by the 3D assembly of interconnected carbon nanotubes with hierarchical and open pores [90].

2.3.2.1 Carbon nanotubes aerogel

There are many techniques adopted for the synthesis of multifunctional and hybrid carbon nanotube aerogels (Figure 4). A liquid containing graphene or carbon nanotubes is extracted from the gel structure at critical point liquid used to produce the desired colloidal particles. This is one of the most commonly used processes. Jung et al. [91], thus, synthesizes single-walled carbon nanotube (CN) aerogels in which the homogeneous distribution that has a solid phase with a dilute concentration is taken first. Subsequently, the drying of the supercritical solvent is conducted to get an aerogel structure. A process which has three stages was applied to form aerogels. In the first step, a solution containing carbon nanotubes in a gel is created with the help of the crosslinker, in the second step, the thermal annealing of the aerogel is carried out, and in the final step, electrical and mechanical properties are enhanced [92].

2.3.2.2 Carbon nanotubes foam

The metal foams can be produced directly on the chemical vapor deposition techniques followed by the creation of 3-dimensional monolith structures of carbon nanotube foams, separating a substrate in a formed metal from carbon-nanotube having a 3D structure. Carbon nanotube-graphene foam was synthesized by Dong et al. [93–95] by a two-step chemical vapor deposition procedure. Firstly, graphene was obtained using nickel foam, and then, this graphene was kept at 750 °C for 40 minutes to reach the desired size.

Microbial Fuel Cells: Materials and Applications
Materials Research Foundation LLC
Materials Research Foundations **46** (2019) 151-176
doi: http://dx.doi.org/10.21741/9781644900116-7

Finally, the resulting pattern was incubated overnight in hydrochloric acid (HCl) to remove the substrate from the metal foam. The obtained pattern exhibits superhydrophobic properties.

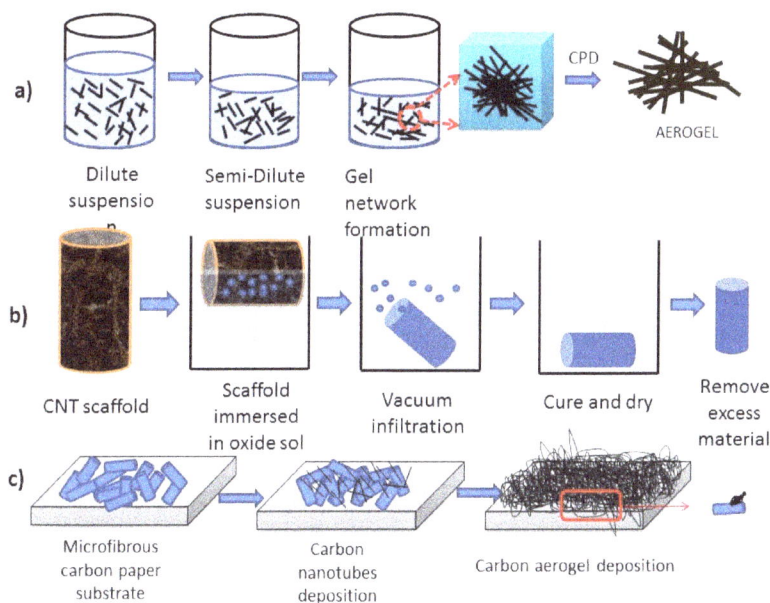

Figure 4 shows the carbon nanotube-based aerogel synthesis process schematically. (a) A new method of synthesis conducted according to mentioned above is used to increase the concentration of colloids for obtaining gel to extract the solvent produced with supercritical drying. (b) formation of a composite containing metal oxide of aerogels or CN synthesized with the sol-gel method. (c) It is an aerogel synthesis method on carbon nanotubes grown in microfibrous carbon paper.

3.   Anode modification

An anode compartment plays an important role in microbial fuel cells because microbial fuel cell acts as an electron generator in power generation. In this section, bacterial/electrode communication, which is the basis for understanding the anode development pathways, was detected. We then, provide an overview of the anode materials used to improve the overall performance of microbial fuel cells [71].

3.1   Bacteria/electrode interaction

The self-reproducible polymeric materials used on the electrode surface are called as biofilms. The complex nature of the biofilms, metabolic activity [96], formation, growth, and environmental effects [97] was examined in different aspects. The backbone of the bacterial/electrode interaction that develops with sequential steps is the bacterial biofilms. Production of three-dimensional carbon nanotubes starts with the adhesion of bacteria and a monolayer form. Followed by the formation of a multi-layered structure, a mature polymeric biofilm having a 3D structure is formed. This biofilm formation process takes place in two stages as starting and ending maturation of biofilm. The electrode surfaces are optimized for the maturation of the electrodes. For this purpose, the electrode surfaces are made suitable for the adhesion of the bacteria before the biofilm maturation. In the case of a low bacterial density on the electrode surface during the application of this process, the microbial fuel cell performs a low energy production. Namely, low energy efficiency is produced as a result of a low electrocatalytic activity between electrodes and bacteria. In the second phase of this function, a high level of catalytic activity takes place as a result of the mass transfer of bacteria by the electrode and the transmission of the electron on the 3D structured biofilm. There are three transitions of electron transport in the multilayered structure of the anode as shown in Figure 1. (i) The structures that provide the transport of electrons occur as a result of reduction and oxidation by the bacteria, (ii) bacterial cell membranes in the outer structure of the c-type cytochromes and the charged enzyme forming from redox reaction, (iii) electron transfer using a self-generated electrical conductive battery called a bacterial nanowire [71].

3.2   Carbon nanotube-based anode materials

Usually, natural polymeric matrices, such as carbon nanotubes, gelatin [98], collagen [99], alginate [100] or chitosan [101], can be synthesized with synthetic polymers such as polycarbosilane [102], polyurethane [103], polycarbonate [104], PLGA [14], PMMA [105] or by metal oxide/catalyst nanoparticles.

3.2.1 Metal/carbon nanotubes

The fixation of metallic nanoparticles on carbon nanotubes can be used as skeletons for electronics [106], fuel cells [107–113] and sensors [114]. Metal oxides including TiO_2, SnO_2, MgO_2 for integration into carbon nanotubes or carbon nanotube/polymer hybrid materials should have structural stability, low cost, and abundant availability. The structure of the microbial fuel cell anodes using carbon nanotube-based materials decorated with metal oxide particles shows better performance compared to bare carbon nanotubes [115–116]. This nanohybrid material has better performance than carbon nanotube and TiO_2 based electrodes. The nanoparticles of tin oxide (SnO_2) bonded to carbon nanotubes were also applied to the microbial fuel cells as an electrode material.

3.3   Carbon nanotube-based anode assembly

The performance of microbial fuel cell is primarily regulated by the chemical properties of carbon nanotube-based materials. As described in the following sections, the microbial fuel cell anodes are constructed as self-assembled carbon nanotubes, carbon nanotube-based monoliths/free-standing structures alone. Carbon nanotube alignment is regulated during the expression of carbon nanotubes thin films, such as porosity and layer resistance, which significantly affect electrode biocompatibility. For example, porosity allows the distribution of bacteria to the biofilm layers. In this context, a membrane-free microbial fuel cell using a modified anode according to conventional gold and nickel electrodes is proven to be a high-strength energy harvesting machine that reaches 6 and 20 times higher current and power density. For the first time for power generation on a sustainable chip with a microbial fuel cell, these findings have enabled applications in electronics [71]. The stay of bacteria within the carbon nanotube network affects the pore size of biofilm formation, cell viability and various surface alignments of carbon nanotubes. Ren-sputtered (SSLbL) carbon nanotube films on the gold substrate, randomly aligned carbon nanotubes (RACNTs), vertically aligned carbon nanotubes (VACNTs) or carbon nanotube forests, and high-pore characteristics were added in the study by Ren et al. [117] As mentioned by Ren et al., the vertically aligned carbon nanotube results in the formation of fine biofilm. On the contrary, randomly aligned carbon nanotube opens thicker biofilms for the easy localization of bacteria within the randomly aligned carbon nanotube network, and thus, higher electricity production can be achieved. The potential for increasing contact-based electron transfer occurs and expands the scope of the bacterial biomass feed support significantly. To create a 3D conductor network, the carbon nanotubes are connected to form a network and determine the volume fraction in a solution. The resulting entangled conductor, carbon nanotube network, is mounted on a carbon-based anode. A membrane is used which adversely affects the sustainability of energy production, power output, and equipment costs [71].

4.   Cathode modification

The power generation cycle in a microbial fuel cell is completed as a result of the generation of water. The oxygen reduction reaction kinetics is very slow. This slow reaction is a significant problem for industrial application of lithium-air batteries and acidic and alkaline fuel cells. The integration of enzymes into microbial fuel cell cathodes is discussed and shows outstanding performance against oxygen reduction reaction, and viability requiring bacterial metabolism [71]. To evaluate the catalytic activity of the cathode, two general oxygen reduction pathways (H_2O) are frequently studied: direct reduction to H_2O by four-electron path and indirect reduction by 2-electron pathways in which hydrogen peroxide (H_2O_2) is the intermediate product and higher energy conversion to the first pathway. Efficacy is the main criteria to decide which pathway to prefer. Catalytic activities for oxygen reduction reaction in carbon materials are linked to the presence of quinone functional groups attributed to the reduction mechanism on the surface of carbon nanotubes. On complexly designed nano-surfaces, the oxygen reduction reaction mechanism varies greatly depending on the type of carbonaceous material and chemical functional groups leading to the formation of a mixture of H_2O_2, OH^-. Rao et al. [118] examined the structure of the oxygen reduction reaction regions on vertically aligned carbon nanotubes supplemented. Developments in microbial fuel cell cathode modification have a wide range of use in many materials. The oxidation of carbon nanotubes with oxygen-carrying functions added to the carbon nanotube structure allows the carbon nanotubes to be more readily bonded to other materials, creating a surface problem. The oxidative treatment is accomplished by boiling the carbon nanotubes in solution, which is a mixture of sulfuric acid (H_2SO_4), nitric acid (HNO_3). Different substrates such as carbon mat, glassy carbon electrode, and metal coated silicon wafers can then be coated with functionalized carbon nanotubes and used as the microbial fuel cell cathode. Zhu et al. [119] studied acid-treated multi-walled carbon nanotubes which were electrophoretically deposited on the graphite felt to be used in the precipitated microbial fuel cell, and a power density of about 60 % higher than that of the naked graphite felt cathode is reached. Functionalized carbon nanotubes tend to remain homogeneous over a more extended period compared to pristine (untreated) carbon nanotubes dispersed in ethanol. Physical adsorption of polymers establishes those types of materials. In situ polymerization is a highly flexible way to inoculate polymers to high-density carbon nanotubes and is provided by initiators in the surface of oxidized carbon nanotubes. The monomers are polymerized to macromolecules and bonded to carbon nanotube surfaces. This method allows the use of the commercial polymer while the steric effect of the macromolecules and the polymer chains are inhibited by low reactivity to anchor the carbon nanotube surface.

5.   Challenges and promises

Carbon nanotube provides excellent growth in nano-engineering materials, energy storage/conversion devices, and production and synthesis techniques for application-oriented nanocomposites. Instead of laser-ablation and arc-discharge techniques, chemical vapor deposition method facilitates time and costs in laboratory-scale investigations on carbon nanotube-based composites. The studies in the carbon nanotube industry have reduced the price of multi-walled carbon nanotubes from 45,000 to 100 $/ kg [120] and if this trend continues, for nano-engineered carbon nanotube materials for commercial use, it is only a matter of time to replace other carbonaceous or valuable materials. The microbial fuel cells need to be scaled, and these systems have been tried to grow and stack in both laboratory and pilot scale tests. Scaling techniques require quick, easy and cost-effective production ways to make bio-source energy recovery more attractive to the industry. In this chapter, we use the carbon nanotube and carbon nanotube composites which have unique properties and turn them into the modification of microbial fuel cell electrodes. The techniques discussed in this article relate to electrode modification of microbial fuel cells. Carbon nanotube composite membranes can replace with ordinary membranes. Carbon nanotube-based membranes are examined for their ionic selectivity and exchangeability and evaluated against high-yield Nafion 117 and anion exchange membranes. Venkatesan et al. [121] reported that the chitosan synthesized by multi-walled carbon nanotube was used as a separator in a patch mode in which the serial was added, and research in this area should be continued. The final step in the rapid and easy modification of carbon nanotube-based microbial fuel cells is the electrode preparation methods. As regards the reduction mechanism in the cathode, the device must limit the power output and depends on the anode performance. The study of biocompatible and efficient microbial anodes is expected to address ongoing problems in slow oxygen reduction reaction kinetics. The rate of growth in nano-engineering material for energy storage will record great improvement in all microbial fuel cells fields.

Conclusions

As a result, carbon nanotubes and carbon nanotube-based materials are extensively used in microbial fuel cells which provide zero emission in energy supply. They have good electrochemical properties for the microbial fuel cells. Furthermore, the use of carbon nanotube-based composites is very important in a large area of application from medical devices to scaled power generation. For this reason, in this chapter, microbial fuel cells formed by carbon nanotube-based composites were evaluated by using microbial fuel cells and the effects of microbial fuel cells.

References

[1]   S. Lijima, Helical Microtubules of Graphitic Varbon. Nature, 354 (1991) 56–58.

[2]   W. Hoenlein, F. Kreupl, G. S. Duesberg, A. P. Graham, M. Liebau, R. V. Seidel, & E. Unger, Carbon Nanotube Applications in Microelectronics. IEEE

Transactions on Components and Packaging Technologies, 27 (2004) 629–634. https://doi.org/10.1109/TCAPT.2004.838876.

[3]   T.-W. C. Erik T. Thostensona, Zhifeng Renb, Advances in The Science and Technology of Carbon Nanotubes and Their Composites: A Review. Computer Language Magazine, 61 (2001) 1899–1912. https://doi.org/10.1016/s0266-3538(01)00094-x.

[4]   P. D. Bradford & A. E. Bogdanovich, Electrical Conductivity Study of Carbon Nanotube Yarns, 3-D Hybrid Braids and Their Composites. Journal of Composite Materials, 42 (2008) 1533–1545. https://doi.org/10.1177/0021998308092206.

[5]   D. T. Savas Berber, Young-Kyun Kwon, Unusually High Thermal Conductivity of Carbon Nanotubes. 84 (2000) 4613–4616. https://doi.org/10.1103/PhysRevLett.84.4613.

[6]   M. F. Yu, O. Lourie, M. J. Dyer, K. Moloni, T. F. Kelly, & R. S. Ruoff, Strength and Breaking Mechanism of Multiwalled Carbon Nanotubes under Tensile Load. Science, 287 (2000) 637–640. https://doi.org/10.1126/science.287.5453.637.

[7]   E. L. Hopley, S. Salmasi, D. M. Kalaskar, & A. M. Seifalian, Carbon Nanotubes Leading The Way Forward in New Generation 3D Tissue Engineering. Biotechnology Advances, 32 (2014) 1000–1014. https://doi.org/10.1016/j.biotechadv.2014.05.003.

[8]   Z. Chen, Y. Yuan, H. Zhou, X. Wang, Z. Gan, F. Wang, & Y. Lu, 3D Nanocomposite Architectures from Carbon-Nanotube-Threaded Nanocrystals for High-Performance Electrochemical Energy Storage. Advanced Materials, 26 (2014) 339–345. https://doi.org/10.1002/adma.201303317.

[9]   D. Yu, K. Goh, H. Wang, L. Wei, W. Jiang, Q. Zhang, L. Dai, & Y. Chen, Scalable synthesis of Hierarchically Structured Carbon Nanotube-Graphene Fibres for Capacitive Energy Storage. Nature Nanotechnology, 9 (2014) 555–562. https://doi.org/10.1038/nnano.2014.93.

[10]   B. Esser, J. M. Schnorr, & T. M. Swager, Selective Detection of Ethylene Gas Using Carbon Nanotube-Based Devices: Utility in Determination of Fruit Ripeness. Angewandte Chemie - International Edition, 51 (2012) 5752–5756. https://doi.org/10.1002/anie.201201042.

[11]   T. Kurkina, A. Vlandas, A. Ahmad, K. Kern, & K. Balasubramanian, Label-free Detection of Few Copies of DNA with Carbon Nanotube Impedance Biosensors. Angewandte Chemie - International Edition, 50 (2011) 3710–3714. https://doi.org/10.1002/anie.201006806.

[12]   C. M. Rivera, H. J. Kwon, A. Hashmi, G. Yu, J. Zhao, J. Gao, J. Xu, W. Xue, &
       A. G. Dimitrov, Towards A Dynamic Clamp for Neurochemical Modalities.
       Sensors (Switzerland), 15 (2015) 10465–10480.
       https://doi.org/10.3390/s150510465.

[13]   X. Shi, A. Von Dem Bussche, R. H. Hurt, A. B. Kane, & H. Gao, Cell Entry of
       One-Dimensional Nanomaterials occurs by Tip Recognition and Rotation. Nature
       Nanotechnology, 6 (2011) 714–719. https://doi.org/10.1038/nnano.2011.151.

[14]   Y. Zhao, S. Wang, Q. Guo, M. Shen, & X. Shi, Hemocompatibility of Electrospun
       Halloysite Nanotube- and Carbon Nanotube-Doped Composite Poly(Lactic- Co -
       Glycolic Acid) Nanofibers. Journal of Applied Polymer Science, 127 (2013)
       4825–4832. https://doi.org/10.1002/app.38054.

[15]   A. Faraj & published by Dove Press, Magnetic Single-Walled Carbon Nanotubes
       As Efficient Drug Delivery Nanocarriers in Breast Cancer Murine Model:
       Noninvasive Monitoring Using Diffusion-Weighted Magnetic Resonance İmaging
       as Sensitive Imaging Biomarker. International Journal of Nanomedicine, 10 (2015)
       157–168. https://doi.org/10.2147/IJN.S75074.

[16]   D. Cai, J. M. Mataraza, Z.-H. Qin, Z. Huang, J. Huang, T. C. Chiles, D. Carnahan,
       K. Kempa, & Z. Ren, Highly Efficient Molecular Delivery into Mammalian Cells
       Using Carbon Nanotube Spearing. Nature Methods, 2 (2005) 449–454.
       https://doi.org/10.1038/nmeth761.

[17]   N. W. Shi Kam, M. O'Connell, J. A. Wisdom, & H. Dai, Carbon Nanotubes as
       Multifunctional Biological Transporters and Near-Infrared Agents for Selective
       Cancer Cell Destruction. Proceedings of the National Academy of Sciences, 102
       (2005) 11600–11605. https://doi.org/10.1073/pnas.0502680102.

[18]   F. Bottacchi, L. Petti, F. Späth, I. Namal, G. Tröster, T. Hertel, & T. D.
       Anthopoulos, Polymer-sorted (6,5) Single-Walled Carbon Nanotubes for Solution-
       Processed Low-Voltage Flexible Microelectronics. Applied Physics Letters, 106
       (2015) 193302–193306. https://doi.org/10.1063/1.4921078.

[19]   A. D. Franklin, M. Luisier, S. J. Han, G. Tulevski, C. M. Breslin, L. Gignac, M. S.
       Lundstrom, & W. Haensch, Sub-10 nm Carbon Nanotube Transistor. Nano
       Letters, 12 (2012) 758–762. https://doi.org/10.1021/nl203701g.

[20]   M. Held, S. P. Schießl, D. Miehler, F. Gannott, & J. Zaumseil, Polymer/Metal
       Oxide Hybrid Dielectrics for Low Voltage Field-Effect Transistors with Solution-
       Processed, High-Mobility Semiconductors. Applied Physics Letters, 107 (2015) 1–
       5. https://doi.org/10.1063/1.4929461.

[21] B. E. Logan, Logan 2009. Exoelectrogenic Bacteria That Power Microbial Fuel
      Cells. Nature Reviews Microbiology, 7 (2009) 375–381.
      https://doi.org/10.1038/nrmicro2113.

[22] S. Choi, Microscale Microbial Fuel Cells: Advances and Challenges. Biosensors
      and Bioelectronics, 69 (2015) 8–25. https://doi.org/10.1016/j.bios.2015.02.021.

[23] J. E. Mink, R. M. Qaisi, B. E. Logan, & M. M. Hussain, Energy Harvesting from
      Organic Liquids in Micro-Sized Microbial Fuel Cells. NPG Asia Materials, 6
      (2014) e89–e89. https://doi.org/10.1038/am.2014.1.

[24] H. Ren, C. I. Torres, P. Parameswaran, B. E. Rittmann, & J. Chae, Improved
      Current and Power Density with A Micro-Scale Microbial Fuel Cell due to A
      Small Characteristic Length. Biosensors and Bioelectronics, 61 (2014) 587–592.
      https://doi.org/10.1016/j.bios.2014.05.037.

[25] D. Vigolo, T. T. Al-Housseiny, Y. Shen, F. O. Akinlawon, S. T. Al-Housseiny, R.
      K. Hobson, A. Sahu, K. I. Bedkowski, T. J. Dichristina, & H. A. Stone, Flow
      Dependent Performance of Microfluidic Microbial Fuel Cells. Physical Chemistry
      Chemical Physics, 16 (2014) 12535–12543. https://doi.org/10.1039/c4cp01086h.

[26] E. Baranitharan, M. R. Khan, D. M. R. Prasad, W. F. A. Teo, G. Y. A. Tan, & R.
      Jose, Effect of Biofilm Formation on The Performance of Microbial Fuel Cell for
      The Treatment of Palm Oil Mill Effluent. Bioprocess and Biosystems Engineering,
      38 (2015) 15–24. https://doi.org/10.1007/s00449-014-1239-9.

[27] F. J. Hernández-Fernández, A. Pérez De Los Ríos, M. J. Salar-García, V. M.
      Ortiz-Martínez, L. J. Lozano-Blanco, C. Godínez, F. Tomás-Alonso, & J.
      Quesada-Medina, Recent Progress and Perspectives in Microbial Fuel Cells For
      Bioenergy Generation and Wastewater Treatment. Fuel Processing Technology,
      138 (2015) 284–297. https://doi.org/10.1016/j.fuproc.2015.05.022.

[28] H. Liu, R. Ramnarayanan, & B. E. Logan, Production of Electricity during
      Wastewater Treatment Using a Single Chamber Microbial Fuel Cell.
      Environmental Science & Technology, 38 (2004) 2281–2285.
      https://doi.org/10.1021/es034923g.

[29] B. E. Logan, M. J. Wallack, K. Y, Kim, W. He, Y. Feng, & P. E. Saikaly,
      Assessment of Microbial Fuel Cell Configurations and Power Densities.
      Environmental Science & Technology Letters, 2 (2015) 206–214.
      https://doi.org/10.1021/acs.estlett.5b00180.

[30] P. F. Tee, M. O. Abdullah, I. A. W. Tan, N. K. A. Rashid, M. A. M. Amin, C.
      Nolasco-Hipolito, & K. Bujang, Review on Hybrid Energy Systems for
      Wastewater Treatment and Bio-Energy Production. Renewable and Sustainable
      Energy Reviews, 54 (2016) 235–246. https://doi.org/10.1016/j.rser.2015.10.011.

[31] A. A. Carmona-Martínez, E. Trably, K. Milferstedt, R. Lacroix, L. Etcheverry, &
      N. Bernet, Long-Term Continuous Production of H_2 in A Microbial Electrolysis
      Cell (MEC) Treating Saline Wastewater. Water Research, 81 (2015) 149–156.
      https://doi.org/10.1016/j.watres.2015.05.041.

[32] T. Catal, K. L. Lesnik, & H. Liu, Suppression of Methanogenesis for Hydrogen
      Production in Single-Chamber Microbial Electrolysis Cells Using Various
      Antibiotics. Bioresource Technology, 187 (2015) 77–83.
      https://doi.org/10.1016/j.biortech.2015.03.099.

[33] P. Kuntke, T. H. J. A. Sleutels, M. Saakes, & C. J. N. Buisman, Hydrogen
      Production and Ammonium Recovery From Urine by A Microbial Electrolysis
      Cell. International Journal of Hydrogen Energy, 39 (2014) 4771–4778.
      https://doi.org/10.1016/j.ijhydene.2013.10.089.

[34] S. Cheng, H. V. M. Hamelers, B. E. Logan, D. Call, S. Cheng, H. V. M. Hamelers,
      T. H. J. A. Sleutels, A. W. Jeremiasse, & R. A. Rozendal, Microbial Electrolysis
      Cells for High Yield Hydrogen Gas Production from Organic Matter.
      Environmental Science and Technology, 42 (2008) 8630–8640.

[35] Y. Zhang & I. Angelidaki, Microbial Electrolysis Cells Turning to be Versatile
      Technology: Recent Advances and Future Challenges. Water Research, 56 (2014)
      11–25. https://doi.org/10.1016/j.watres.2014.02.031.

[36] N. Jayasinghe, A. Franks, K. P. Nevin, & R. Mahadevan, Metabolic Modeling of
      Spatial Heterogeneity of Biofilms in Microbial Fuel Cells Reveals Substrate
      Limitations in Electrical Current Generation. Biotechnology Journal, 9 (2014)
      1350–1361. https://doi.org/10.1002/biot.201400068.

[37] Y. Jiang, P. Liang, C. Zhang, Y. Bian, X. Yang, X. Huang, & P. R. Girguis,
      Enhancing The Response of Microbial Fuel Cell Based Toxicity Sensors to Cu(II)
      with The Applying of Flow-Through Electrodes And Controlled Anode Potentials.
      Bioresource Technology, 190 (2015) 367–372.
      https://doi.org/10.1016/j.biortech.2015.04.127.

[38] K. P. Katuri, A. M. Enright, V. O'Flaherty, & D. Leech, Microbial Analysis of
      Anodic Biofilm in A Microbial Fuel Cell Using Slaughterhouse Wastewater.
      Bioelectrochemistry, 87 (2012) 164–171.
      https://doi.org/10.1016/j.bioelechem.2011.12.002.

[39] L. Lu, D. Xing, & Z. J. Ren, Microbial Community Structure Accompanied with
      Electricity Production in A Constructed Wetland Plant Microbial Fuel Cell.
      Bioresource Technology, 195 (2015) 115–121.
      https://doi.org/10.1016/j.biortech.2015.05.098.

[40]   N. S. Malvankar, M. T. Tuominen, & D. R. Lovley, Biofilm Conductivity is a
       Decisive Variable for High-Current-Density Geobacter Sulfurreducens Microbial
       Fuel Cells. Energy and Environmental Science, 5 (2012) 5790–5797.
       https://doi.org/10.1039/c2ee03388g.

[41]   Y. Li, Photosynthetic Conversion of CO2 To Acetic Acid by An Inorganic-
       Biological Hybrid System. Science China Materials, 59 (2016) 93–94.
       https://doi.org/10.1007/s40843-016-0116-z.

[42]   K. K. Sakimoto, A. B. Wong, & P. Yang, Self-Photosensitization of
       Nonphotosynthetic Bacteria for Solar-to-Chemical Production. 351 (2016) 74–77.
       https://doi.org/10.4271/2011-01-2699.

[43]   Z. Wang, Y. Zheng, Y. Xiao, S. Wu, S. Wu, Z. Yang, & F. Zhao, Analysis of
       Oxygen Reduction and Microbial Community of Air-Diffusion Biocathode in
       Microbial Fuel Cells. Bioresource Technology, 144 (2013) 74–79.
       https://doi.org/10.1016/j.biortech.2013.06.093.

[44]   X. Xia, J. C. Tokash, F. Zhang, P. Liang, X. Huang, & B. E. Logan, Oxygen-
       Reducing Biocathodes Operating with Passive Oxygen Transfer in Microbial Fuel
       Cells. Environmental Science & Technology, 47 (2013) 2085–2091.
       https://doi.org/10.1021/es3027659.

[45]   S. Ci, P. Cai, Z. Wen, & J. Li, Graphene-Based Electrode Materials for Microbial
       Fuel Cells. Science China Materials, 58 (2015) 496–509.
       https://doi.org/10.1007/s40843-015-0061-2.

[46]   E. Antolini, Composite Materials for Polymer Electrolyte Membrane Microbial
       Fuel Cells. Biosensors and Bioelectronics, 69 (2015) 54–70.
       https://doi.org/10.1016/j.bios.2015.02.013.

[47]   Mustakeem, Electrode Materials for Microbial Fuel Cells: Nanomaterial
       Approach. Materials for Renewable and Sustainable Energy, 4 (2015) 1–11.
       https://doi.org/10.1007/s40243-015-0063-8.

[48]   A. Thess, R. Lee, P. Nikolaev, H. Dai, P. Petit, J. Robert, C. Xu, Y. H. Lee, S. G.
       Kim, A. G. Rinzler, D. T. Colbert, G. E. Scuseria, D. Tomanek, J. E. Fischer, & R.
       E. Smalley, Crystalline Ropes of Metallic Carbon Nanotubes. Science, 273 (1996)
       483–487. https://doi.org/10.1126/science.273.5274.483.

[49]   C. Journet, W. K. Maser, P. Bernier, A. Loiseau, M. L. de la Chapelle, S. Lefrant,
       P. Deniard, R. Lee, & J. E. Fischer, Large-Scale Production of Single-Walled
       Carbon Nanotubes by The Electric-Arc Technique. Nature, 388 (1997) 756–758.
       https://doi.org/10.1038/41972.

[50]   A. M. Cassell, J. A. Raymakers, J. Kong, & H. Dai, Large Scale CVD Synthesis of
        Single-Walled Carbon Nanotubes. The Journal of Physical Chemistry B, 103
        (1999) 6484–6492. https://doi.org/10.1021/jp990957s.

[51]   R. T. K. Baker, Catalytic Growth of Carbon Filaments. Carbon, 27 (1989) 315–
        323. https://doi.org/10.1016/0008-6223(89)90062-6.

[52]   C. Journet & P. Bernier, Production of Carbon Nanotubes. Applied Physics A:
        Materials Science & Processing, 67 (1998) 1–9.
        https://doi.org/10.1007/s003390050731.

[53]   Y. Yildiz, H. Pamuk, Ö. Karatepe, Z. Dasdelen, & F. Sen, Carbon Black Hybrid
        Material Furnished Monodisperse Platinum Nanoparticles as Highly Efficient and
        Reusable Electrocatalysts for Formic Acid Electro-Oxidation. RSC Advances, 6
        (2016) 32858–32862. https://doi.org/10.1039/c6ra00232c.

[54]   B. Çelik, G. Başkaya, H. Sert, Ö. Karatepe, E. Erken, & F. Şen, Monodisperse
        Pt(0)/DPA@GO Nanoparticles as Highly Active Catalysts for Alcohol Oxidation
        and Dehydrogenation of DMAB. International Journal of Hydrogen Energy, 41
        (2016) 5661–5669. https://doi.org/10.1016/j.ijhydene.2016.02.061.

[55]   E. Erken, Y. Yıldız, B. Kilbaş, & F. Şen, Synthesis and Characterization of Nearly
        Monodisperse Pt Nanoparticles for C_1 to C_3 Alcohol Oxidation and
        Dehydrogenation of Dimethylamine-borane (DMAB). Journal of Nanoscience and
        Nanotechnology, 16 (2016) 5944–5950. https://doi.org/10.1166/jnn.2016.11683.

[56]   G. Baskaya, I. Esirden, E. Erken, F. Sen, & M. Kaya, Synthesis of 5-Substituted-
        1H-Tetrazole Derivatives Using Monodisperse Carbon Black Decorated Pt
        Nanoparticles as Heterogeneous Nanocatalysts. Journal of Nanoscience and
        Nanotechnology, 17 (2017) 1992–1999. https://doi.org/10.1166/jnn.2017.12867.

[57]   B. Şen, A. Aygün, T. O. Okyay, A. Şavk, R. Kartop, & F. Şen, Monodisperse
        Palladium Nanoparticles Assembled on Graphene Oxide With The High Catalytic
        Activity And Reusability in The Dehydrogenation of Dimethylamine-borane.
        International Journal of Hydrogen Energy, 3 (2018) 2–8.
        https://doi.org/10.1016/j.ijhydene.2018.03.175.

[58]   B. Sen, S. Kuzu, E. Demir, S. Akocak, & F. Sen, Polymer-graphene Hybride
        Decorated Pt Nanoparticles as Highly Efficient and Reusable Catalyst for The
        Dehydrogenation of Dimethylamine–Borane at Room Temperature. International
        Journal of Hydrogen Energy, 42 (2017) 23284–23291.
        https://doi.org/10.1016/j.ijhydene.2017.05.112.

[59]   B. Sen, S. Kuzu, E. Demir, S. Akocak, & F. Sen, Highly Monodisperse RuCo
        Nanoparticles Decorated on Functionalized Multiwalled Carbon Nanotube with
        The Highest Observed Catalytic Activity in The Dehydrogenation of

Dimethylamine−borane. International Journal of Hydrogen Energy, 42 (2017) 23292–23298. https://doi.org/10.1016/j.ijhydene.2017.06.032.

[60] E. Erken, H. Pamuk, Ö. Karatepe, G. Başkaya, H. Sert, O. M. Kalfa, & F. Şen, New Pt(0) Nanoparticles as Highly Active and Reusable Catalysts in the C_1–C_3 Alcohol Oxidation and the Room Temperature Dehydrocoupling of Dimethylamine-Borane (DMAB). Journal of Cluster Science, 27 (2016) 9–23. https://doi.org/10.1007/s10876-015-0892-8.

[61] B. Çelik, Y. Yildiz, H. Sert, E. Erken, Y. Koşkun, & F. Şen, Monodispersed Palladium-Cobalt Alloy Nanoparticles Assembled on Poly(N-Vinyl-Pyrrolidone) (PVP) as A Highly Effective Catalyst for Dimethylamine Borane (DMAB) Dehydrocoupling. RSC Advances, 6 (2016) 24097–24102. https://doi.org/10.1039/c6ra00536e.

[62] N. Karousis, N. Tagmatarchis, & D. Tasis, Current Progress on the Chemical Modification of Carbon Nanotubes. Chemical Reviews, 110 (2010) 5366–5397. https://doi.org/10.1021/cr100018g.

[63] V. Georgakilas, D. Gournis, V. Tzitzios, L. Pasquato, D. M. Guldi, & M. Prato, Decorating Carbon Nanotubes With Metal or Semiconductor Nanoparticles. Journal of Materials Chemistry, 17 (2007) 2679–2694. https://doi.org/10.1039/b700857k.

[64] D. P. Dubal, G. S. Gund, C. D. Lokhande, & R. Holze, Decoration of Spongelike Ni(OH)$_2$ Nanoparticles onto Mwcnts Using An Easily Manipulated Chemical Protocol for Supercapacitors. ACS Applied Materials and Interfaces, 5 (2013) 2446–2454. https://doi.org/10.1021/am3026486.

[65] M. T. Byrne & Y. K. Guin'Ko, Recent advances in Research on Carbon Nanotube-Polymer Composites. Advanced Materials, 22 (2010) 1672–1688. https://doi.org/10.1002/adma.200901545.

[66] S. W. Lee, N. Yabuuchi, B. M. Gallant, S. Chen, B. S. Kim, P. T. Hammond, & Y. Shao-Horn, High-power Lithium Batteries from Functionalized Carbon-Nanotube Electrodes. Nature Nanotechnology, 5 (2010) 531–537. https://doi.org/10.1038/nnano.2010.116.

[67] M. N. Hyder, S. W. Lee, F. Ç. Cebeci, D. J. Schmidt, Y. Shao-Horn, & P. T. Hammond, Layer-by-Layer Assembled Polyaniline Nanofiber/Multiwall Carbon Nanotube Thin Film Electrodes for High-Power and High-Energy Storage Applications. ACS Nano, 5 (2011) 8552–8561. https://doi.org/10.1021/nn2029617.

[68] H. Paloniemi, M. Lukkarinen, T. Ääritalo, S. Areva, J. Leiro, M. Heinonen, K. Haapakka, & J. Lukkari, Layer-by-Layer Electrostatic Self-Assembly of Single-

Wall Carbon Nanotube Polyelectrolytes. Langmuir, 22 (2006) 74–83.
https://doi.org/10.1021/la051736i.

[69]   M. H. Merrill & C. T. Sun, Fast, Simple And Efficient Assembly of Nanolayered
       Materials and Devices. Nanotechnology, 20 (2009) 075606-13.
       https://doi.org/10.1088/0957-4484/20/7/075606.

[70]   F. S. Gittleson, D. J. Kohn, X. Li, & A. D. Taylor, Improving The Assembly
       Speed, Quality, and Tunability of Thin Conductive Multilayers. ACS Nano, 6
       (2012) 3703–3711. https://doi.org/10.1021/nn204384f.

[71]   A. A. Yazdi, L. D'Angelo, N. Omer, G. Windiasti, X. Lu, & J. Xu, Carbon
       Nanotube Modification of Microbial Fuel Cell Electrodes. Biosensors and
       Bioelectronics, 85 (2016) 536–552. https://doi.org/10.1016/j.bios.2016.05.033.

[72]   M. Ghasemi, M. Ismail, S. K. Kamarudin, K. Saeedfar, W. R. W. Daud, S. H. A.
       Hassan, L. Y. Heng, J. Alam, & S.-E. Oh, Carbon Nanotube As an Alternative
       Cathode Support and Catalyst for Microbial Fuel Cells. Applied Energy, 102
       (2013) 1050–1056. https://doi.org/10.1016/j.apenergy.2012.06.003.

[73]   Y. Yıldız, İ. Esirden, E. Erken, E. Demir, M. Kaya, & F. Şen, Microwave (Mw)-
       assisted Synthesis of 5-Substituted 1H-Tetrazoles via [3+2] Cycloaddition
       Catalyzed by MW-Pd/Co Nanoparticles Decorated on Multi-Walled Carbon
       Nanotubes. ChemistrySelect, 1 (2016) 1695–1701.
       https://doi.org/10.1002/slct.201600265.

[74]   E. Erken, I. Esirden, M. Kaya, & F. Sen, A Rapid and Novel Method for The
       Synthesis of 5-Substituted 1H-Tetrazole Catalyzed by Exceptional Reusable
       Monodisperse Pt NPs@AC under The Microwave Irradiation. RSC Advances, 5
       (2015) 68558–68564. https://doi.org/10.1039/c5ra11426h.

[75]   S. Günbatar, A. Aygun, Y. Karataş, M. Gülcan, & F. Şen, Carbon-nanotube-based
       Rhodium Nanoparticles as Highly-Active Catalyst for Hydrolytic
       Dehydrogenation of Dimethylamineborane at Room Temperature. Journal of
       Colloid and Interface Science, 530 (2018) 321–327.
       https://doi.org/10.1016/j.jcis.2018.06.100.

[76]   G. Başkaya, Y. Yıldız, A. Savk, T. O. Okyay, S. Eriş, H. Sert, & F. Şen, Rapid,
       Sensitive, and Reusable Detection of Glucose By Highly Monodisperse Nickel
       Nanoparticles Decorated Functionalized Multi-Walled Carbon Nanotubes.
       Biosensors and Bioelectronics, 91 (2017) 728–733.
       https://doi.org/10.1016/j.bios.2017.01.045.

[77]   R. Ulus, Y. Yıldız, S. Eriş, B. Aday, F. Şen, & M. Kaya, Functionalized Multi-
       Walled Carbon Nanotubes (f-MWCNT) as Highly Efficient and Reusable

Microbial Fuel Cells: Materials and Applications   Materials Research Forum LLC
Materials Research Foundations **46** (2019) 151-176   doi: http://dx.doi.org/10.21741/9781644900116-7

Heterogeneous Catalysts for the Synthesis of Acridinedione Derivatives. ChemistrySelect, 1 (2016) 3861–3865. https://doi.org/10.1002/slct.201600719.

[78] B. Çelik, S. Kuzu, E. Erken, H. Sert, Y. Koşkun, & F. Şen, Nearly Monodisperse Carbon Nanotube Furnished Nanocatalysts as Highly Efficient and Reusable Catalyst for Dehydrocoupling of DMAB and C_1 to C_3 Alcohol Oxidation. International Journal of Hydrogen Energy, 41 (2016) 3093–3101. https://doi.org/10.1016/j.ijhydene.2015.12.138.

[79] S. Sen, F. Sen, A. A. Boghossian, J. Zhang, & M. S. Strano, Effect of Reductive Dithiothreitol And Trolox on Nitric Oxide Quenching of Single-Walled Carbon Nanotubes. Journal of Physical Chemistry C, 117 (2013) 593–602. https://doi.org/10.1021/jp307175f.

[80] F. Sen, A. A. Boghossian, S. Sen, Z. W. Ulissi, J. Zhang, & M. S. Strano, Observation of Oscillatory Surface Reactions of Riboflavin, Trolox, and Singlet Oxygen Using Single Carbon Nanotube Fluorescence Spectroscopy. ACS Nano, 6 (2012) 10632–10645. https://doi.org/10.1021/nn303716n.

[81] N. M. Iverson, P. W. Barone, M. Shandell, L. J. Trudel, S. Sen, F. Sen, V. Ivanov, E. Atolia, E. Farias, T. P. McNicholas, N. Reuel, N. M. A. Parry, G. N. Wogan, & M. S. Strano, In vivo Biosensing via Tissue-Localizable Near-İnfrared-Fluorescent Single-Walled Carbon Nanotubes. Nature Nanotechnology, 8 (2013) 873–880. https://doi.org/10.1038/nnano.2013.222.

[82] Z. W. Ulissi, F. Sen, X. Gong, S. Sen, N. Iverson, A. A. Boghossian, L. C. Godoy, G. N. Wogan, D. Mukhopadhyay, & M. S. Strano, Spatiotemporal Intracellular Nitric Oxide Signaling Captured Using İnternalized, Near-İnfrared Fluorescent Carbon Nanotube Nanosensors. Nano Letters, 14 (2014) 4887–4894. https://doi.org/10.1021/nl502338y.

[83] Z. D. Lin, S. J. Young, & S. J. Chang, CO2 Gas Sensors Based on Carbon Nanotube Thin Films Using a Simple Transfer Method on Flexible Substrate. IEEE Sensors Journal, 15 (2015) 7017–7020. https://doi.org/10.1109/JSEN.2015.2472968.

[84] A. Abdelhalim, A. Abdellah, G. Scarpa, & P. Lugli, Fabrication of Carbon Nanotube Thin Films on Flexible Substrates by Spray Deposition and Transfer Printing. Carbon, 61 (2013) 72–79. https://doi.org/10.1016/j.carbon.2013.04.069.

[85] E. Sunden, J. K. Moon, C. P. Wong, W. P. King, & S. Graham, Microwave Assisted Patterning of Vertically Aligned Carbon Nanotubes onto Polymer Substrates. Journal of Vacuum Science & Technology B: Microelectronics and Nanometer Structures, 24 (2006) 1947–1950. https://doi.org/10.1116/1.2221320.

[86] S. K. Chang-Jian, J. R. Ho, J. W. J. Cheng, & C. K. Sung, Fabrication of Carbon
      Nanotube Field Emission Cathodes in Patterns by A Laser Transfer Method.
      Nanotechnology, 17 (2006) 1184–1187. https://doi.org/10.1088/0957-
      4484/17/5/003.

[87] P. D. Bradford, X. Wang, H. Zhao, & Y. T. Zhu, Tuning The Compressive
      Mechanical Properties of Carbon Nanotube Foam. Carbon, 49 (2011) 2834–2841.
      https://doi.org/10.1016/j.carbon.2011.03.012.

[88] S. Nardecchia, D. Carriazo, M. L. Ferrer, M. C. Gutiérrez, & F. Del Monte, Three
      Dimensional Macroporous Architectures and Aerogels Built of Carbon Nanotubes
      and/or Graphene: Synthesis and Applications. Chemical Society Reviews, 42
      (2013) 794–830. https://doi.org/10.1039/c2cs35353a.

[89] X. Cheng, K. Ye, D. Zhang, K. Cheng, Y. Li, B. Wang, G. Wang, & D. Cao,
      Methanol Electrooxidation on Flexible Multi-Walled Carbon Nanotube-Modified
      Sponge-Based Nickel Electrode. Journal of Solid State Electrochemistry, 19
      (2015) 3027–3034. https://doi.org/10.1007/s10008-015-2897-5.

[90] R. Du, Q. Zhao, N. Zhang, & J. Zhang, Macroscopic Carbon Nanotube-based 3D
      Monoliths. Small, 11 (2015) 3263–3289. https://doi.org/10.1002/smll.201403170.

[91] S. Jung & M. Daniels, Conceptualizing Sex Offender Denial from A Multifaceted
      Framework: Investigating The Psychometric Qualities of A New İnstrument.
      Journal of Addictions and Offender Counseling, 33 (2012) 2–17.
      https://doi.org/10.1002/j.2161-1874.2012.00001.x.

[92] R. R. Kohlmeyer, M. Lor, J. Deng, H. Liu, & J. Chen, Preparation of Stable
      Carbon Nanotube Aerogels with High Electrical Conductivity and Porosity.
      Carbon, 49 (2011) 2352–2361. https://doi.org/10.1016/j.carbon.2011.02.001.

[93] M. A. Worsley, S. O. Kucheyev, J. D. Kuntz, T. Y. Olson, T. Y.-J. Han, A. V.
      Hamza, J. H. Satcher, & T. F. Baumann, Carbon Scaffolds for Stiff and Highly
      Conductive Monolithic Oxide–Carbon Nanotube Composites. Chemistry of
      Materials, 23 (2011) 3054–3061. https://doi.org/10.1021/cm200426k.

[94] T. Bordjiba, M. Mohamedi, & L. H. Dao, New Class of Carbon-Nanotube Aerogel
      Electrodes for Electrochemical Power Sources. Advanced Materials, 20 (2008)
      815–819. https://doi.org/10.1002/adma.200701498.

[95] X. Dong, J. Chen, Y. Ma, J. Wang, M. B. Chan-Park, X. Liu, L. Wang, W. Huang,
      & P. Chen, Superhydrophobic and Superoleophilic Hybrid Foam of Graphene and
      Carbon Nanotube for Selective Removal of Oils or Organic Solvents from The
      Surface of Water. Chemical Communications, 48 (2012) 10660–62.
      https://doi.org/10.1039/c2cc35844a.

[96] A. M. Speers & G. Reguera, Electron Donors Supporting Growth and Electroactivity of Geobacter Sulfurreducens Anode Biofilms. Applied and Environmental Microbiology, 78 (2012) 437–444. https://doi.org/10.1128/AEM.06782-11.

[97] A. Kumar, D. Karig, R. Acharya, S. Neethirajan, P. P. Mukherjee, S. Retterer, & M. J. Doktycz, Microscale Confinement Features Can Affect Biofilm Formation. Microfluidics and Nanofluidics, 14 (2013) 895–902. https://doi.org/10.1007/s10404-012-1120-6.

[98] G. Kavoosi, S. M. M. Dadfar, S. M. A. Dadfar, F. Ahmadi, & M. Niakosari, Investigation of Gelatin/Multi-Walled Carbon Nanotube Nanocomposite Films as Packaging Materials. Food Science & Nutrition, 2 (2014) 65–73. https://doi.org/10.1002/fsn3.81.

[99] J. H. Lee, J. Y. Lee, S. H. Yang, E. J. Lee, & H. W. Kim, Carbon Nanotube-Collagen Three-Dimensional Culture of Mesenchymal Stem Cells Promotes Expression of Neural Phenotypes and Secretion of Neurotrophic Factors. Acta Biomaterialia, 10 (2014) 4425–4436. https://doi.org/10.1016/j.actbio.2014.06.023.

[100] K. Sui, Y. Li, R. Liu, Y. Zhang, X. Zhao, H. Liang, & Y. Xia, Biocomposite Fiber of Calcium Alginate/Multi-Walled Carbon Nanotubes with Enhanced Adsorption Properties for Ionic Dyes. Carbohydrate Polymers, 90 (2012) 399–406. https://doi.org/10.1016/j.carbpol.2012.05.057.

[101] P. E. Canavar, E. Ekşin, & A. Erdem, Electrochemical Monitoring of The Interaction between Mitomycin C and DNA at Chitosan-Carbon Nanotube Composite Modified Electrodes. Turkish Journal Of Chemistry, 39 (2015) 1–12. https://doi.org/10.3906/kim-1402-11.

[102] F. Dalcanale, J. Grossenbacher, G. Blugan, M. R. Gullo, J. Brugger, H. Tevaearai, T. Graule, & J. Kuebler, CNT and PDCs: A Fruitful Association? Study of a Polycarbosilane-MWCNT Composite. Journal of the European Ceramic Society, 35 (2015) 2215–2224. https://doi.org/10.1016/j.jeurceramsoc.2015.02.016.

[103] L. D. Tijing, C. H. Park, W. L. Choi, M. T. G. Ruelo, A. Amarjargal, H. R. Pant, I. T. Im, & C. S. Kim, Characterization and Mechanical Performance Comparison of Multiwalled Carbon Nanotube/Polyurethane Composites Fabricated by Electrospinning and Solution Casting. Composites Part B: Engineering, 44 (2013) 613–619. https://doi.org/10.1016/j.compositesb.2012.02.015.

[104] J. R. Bautista-Quijano, P. Pötschke, H. Brünig, & G. Heinrich, Strain Sensing, Electrical and Mechanical Properties of Polycarbonate/Multiwall Carbon Nanotube Monofilament Fibers Fabricated by Melt Spinning. Polymer, 82 (2016) 181–189. https://doi.org/10.1016/j.polymer.2015.11.030.

[105] R. Ormsby, T. McNally, P. O'Hare, G. Burke, C. Mitchell, & N. Dunne, Fatigue and Biocompatibility Properties of A Poly(Methyl Methacrylate) Bone Cement With Multi-Walled Carbon Nanotubes. Acta Biomaterialia, 8 (2012) 1201–1212. https://doi.org/10.1016/j.actbio.2011.10.010.

[106] V. Vinoth, J. J. Wu, A. M. Asiri, T. Lana-Villarreal, P. Bonete, & S. Anandan, SnO2-Decorated Multiwalled Carbon Nanotubes and Vulcan Carbon Through A Sonochemical Approach for Supercapacitor Applications. Ultrasonics Sonochemistry, 29 (2016) 205–212. https://doi.org/10.1016/j.ultsonch.2015.09.013.

[107] S. Eris, Z. Daşdelen, Y. Yıldız, & F. Sen, Nanostructured Polyaniline-rGO Decorated Platinum Catalyst with Enhanced Activity and Durability for Methanol Oxidation. International Journal of Hydrogen Energy, 43 (2018) 1337–1343. https://doi.org/10.1016/j.ijhydene.2017.11.051.

[108] S. Eris, Z. Daşdelen, & F. Sen, Enhanced Electrocatalytic Activity and Stability of Monodisperse Pt Nanocomposites for Direct Methanol Fuel Cells. Journal of Colloid and Interface Science, 513 (2018) 767–773. https://doi.org/10.1016/j.jcis.2017.11.085.

[109] S. Eris, Z. Daşdelen, & F. Sen, Investigation of Electrocatalytic Activity and Stability Of Pt@F-VC Catalyst Prepared by In-situ Synthesis for Methanol Electrooxidation. International Journal of Hydrogen Energy, 43 (2018) 385–390. https://doi.org/10.1016/j.ijhydene.2017.11.063.

[110] Z. Daşdelen, Y. Yıldız, S. Eriş, & F. Şen, Enhanced Electrocatalytic Activity and Durability of Pt Nanoparticles Decorated on GO-PVP Hybride Material for Methanol Oxidation Reaction. Applied Catalysis B: Environmental, 219 (2017) 511–516. https://doi.org/10.1016/j.apcatb.2017.08.014.

[111] Y. Yıldız, S. Kuzu, B. Sen, A. Savk, S. Akocak, & F. Şen, Different Ligand Based Monodispersed Pt Nanoparticles Decorated with rGO As Highly Active and Reusable Catalysts for The Methanol Oxidation. International Journal of Hydrogen Energy, 42 (2017) 13061–13069. https://doi.org/10.1016/j.ijhydene.2017.03.230.

[112] Ö. Karatepe, Y. Yıldız, H. Pamuk, S. Eris, Z. Dasdelen, & F. Sen, Enhanced Electrocatalytic Activity And Durability of Highly Monodisperse Pt@PPy–PANI Nanocomposites as A Novel Catalyst for The Electro-Oxidation of Methanol. RSC Advances, 6 (2016) 50851–50857. https://doi.org/10.1039/C6RA06210E.

[113] Y. Yıldız, E. Erken, H. Pamuk, H. Sert, & F. Şen, Monodisperse Pt Nanoparticles Assembled on Reduced Graphene Oxide: Highly Efficient and Reusable Catalyst for Methanol Oxidation and Dehydrocoupling of Dimethylamine-Borane

(DMAB). Journal of Nanoscience and Nanotechnology, 16 (2016) 5951–5958. https://doi.org/10.1166/jnn.2016.11710.

[114] P. Nayak, B. Anbarasan, & S. Ramaprabhu, Fabrication of Organophosphorus Biosensor Using Zno Nanoparticle-Decorated Carbon Nanotube-Graphene Hybrid Composite Prepared by A Novel Green Technique. Journal of Physical Chemistry C, 117 (2013) 13202–13209. https://doi.org/10.1021/jp312824b.

[115] Z. Wen, S. Ci, S. Mao, S. Cui, G. Lu, K. Yu, S. Luo, Z. He, & J. Chen, TiO2 Nanoparticles-Decorated Carbon Nanotubes for Significantly Improved Bioelectricity Generation in Microbial Fuel Cells. Journal of Power Sources, 234 (2013) 100–106. https://doi.org/10.1016/j.jpowsour.2013.01.146.

[116] A. Mehdinia, E. Ziaei, & A. Jabbari, Multi-Walled Carbon Nanotube/SnO2 Nanocomposite: A Novel Anode Material for Microbial Fuel Cells. Electrochimica Acta, 130 (2014) 512–518. https://doi.org/10.1016/j.electacta.2014.03.011.

[117] H. Ren, S. Pyo, J. I. Lee, T. J. Park, F. S. Gittleson, F. C. C. Leung, J. Kim, A. D. Taylor, H. S. Lee, & J. Chae, A High Power Density Miniaturized Microbial Fuel Cell Having Carbon Nanotube Anodes. Journal of Power Sources, 273 (2015) 823–830. https://doi.org/10.1016/j.jpowsour.2014.09.165.

[118] C. V. Rao, C. R. Cabrera, & Y. Ishikawa, In Search of The Active Site in Nitrogen-Doped Carbon Nanotube Electrodes for The Oxygen Reduction Reaction. Journal of Physical Chemistry Letters, 1 (2010) 2622–2627. https://doi.org/10.1021/jz100971v.

[119] D. Zhu, D. Bin Wang, T. shun Song, T. Guo, P. Ouyang, P. Wei, & J. Xie, Effect of Carbon Nanotube Modified Cathode by Electrophoretic Deposition Method on The Performance of Sediment Microbial Fuel Cells. Biotechnology Letters, 37 (2015) 101–107. https://doi.org/10.1007/s10529-014-1671-6.

[120] Q. Zhang, J. Q. Huang, W. Z. Qian, Y. Y. Zhang, & F. Wei, The Road for Nanomaterials Industry: A Review of Carbon Nanotube Production, Post-Treatment, and Bulk Applications for Composites and Energy Storage. Small, 9 (2013) 1237–1265. https://doi.org/10.1002/smll.201203252.

[121] P. Narayanaswamy Venkatesan & S. Dharmalingam, Characterization and Performance Study on Chitosan-Functionalized Multi Walled Carbon Nano Tube as Separator in Microbial Fuel Cell. Journal of Membrane Science, 435 (2013) 92–98. https://doi.org/10.1016/j.memsci.2013.01.064.

Microbial Fuel Cells: Materials and Applications   Materials Research Forum LLC
Materials Research Foundations **46** (2019) 177-220   doi: http://dx.doi.org/10.21741/9781644900116-8

Chapter 8

Biofuel Production from Food Processing Waste

Nivedita Sharma*, Kanika Sharma, Neha Kaushal and Ranjana Sharma

[1] Microbiology Research Laboratory, Department of Basic Sciences, Dr. Y. S. Parmar University of Horticulture and Forestry, Nauni (Solan) Himachal Pradesh-173230, India

*niveditashaarma@yahoo.co.in

Abstract

Food waste is simply disposed off in land fillings/incinerators worldwide without making much use of it. However this food waste is rich in many nutrients and by employing suitable technology, it can be converted to value added products like biofuel. This type of food waste management not only resolves the serious pollution problem but also helps to reduce partially the dependency of the energy sector on crude oil. The important aspects like a critical evaluation of types of food waste, current food waste scenario and its strategic way of management and bioconversion to biofuel have thoroughly been explored in this chapter to discover the utility of this waste in real terms.

Keywords

Food Processing Waste, Biofuel, Fermentation, Saccharification, Renewable Energy

List of abbreviations

EJ: Exajoule

GHGs: Greenhouse gases

FAME: Fatty acid methyl ester

FFA: Free fatty acids

WCO: Waste cooking oil

CNG: Compressed natural gas

LBM: Liquefied biomethane

LNG   Liquefied natural gas

C/N ratio: Carbon/Nitrogen ratio

Mo: Molar

ASTM: American society for testing and materials

ABE: Acetone-butanol-ethanol

SHF: Separate hydrolysis and fermentation

SSF: Simultaneous saccharification and fermentation

FW: Food waste

AD: Anaerobic digestion

Contents

1.   Introduction

The fast economic growth and industrialization in the world are characterized by a significant gap between energy demand and supply. The significant increase in world population and economic growth in emerging economies have substantially resulted in increased energy consumption. This high energy demand is further expected to develop more rapidly in the coming years which will have an intense impact on the global energy market. Considering this, different biofuels such as bioethanol and biodiesel are being used as a replacement for current fuels such as diesel and gasoline in many countries. Thus biofuels can be used as a renewable alternative energy source [1,2]. Approximately

1.3 billion tons of food waste which consits of one-third of the worldwide food produced is discarded around the world without any further use and due to rapid economic expansion and continuous population growth in developing Asian countries, generated food waste is increasing consistently. For instance, 1300 million tonnes of food waste is generated worldwide and 82.8 million tonnes of food waste is produced by China alone. [1,2,3].

Table1: Usual food waste produced in different countries [6,7,8]

Waste (KT)	World	Asia	Australia	Cambodia	China	Indonesia	Japan	Malaysia	Vietnam
Cereal	95,245	52,374	1,380	506.1	18,990	4,588	413.4	183.4	2,706
Rice	26,738	22,668	0.4	506.0	6,046	3,307	139.4	50.2	2,47
Sugar	459.9	188.9	93.6	-	0.4	-	20.8	-	-
Pulses	2,735	1,134	36.0	0.9	142.3	38.0	7.1	-	8.6
Oil crops	18,424	13,590	3.9	3.8	9,017	2,238	69.6	1.4	30.5
Beans	1,049	447.3	1.1	0.9	49.1	37.2	6.5	-	5.2
Onions	5,891	3,887	14.6	-	2,107	99.9	68.1	-	22.7
Peas	412.7	145.1	7.2	-	39.9	-	0.4	-	-
Tomatoes	12,874	7,415	-	-	3,181	85.3	100.7	1.6	-
Potatoes	62,229	12,912	23.6	-	7,501	250.0	177	-	83.3
Fruits	53,796	28,238	30.9	30.5	8,323	2,706	749	89.1	531.0
Apples	5,742	4,116	5.9	-	3,192	3.1	84.6	-	5.1
Banana	13,532	8,544	5.4	7.8	949.3	637.4	213.0	56.1	137
Coconut	3,038	2,488	-	-	20.5	2,066	-	1.3	0.9
Pineapple	1,829	579	-	2.2	97.7	-	15.4	-	50
Coffee	105.0	33.3	-	-	0.033	20.9	-	0.6	-
Milk	16,560	10,887	-	1.6	1,447	45	-	3.8	9.5
Cream	33.9	0.1	-	-	0.1	-	-	-	-
Butter	84.0	1.7	-	-	-	-	-	-	-
Animal fats	174.1	1.8	-	-	0.1	-	-	-	-
Meat	1,184	183.2	-	-	-	-	107.2	-	-
Offal	63.0	19.6	5.7	-	-	-	-	-	-
Poultry meat	97.5	61.2	-	-	-	-	34.5	-	-
Annual waste production	0.184	-	0.277	0.173	0.061	0.130	0.129	0.113	0.130

KT: Kilotons;

In order to maintain the current levels of energy use, there is a pressing need to increase the use of renewable energy and natural resources more efficiently. Bioenergy obtained from biomass resources such as crops, trees, food or agricultural waste can be a key to these problems and may be used to generate heat, electricity or transport fuels with a lower level of greenhouse gas emissions than fossil fuel sources [1,2]. A **biofuel** is an outcome of ultimate fermentation of carbohydrates using aerobic/ anaerobic microorganisms instead of a conventional fossil fuel synthesized from geological upheavals such as petroleum and coal, from primitive biological matter. People are turning to the novel and new methods of energy production, as the need for alternatives to fossil fuels becomes more requisite. The carbon dioxide emitted by the biofuel approximately correlated to the amount absorbed by the biomass during the growth period which is driven by photosynthesis with sunlight being the ideal source of energy leading to the minimal carbon footprint. Liquid biofuels *viz.* bioethanol, biodiesel, methanol, ethers and biobutanol with properties similar to diesel fuels or gasoline are different forms that can be produced from these renewable resources. The biomass of food processing residues (Table 1) alone constitutes about 30 EJ/year energy i.e. an abundant alternative source of organic raw material to petroleum, and a significant amount of the global primary energy demand of more than 400 EJ/year could be covered by utilizing dedicated agricultural crops in addition to the waste/residues of different sorts [5].

2. Biofuels and their generations

People are turning to the novel and new methods of energy production, as the need for alternatives to fossil fuels becomes more requisite. Many types of liquid fuels along with gaseous fuels such as hydrogen and methane can be extracted and developed from a wide variety of biomass. The main use of biofuels now-days is to provide power to vehicles but these can be used to fuel cells or fuel engines for the generation of electricity. Some of the major producers and users of biofuels are Europe, Asia and America [3,9].

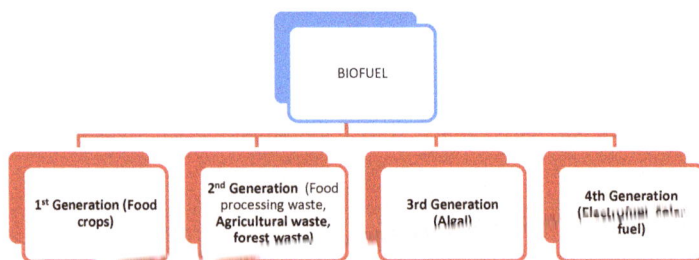

Figure 1. Different generations of Biofuel

There are four generations of biofuel, out of these, sugar, lipid or starch are the major sources of carbon for production of first-generation biofuels (Fig. 2), while in second-generation biofuel, carbon is obtained from cellulose, lignin or pectin substrates including food processing waste, agricultural and forestry residues (Fig. 3). Third generation biofuels are mainly produced from aquatic autotrophic organisms such as algae (Fig. 4) and fourth generation biofuels include electro and solar fuels.

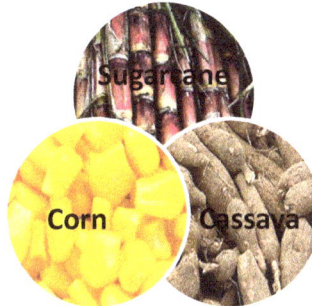

Figure 2. 1st Generation biofuel [10]

Figure 3. 2nd Generation biofuel

Figure 4. 3rd generation biofuel

Biofuels offer many benefits such as local availability, easy accessibility, sustainability as renewable fuels, reduced greenhouse gas emissions and security of supply [11]. The only sustainable alternatives to transport fuels are liquid biofuels from non-food crops as well as waste. Thermochemical and biological methods have been used to convert biomass into liquid and gaseous fuels [12]. Biofuels fall into two categories: (1) Gaseous biofuels (biomethane, biohydrogen and biohythane) and (2) Liquid biofuels (biodiesels, bioethanol, biobutanol, biomethanol) (Fig. 5).

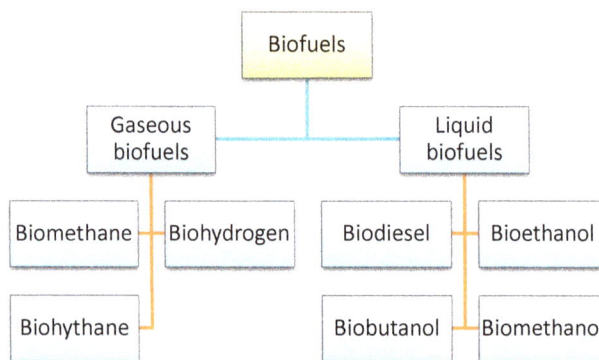

Figure 5. Types of biofuels

Organic waste released from different sources viz. domestic and commercial kitchens, food-processing plants, restaurants and cafeterias is known as Food Waste (FW). Food waste comprises of by-products emerging out from different food processing industries which have low budgetary value than the price of collection and recovery and which have not been reused for further purposes [13]. In accordance with FAO 2007, approximately 1.3-1.6 billion tonnes of food from food processing industries, kitchens, cafeterias and restaurants are discarded. These food wastes include fruits, fresh vegetables, meat, dairy products, bakery products, etc. which constitute about one third of the food prepared worldwide for human usage out of which 95% of food waste is converted into methane, carbon dioxide and other greenhouse gases (GHGs) by anaerobic digestion (AD) at landfill sites and greatest loss faced by low- income / developing countries during food production. Also, the high energy value of food waste and carbon-rich feedstock are not derived when food waste is buried at landfill sites [14,15].

With the continuous increase in population, the demand for energy is also increasing constantly which cannot be fulfilled with non-renewable and limited energy sources such as petroleum, natural gas and coal. According to Intergovernmental Panel on climate change [14], it is becoming more and more important to search for new sustainable, renewable, environmentally friendly energy sources as incineration of fossil-fuel results in 57% of the emissions to blame for global warming. Rapid economic development and dwindling fossil fuel resources have attracted researchers to use food processing waste as a renewable and sustainable resource of energy and fuel in industries as it is rich in carbohydrates, proteins, lipids and organic acids, etc. Food processing waste can be used as a microbial feedstock, for production of methane, biohydrogen, bioethanol, biopolymers, biodegradable plastics and enzymes, etc.

Food processing waste may be generated in solid, liquid or semi-solid form. Liquid waste comprises of suspended solids, fats, oils, organic acids and inorganic matter (Fig.6). Solid food wastes are rich in starch, lignin and cellulose. Food processing wastes such as stem, straw, stalk, husk, shell, peel, pulp, leaves, stubble, etc. which come from cereals (rice, wheat, maize or corn, sorghum, millet, barley), cotton, jute, groundnut, legumes (bean, tomato, soy) tea, coffee, cacao, fruits (mango, banana, coco, cashew), cooking oil and palm oil have substantial ability for transformation to alcohols or other renewable fuels (Fig. 7). Horticulture industries generate a huge quantity of wastes are beneficially containing a large number of polysaccharides for production of bioethanol [15].

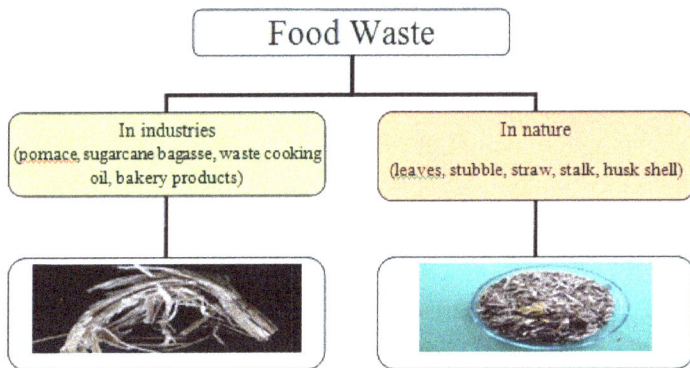

Figure 6. Types of food waste

A large amount of waste is produced by food and vegetable industries such as peels, seeds and pulps that corresponds to a global problem for the environment [16]. Wastes generated by food processing industries contain 30-50% of input materials, which generate a critical issue for food processing industries to manage such tonnes of feedstock as waste every year. Conversion of these wastes into more resourceful and sustainable sources through modern eco-compatible technologies can be a remarkable twist for removal of multiple sugars which can be converted into alcohol instead of disposing of food waste in landfills by industries. However, the primary obstacle for the adoption of food processing waste for biofuels production is the high moisture content. [16].

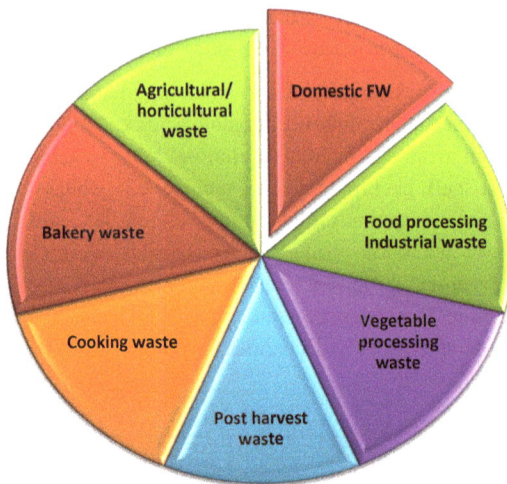

Figure 7. Major chunks of food waste

There is an urgent need of appropriate management of food waste to convert it into commercially useful products because the incineration of food waste with other municipal wastes generates dioxins causing air pollution [18,19]. Food processing waste is mainly composed of carbohydrate polymers, lignin, proteins, lipids and organic acids, etc. (Table 2). For the production of different value-added products such an methane, biohydrogen, bioethanol, etc., food waste has been used as a microbial feedstock because the hydrolysis of carbohydrates present in food waste results in oligosaccharides and monosaccharide which can be easily converted into biofuel through fermentation [20-30]. The value of biofuels ($200-400/ ton biomass) is higher than electricity ($60-150/ton biomass) and animal feed ($70-200/ton biomass) (Figs. 8 and 9). Utilization of these wastes could solve the disposal problem and reduce the cost of waste treatment (Fig. 10).

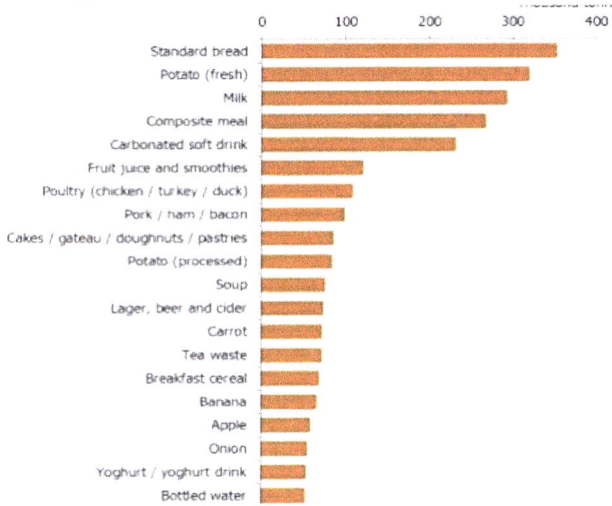

Figure 8. Tons of avoidable food processing waste [17]

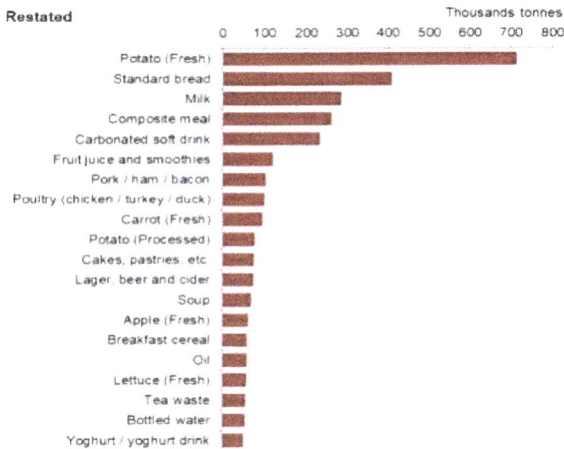

*Figure 9. Food processing industrial waste across worldwide
(Source: Report-Waste & resource action programme, UK [31])*

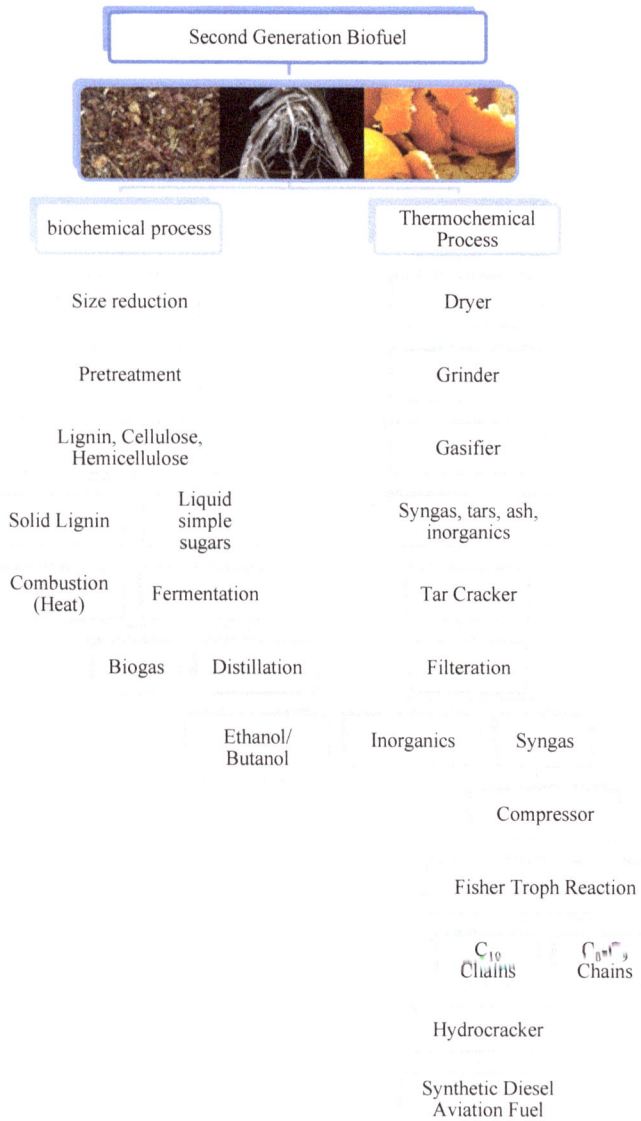

Figure 10. Production of biofuel

2.2   Liquid biofuels

Liquid biofuels from biomass are great optional sources of energy to substitute traditional liquid fuels (diesel and petrol) having similar qualities of gasoline, diesel or other petroleum-derived fuels.

2.2.1 Bioethanol

Ethanol is one of the essential renewable and sustainable fuels and due to potential and wider applications in industries, increasing pollution produced by use of fossil fuels, its demand has increased globally.   Therefore, production of bioethanol from cheap feedstock such as food processing waste is useful to reduce worldwide consumption of dwindling fossil fuels [32]. Usually, bioethanol has been produced from starch-rich crops belonging to direct food chain [33].

Development of technology for the production of bioethanol from waste materials obtained from agricultural and food industries is a feasible alternative to minimize the cost of ethanol production and aid to wipe out these wastes from the direct trade. Broad range of food wastes from industries [34, 35, 36] which include banana peels waste [37], waste potato peels waste [38], municipal solid waste [39], domestic food waste [40], cassava grate waste [41], and many others can be used in production of bioethanol. Most of these wastes contain 50 to 70% of carbohydrates. [42]. In the bakery industry, bread waste can be used as a raw material for bioethanol production and its yield, depending on the processing technology i.e. 350–366 g ethanol per kilogram of the resource. Its low price, maximum accessibility and, gainful chemical composition facilitate its use.

Fruit waste is another inexpensive and easily suitable potential energy source for the production of bioethanol. *Carica papaya* (pawpaw) and grapefruit waste with the highest source of energy and invert sugar are available . for the production of bioethanol by fermentation process with a yield of ethanol greater than 80% of the fermentable sugar consumed [43]. The waste from bananas (*Musa paradisiaca*), commonly discarded as waste represent a potential energy feedstock which may be especially suited for ethanol production. Apple (*Malus pumila)* occupies a prominent position in India and apple processing industries are one of the superior industries of Himachal Pradesh, Jammu and Kashmir, and Uttaranchal. These industries produce juices, concentrates, wines, cider, etc. After the extraction of the juice, apple pomace remains unconsumed which contains peels, seeds and other waste solid parts [44] and contributes about 20–35% of the total fruit production.   Apple pomace is a waste and its disposal is challenging. 'Biodegradation' and 'Bioconversion' of this waste will not only solve the serious pollution problem but at the same time, it could be utilized to meet partially the global challenge of the modern world i.e. fuel depletion. Bioconversion of sugars produced from

Microbial Fuel Cells: Materials and Applications   Materials Research Forum LLC
Materials Research Foundations **46** (2019) 177-220   doi: http://dx.doi.org/10.21741/9781644900116-8

complex cellulose, hemicellulose and pectin present in this postharvest waste will provide the commercially high-value compounds of high utility viz. ethanol, enzymes and single cell protein [44]. By employing an appropriate technology it can be utilized efficiently for bioethanol production and thereby reducing the environmental pollution as well as generating clean technology. Pathania et al., [45] described the utilization of this waste for multiple carbohydrase productions from *Rhizopus delemar* F_2 under solid state fermentation and consequently ethanol production.

Many studies have been performed for the exploration of suitable methods with high efficiency to produce ethanol from food processing waste [46]. Table 2 displays the findings obtained on the processes and yields associated with producing ethanol from different types of food wastes such as lignocellulosic waste, municipal waste, food processing waste such as banana peel, apple pomace, bakeries, cheese whey, etc. (Fig. 11). Commercial bioethanol can be produced from many types of food processing wastes such as sugarcane, corn, wheat, potatoes, apple pomace, banana peels, dairy by-products, bakeries waste etc. Lignocellulose is a complex carbohydrate polymer of cellulose, hemicellulose and lignin (Table 2). Cellulose is linear and crystalline. It is a homopolymer of repeating sugar units of glucose linked by b-1,4glycosidic bonds. Lignocellulose is hydrophobic with complex structure.

Table 2. Chemical composition of agricultural food waste

Substrate	Cellulose (%)	Hemicellulose (%)	Lignin (%)	Ash (%)
Wheat straw	34-46	20-30	8-15	10.1
Rice straw	32-47	18-27	5-23	12.4
Corn straw	42.6	21.0	8.5	4.3
Bagasse	35-50	25-35	18.4	2.4

Food waste residues are processed for bioethanol production through three major operations (Fig. 9):

i. pretreatment for delignification is necessary to release cellulose and hemicellulose before hydrolysis;

ii. hydrolysis of complex sugars (cellulose and hemicellulose) to produce fermentable sugars including glucose, xylose, mannose, arabinose, galactose and fermentation of reducing sugars.

iii. Lignin (non-carbohydrate components) also have value

Pretreatment methods refer to the solubilization and separation of components of biomass thus making the remaining biomass more accessible for further processes (Table 3, Fig. 11). The pretreatment method is utilized for breaking the matrix made up of cellulose and lignin for making biomass more vulnerable for enzymatic attack thus results in the production of a large number of monomeric sugars [47,48]. The goals of an effectual pretreatment process are:

  i. Direct formation of sugars by hydrolysis

 ii. Deterioration of resulted sugars

iii. Decrease formation of inhibitory products

 iv. By decreasing the need for energy to reduce operational costs.

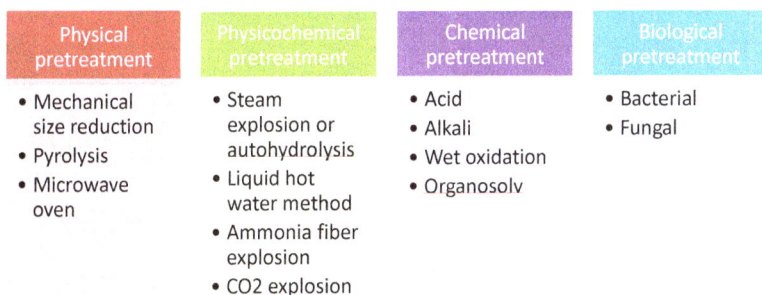

Physical pretreatment	Physicochemical pretreatment	Chemical pretreatment	Biological pretreatment
• Mechanical size reduction • Pyrolysis • Microwave oven	• Steam explosion or autohydrolysis • Liquid hot water method • Ammonia fiber explosion • CO2 explosion	• Acid • Alkali • Wet oxidation • Organosolv	• Bacterial • Fungal

Figure 11. Different pretreatment methods for degradation of agricultural food waste

Table 3. Exploitation of industrial food processing waste for sugar productions

Waste Source	Pretreatment method
Apple pomace	Chemical- acidic Physical-autoclave
Orange peels	Physical- microwave, drying
Citrus peels	Physico-chemical (Sequential extraction)
Banana peels	Physico-chemical (Sequential extraction)
Peanut cakes	Chemical (Alkali)
Agro-waste materials	Physical, chemical, physicochemical and biological
Soy Hull	Chemical (Acidic- 0.1M HCl)

Harsh pre-treatment before enzymatic hydrolysis is not essential as autoclaving of food waste can be done to improve purity and yield of products which result in fragmentary degradation of sugars [22,49,50]. Moreover, fresh and wet food wastes are more potent than rewetted dried food waste as reduced specific surface area of dried substrate lowers the reaction effectiveness [51]. The acidic condition is needed to prevent microbial contamination and putrefaction [26, 52]. As such, acid-tolerant ethanol producing microorganisms such as *Zymomonas mobilis*, have utilized for the fermentation of food waste [53, 54].

Saccharification

A key step for ethanol fermentation is basically saccharification because the amount of reducing sugars produced in turn will affect the ethanol yield. The depolymerization of enzymes like amylase, cellulase, xylanase, pectinase, etc. plays a vital role in solubilizing the complex carbohydrates to simple fermentable sugars [55,56,57]. The highest glucose concentration of about 65 g reducing sugar /100 g food waste was obtained with α-amylase [58]. In a study of Hong and Yoon [59], a mixture of commercial enzymes consisting of glucoamylase, α-amylase, and protease resulted in 60 g reducing sugars /100 g food waste.

Use of enzymatic saccharification is not only an eco-friendly approach but also commercially more viable over alkali or acid hydrolysis because of favourable and optimally low reaction conditions *viz.* temperature, pH etc. and minimization of inhibitor generations, toxicity and correlation of equipment, etc. However, cellulase and xylanase enzymes are highly substrate specific enzymes that are used for enzymatic hydrolysis [60,61]. Here the bonds of cellulose and hemicellulose are cleaved by cellulase and hemicellulase respectively into simple sugars. Many factors that control the outputs of monomeric sugars from lignocelluloses include pH, temperature, mixing rate and all other factors of enzymatic hydrolysis of lignocellulosic material. Some factors that influence outputs are cellulase enzyme loading, substrate concentration and surfactant supplement [62].

Fermentation

The saccharified/ hydrolyzed biomass is used for fermentation by many microorganisms but the industrial utilization of agricultural lignocelluloses for bioethanol production (Table 4, Fig. 3) is impeded by the scarcity of ideal microorganisms which can expeditiously ferment both pentose and hexose sugars [63]. The lignocellulosic hydrolysate fermentation is normally employed by two different approaches i.e. i) separate hydrolysis and fermentation (SHF) and ii) simultaneous saccharification and fermentation (SSF) (Table 5). The SSF is important for the production of bioethanol because it improves the yield of ethanol and removes the end product inhibitors [68,69].

Table 4. Production of ethanol from FW [64]

Waste	Method	Vessel type	Pretreatment	Microorganism	Duration(h)	Reference
Bakery waste	Simultaneous	14L fermenter	-	As specified	14	[52]
FW	Repeated batch Simultaneous	1L fermenter.	-	As specified	264	[65]
Mandarin waste, banana peel	Simultaneous	500 ml vol flask	Explosion, Steam drying	As specified	24	[66]
FW	Separate	500 ml flask	-	As specified	16	[39]
FW	Simultaneous	Flask with 100g FW	-	As specified	48	[67]
FW	Separate Continuous	0.45L fermenter	spraying of LAB	As specified	15	[49]
FW	Simultaneous	Flask with 100g FW	-	As specified	67.6	[53]
FW	Continuous Simultaneous	Fermenter (4.3Kg)	spraying of LAB	As specified	25	[22]
FW	Simultaneous	Fermenter (1L)	-	As specified	48	[19]
FW	Repeated batch Simultaneous	250 ml flask150 ml working vol.	-	As specified	14	[19]

Table 5. Comparison between the two main fermentation processes

Fermentation process	Features and advantages	Limitations
Separate hydrolysis and fermentation	• Separate steps minimize interaction between the steps at optimized conditions	• Feedback inhibition. • Contamination during fermentation
Simultaneous saccharification and fermentation	• Low costs • Simplified cost effective process with high yield of alcohol	• To devise precision between hydrolysis and fermentation temperature.

Co-culture of ethanologens or fermenting microorganisms is commonly utilized to ferment cellulose into bioethanol. Some native or wild-type fermenting microorganisms used in the fermentation are *S. cerevisiae, Escherichia coli, Zymomonas mobilis, Pichia stipitis, Candida shehatae, Pachy solentannophilus, Candida brassicae* and *Mucor indicus etc.* Ethanol production from food processing waste material is a widely explored alternative fuel. Cleaner and greener production of compounds like ethanol is important for sustainable growth. These environment eco-friendly processes can be made more economical by optimizing the process parameters and finding more effective techniques for the conversion of these agricultural and other low-cost raw materials into usable products (Fig 12).

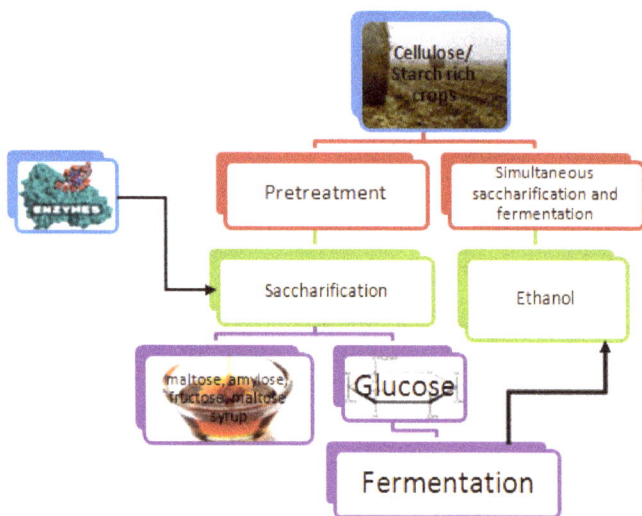

Figure 12. The process of ethanol production from FW

As compared to batch culture, Yan [70] found that by using fed-batch configuration, saccharification and consecutive fermentation of ethanol were both enhanced e.g. the yield of glucose bioconversion was found to be 92% compared to the theoretical value. Hence, SSF can be excluded from the low danger of catabolite repression. This associates fermentation of ethanol and hydrolysis of enzymes into a sole operation for holding the concentration of glucose produced by enzymatic reaction at a low level so as to eliminate inhibition due to hydrolysis of enzymes [72]. This mixed process can be executed in a single tank with more ethanol production and less consumption of energy in a short time of processing using a small quantity of enzyme [71]. For successful completion of the SSF process fermentation conditions should be optimized critically because the fermenting microorganisms and enzymes can have various optimum temperatures and pH [59]. Koike et al. [27] have reported ethanol production from non-diluted FW i.e. garbage in a continuous SSF operation with 17.7g/L ethanol production. Ma and coworkers have [73] evaluated the performance of the SSF process using *Zymomonas mobilis* (acid tolerant) from kitchen garbage without sterilization. The results obtained showed 15.4 g sugar/100 g of garbage and 0.49 g ethanol/g of sugar in 14 hours, yielding 10.08 g/L ethanol.

2.2.2 Biobutanol

Biobutanol is a second generation transport alcoholic fuel which has an advantage over ethanol as biobased butanol fuel because of more energy density, less volatility, adequate hygroscopicity. Biobutanol contains a series of straight-chain alcohols with each molecule of butanol ($C_4H_{10}O$) containing four carbon atoms as compared to two carbon atoms present in ethanol [74] (Table 6, Fig. 13). Biobutanol can be produced from cellulosic raw materials, cereal crops, sugar cane and sugar beet, etc. Production of n-butanol from food processing waste is an encouraging choice for market processes due to its low cost. The acetone-butanol-ethanol (ABE) processes generally produce n-butanol with 15g/L concentration because of the toxicity of n-butanol effects the development of cell membranes and regular physiological functions of microbial cells [75]. Genetic and metabolic engineering has been used to construct *Escherichia coli* and *S. cerevisiae* strains for the improvement of n- butanol yield [76,77].

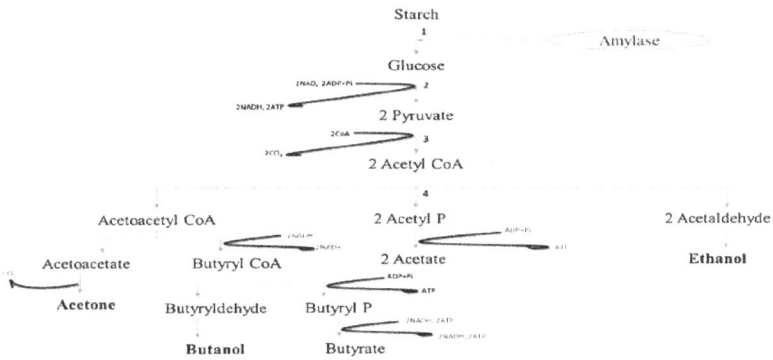

Figure 13. General metabolic pathway of biobutanol production

Table 6: First generation biobutanol from food source [64]

First generation feedstock	Fermentation conditions T°C and pH	*Clostridium* species	Yields		Productivity ($gL^{-1} h^{-1}$)	Product distribution
			gL^{-1}	gg^{-1}		
Casava starch	37 and 5	*C. beijerincki*	6.66	0.18	0.96	Butanol
Glucose	37 and over 4	*C. acetobutylicum* CLCC8012			0.13	Butanol
Casava flour	37 and pH controlled	*C. acetobutylicum* DP 217	574.3		0.76	ABE
Oil palm sap	37 and 6	*C. acetobutylicum* DSM 1731	14.4	0.35		Butanol
Maize meal	37 and 6	*C. beijerincki* BA101	26			Acetone and Butanol

2.2.3 Biodiesel

The American Society for Testing and Materials (ASTM) defined "Biodiesel" as a mono-alkyl fatty acyl ester originated from inexhaustible feedstocks (i.e. vegetable oils) [78]. Biodiesel is water-insoluble light-to dark-yellow coloured with stable reactivity fine liquid. It has a turning point between 145–175°C, a boiling point above 200°C, 195–325°C range of distillation and less than 5 vapour pressure (mm Hg at 22°C). [78]. Biodiesel being recyclable fuel is very useful as replacement of diesel and petroleum. Biodiesel can be characterized as 'carbon neutral' due to nil net release of carbon production by the fuel in the mode of carbon dioxide and thus has a great advantage. Chemically reacting lipids such as vegetable oil, soybean oil and animal fat having alcohol producing fatty acid esters are typically used for making biodiesel. Crops containing oil (palm, rapeseed or soyabean) are the most favourable for the production of biodiesel by using base-catalyzed transesterification which only requires low temperature and pressure and gives 98% conversion yield [79]. To treat waste cooking oil (WCO), one of the alternate is by converting it into biodiesel which is also an eco-friendly and economically viable process. WCO is an Inexpensive waste which can be utilized for production of biodiesel hence saving the treatment and disposal cost of waste. Other than this, it decreases or minimizes the requirement of land for the production of biodiesel crops. But, as compared to refined and crude vegetable oils, waste cooking oils (used frying oils) have different properties. The optimal requirements for the production of biodiesel (concentration of catalyst and methanol/oils ratio) are incongruous. These conditions are strongly dependent on the properties of WCO. Transesterification reaction catalyzes a chemical reaction that involves the conversion of alcohol/oil/vegetable oil into biodiesel and glycerol. To analyze and establish substitute energy sources for degradation of non-recyclable energy sources, biodiesel as a fuel has been utilized [80]. Due to the detrimental effects on the environment from discharge through combustion of conventional fossil fuels and a decrease in the production of domestic oil, interest in biodiesel is growing among the people [81].

Biodiesel generated from a renewable lipid feedstock i.e. animal fat or vegetable oil consists mainly of the combination of fatty acid alkyl esters i.e. carbon chain length C14-C22. Due to low cost and easy availability, methanol is most commonly used in biodiesel production. Other alcohols such as butyl and isopropanol may also be used for this purpose. Biodiesel made from renewable resources and waste lipid, which is a low-emission diesel substitute fuel. Transesterification is the most common way to produce biodiesel (Fig. 14), especially alkali-catalyzed transesterification [81]. FAME from animal fats and vegetable oils had shown that these can be utilized without modifications in conventional diesel engines for biodiesel production due to their volatility, enhanced viscosity, combustion nature towards triacylglycerols [82]. Most common animal fat origins are lard, poultry fat, fish oils and beef tallow. Yellow greases can be a combination of animal fats and vegetable oils. Lard means hog fat whereas chicken fat indicates poultry while tallow is beef fat produced by the slaughterhouse. Rendered animal fats also other interesting raw materials for the production of biodiesel in the industry and these are subsequently present in large amounts at relatively low rates in areas with intensive livestock.

Transesterification catalyzed by homogeneous alkali is the most commonly used industrial process for the production of biodiesel [82,84,85]. This investigation exists just because the base-catalyzed reaction is resulting in a fuel-grade biodiesel which is more rapid than the acid one under mild conditions [86] ending in a fuel-grade biodiesel.

Biodiesel, as an alternative fuel has shown to have great potential and may be produced from FW by transesterification of microbial oils by various oleaginous microorganisms or either by direct transesterification or using alkaline or acid catalysts (Fig. 14). Globally, biodiesel is generated by transesterification of alcohol in the presence of a catalyst. This transesterification process involves the conversion of oil (triglycerides) to biodiesel (methyl esters) and glycerol [87]. Biodiesel is non-toxic, biodegradable, produce little sulphur when burned and has an aromatic-free emission profile. By means of recycling spent fats and oils, it can even benefit the environment [88]. Waste cooking oil (WCO) consists of residues from restaurants, kitchens, food industries and even animal and human wastes which are not only harmful to the human health but cause pollution in the environment. The twin problems of environmental pollution and energy shortage can be reduced by the formulation of biodiesel from FW as a partial alternative for petroleum diesel. Cutting of forests will affect plants and animals ecological communities and thus the utilization of WCO as a biodiesel feedstock/biomass is important to address issues of blockages in water drainage systems and water pollution (Table 7). However, there is still an impact on the environment due to the enhanced production of WCO in homes and restaurants [89].

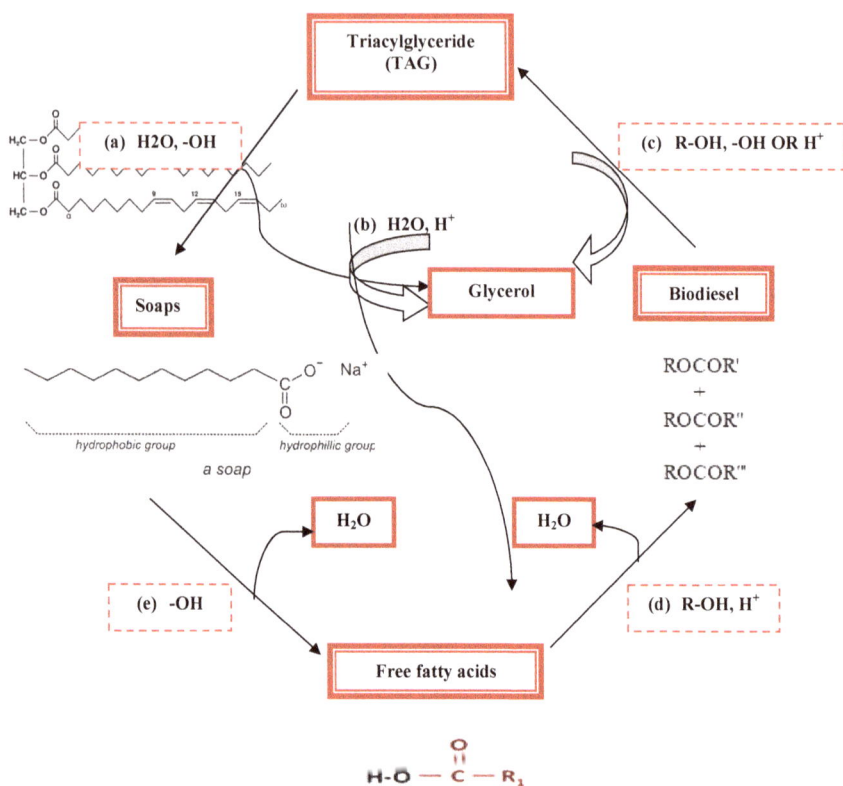

Figure 14. Reactions involved in conventional biodiesel production: (a) alkali-catalyzed transesterification (expected route); (b) acid-catalyzed hydrolysis; (c) alkali=catalyzed hydrolysis; (d) acid-base neutralization; (e) acid-catalyzed esterification (expected route)

Table 7. Comparison between characteristics of commercial diesel fuel and diesel from WCO [89]

Fuel property	Commercial diesel fuel	Biodiesel from WCO
Flash point	144	212
Pour point	254-260	262
Ash content	0.009	0.003
Carbon residue	0.35-0.39	0.38
Octane number	42	54

Raw substrates for production of biomass are based on plant lipids, such as coconut oil, corn oil, soybean oil, palm oil, olive oil, margarine, grape seed oil and canola oil or lipids from an animal, such as ghee, butter, fish oil and Kermanshah oil [90].

Properties of Biodiesel

Biodiesel is most preferable over diesel obtained from petroleum because it is eco-friendly, renewable, biodegradable and non-toxic. It has high lubricity and produces low emissions, aromatic as well as sulphur contents in addition to showing good ignition properties [91,92]. Furthermore, the WCO is cost effective compared to fresh vegetable oil (Table 8).

Table 8. Production of biodiesel and fatty acids from FW [64]

Waste	Microrganisms	Pretreatment	Time required	Yield (g cell/g waste)	Yield (g lipid/ g cell)	References
Waste cooking oil	*Aspergillus niger* NRRI363	Filteration	8	1.15	0.64	[91]
FW	*Schizochytrium mangrovei*	Fungal hydrolysis	4	-	0.321	[92]
FW	*Chlorella pyrenoidosa*	Fungal hydrolysis, autolysis	4	-	0.208	[92]

Pretreatment of waste cooking oil

The WCO in the production of biodiesel has some major drawbacks i.e. the occupancy of free fatty acids (FFA), impurities and water [93] which can lead to hydrolysis and

Materials Research Forum LLC

doi: http://dx.doi.org/10.21741/9781644900116-8

saponification [94] reactions which decrease the yield of biodiesel. To eliminate this error various techniques have been proposed as follows:

- Esterification of acids with sulphuric acid and methanol,
- Esterification using ion- exchange resins,
- Soap separation through decanter proceeded by neutralisation with alkalis,
- Esterification of acids and FFA distillation with polar liquids extraction
- Heating over 100°C to eliminate water from WCO
- Vacuum distillation (0.05 bar)

Moreover, phospholipids, drooping solids and impurities can be removed by centrifugation, washed off using hot water or filtration through paper [95,96].

Trans-esterification reaction

For biodiesel production, transesterification using alcohol of vegetable/agricultural oils is the main method (Fig. 15). There are two methods of trans-esterification as follows:

(a) without catalyst

(b) with catalyst

The yield and rate of biodiesel production can be enhanced by using various types of catalysts. This transesterification process is changeable and the equilibrium can be shifted to the product side due to excess alcohol. Several different alcohols i.e. methanol, ethanol, propanol, and butanol, can be used in this reaction [97]. The methanol application is more feasible because of its low-cost and physical as well as chemical advantages, such as being polar and having the shortest alcohol chain [96].

$$RCOOR^1 + R^2OH \xrightleftharpoons{\text{Catalyst}} RCOOR^2 + R^1OH$$

| Triglyceride | Alcohol | Alkyl ester | Glycerol |

Figure 15 General Transesterification reaction equation

The major factors in biodiesel production are temperature, agitation, the ratio of alcohol and oil and reactants purity. Conventionally, alkaline and acidic catalysts are used mostly for transesterification [97]. The transesterification process is very important not only in the improvement of yield but also in recovery, environmental hazard safety, availability of feedstock and conservation of resources (Fig 16).

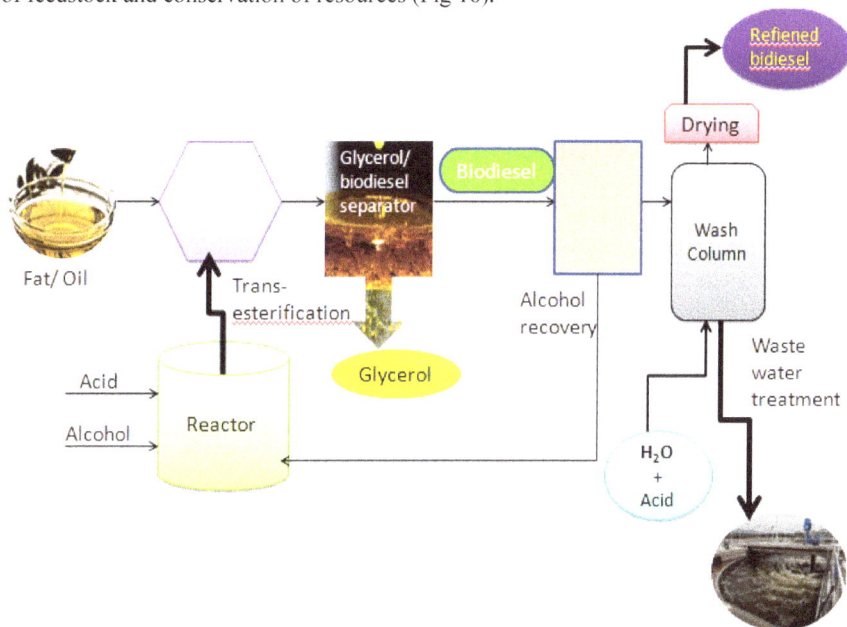

Figure 16. Conventional process of Biodiesel production

2.2.4 Biomethanol

Methanol is also known as "wood alcohol." Generally, methanol is easier to find than ethanol. Feasible methods of methanol production are presently not economically usable. Methanol is produced from biogas or synthetic gas and evaluated as a fuel for internal combustion engines.

Characteristics: Methanol burns with an unseen flame and is toxic. It has a high octane rating like ethanol and can be used as an ignition booster in a diesel engine.

About 2.2 litres methanol can be obtained from 1 litre of gasoline. The supply of oxygen and hydrogen in most of the processes enhances the yield of methanol derived from biomass and produces synthetic gas (H_2 plus CO) from biomass [68].

Agricultural food waste biomass is first converted into bio-oil by fast pyrolysis, after that the bio-oil is converted into hydrogen by catalytic steam reforming [98, 99]. The yield or production of hydrogen that can be produced from biomass is relatively low i.e. 16–18% of dry weight of biomass [100]. Only the bio-oil fraction derived from carbohydrate generated from biomass undergoes reformation (Table 9). Nowadays, methanol is processed from natural gas while it can also be produced via partial oxidation of biomass [102]. Coal and biomass have been used as possible fuels for methanol synthesis and production of syngas [103]. Sufficient amount of hydrogen is added to the gas to be synthesized and that converts the carbon biomass into methanol carbon [104]. The methanol can be partially produced from the waste of lignocellulosic material with the estimated yield as 185 kg of methanol per metric ton of solid waste [105,106]. Agriculture methanol (biomethanol) is currently more expensive than synthetic methanol from natural gas and ethanol from ethylene [107]. Table 4 lists the main production facilities of methanol and biomethanol.

Table 9. A comparison between biomethanol and methanol

Biomethanol	Methanol
Synthesis through catalysis from H_2 and CO	Catalytic synthesis from CO and H_2
Products in gaseous form from gasification of biomass	Petroleum gas
Synthetic gas from coal and biomass	Natural gas
Distillation of liquid from wood pyrolysis	Distillation of liquid from coal pyrolysis

The concurrent bio-methanol production (generated from the hydrogenation of CO_2 formed during the fermentation), similar to bio-ethanol production (Fig. 17), seems more cost effective and promising in areas where hydro-electricity is available at very cheap cost i.e. ~0.01$ Kwh and where lignocellulosic wastes are easily available in huge amount [107].

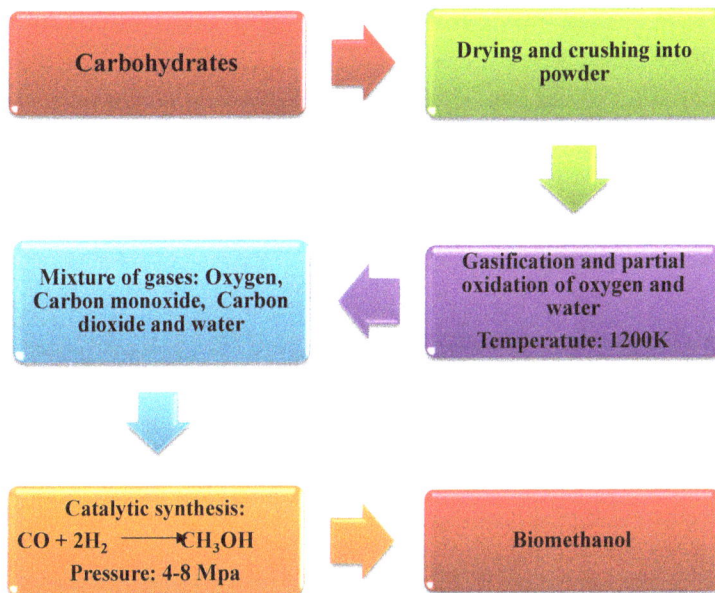

Figure 17. Biomethanol from carbohydrates by gasification and partial oxidation with O_2 and H_2O.

The energy value of agricultural residues generated worldwide and the forest products industry amounts to more than one-third of the total commercial primary energy use at present as well [108]. Supply of bio-energy may be divided into two main categories:

(1) organic domestic waste from the households of food and materials origin,

(2) plantations of energy crops.

Biomass energy from both crops and residues can lead to modern energy transporter [109]. Biomass seems to be a promising feedstock for the main three reasons:

- First, it appears to have beneficial environmental properties resulting in zero net production of carbon dioxide and very less content of sulfur.

- Second, it is a renewable in nature i.e. can be easily sustainably produced in the future.

▪ Third, it seems to be cost-effective unless the prices of fossil fuel increase in the future [110].

Lignocellulosic bio-methanol have fewer emissions due to the carbon content of the alcohol sequestered in the developing of bio-feedstock which is being re-emitted into the atmosphere [111].

2.3   Gaseous biofuels

2.3.1 Biomethane

Biogas a mixture of methane and carbon dioxide gas is produced through a natural process that breaks down organic material in an oxygen-free environment by anaerobic digestion. Anaerobic digester plays an important role as a renewable energy source that helps dairies to generate and capture biogas. Most dairies utilizing anaerobic digesters for production of energy store the biogas and incinerate it as renewable electricity for operations on-farm. Anaerobic digesters can also be used to exclude pathogens, control flies and removing odours.

To collect and transport biomethane, high pressures can be used to compress the gas (CBM) similar to compressed natural gas (CNG), and also very low temperatures can be utilized to generate liquefied biomethane (LBM) similar to liquefied natural gas (LNG). There are three levels for biomethane production. Through upgrading biogas By removal of(1) moisture, (2) hydrogen sulfide, and (3) carbon dioxide. The easy way to take off moisture is by refrigeration whereas H_2S can be removed by

- injecting air into the digester biogas holder
- biological removal on a filter bed
- adding iron chloride into the digester influent
- reacting with hydroxide (iron sponge) or iron oxide
- water scrubbing
- sodium hydroxide or lime scrubbing
- use of an activated-carbon sieve

To scavenge CO_2 and traces of H_2S from biogas of dairy manure the following processes,.shown according to their present availability are considered:

- Water scrubbing

• Pressure swing adsorption

Chemical scrubbing with amines

• Membrane separation

• Chemical glycols scrubbing (e.g. Selexol™)

• Other methods

Some methods are mostly employed for dairy farm operations, because of economic ease, simple operation and possible environmental effects. A layout for a small dairy biogas upgrading plant simply comprises of the following:

• Water scrubber having one or two columns to exclude carbon dioxide

• Iron sponge unit to exclude hydrogen sulfide

• Compressors and storage units

• Final compressor for producing CBM, if desired

• Refrigeration unit to remove water

The reduction of waste generated through biomethanation of FW has received enormous importance. Biomethanation is a low-cost process, generates less residual waste and is a renewable source of energy hence it could play an important role in waste management [112]. As a surplus to producing biogas, biomethanation also leads to a nutrient-rich digestate which may be utilized as a soil conditioner or fertilizer. The optimal amount of methane can only be achieved when microorganisms interact with each other [113]. Misbalance between microbes can lead to process failure and in the reduction of methane production [114], due to the concentration of intermediate products that retards methanogens.

Anaerobic digestion of food processing waste is a well-established method for production of biogases i.e. mainly CO_2 and methane. By 2022, the global installed capacity for power generation is estimated to be 29.5 GW from biogas production [115]. The generation of methane involves four stages: hydrolysis, acidogenesis, acetogenesis, and methanogenesis [83,116,117,118]. The inclusion of food waste is beneficial for methane yield [119,120,121], while the use of food waste as the sole substrate in digestion processes is unsuitable [82,122]. Use of mixed feedstocks can be beneficial as it increases biogas production, enhance degradation rate and digester capacity [123,124]. The beneficial effects of co-digestion are directly related to balanced availability of macro- and micronutrients required by the microbial community, optimal moisture content, buffer capacity and dilution of inhibitory or toxic compounds. Ebner et al. [124] found

Microbial Fuel Cells: Materials and Applications   Materials Research Forum LLC
Materials Research Foundations **46** (2019) 177-220   doi: http://dx.doi.org/10.21741/9781644900116-8

that co-digestion increases the rate of hydrolysis when food waste and manure were co-digested in biomethane potential assay. Methane (50-70%) is the prominent biogas which can be utilized as a source of light energy, water pumps, cooking and electric generators [125]. Benefits of generating biogas from anaerobic digestion in comparison to other waste treatment processes include effective removal of pathogens, less odour leakage and low sludge production from biomass compared to aerobic treatment processes [126]. In India, biomethane production from cattle dung is a well-developed, adopted technology and government had taken considerable steps for the adaptation of biogas technology [127].

2.3.2 Biohydrogen

Hydrogen has been globally recognized as an eco-friendly and renewable energy resource. Hence, it is an ideal substitute to fossil fuels, as it has the greater energy density i.e. 142 kJ/g out of all the known fuels, and generates sole by-product i.e. in the form of water during combustion [128]. H_2 can be generated in two processes i.e. light dependent and light independent bioprocesses: the light dependent process involves photo-fermentation and biophotolysis, while the light-independent reactions involve dark fermentation and bio-electrochemical system) [129]. Integrated photo- dark fermentation process using food processing wastes proves to be a worthwhile technology in terms of economy for production of hydrogen.

Substrate composition

Hydrogen production was observed to be higher than that of protein and fat-based waste i.e. 20 times more which is the potential of carbohydrate-based waste [130]. Kim et al. [131] observed that the biohydrogen yield was cultivated around 0.5 mol H_2/mol hexose at lower than 20 C/N ratio, whereas H_2 yield decreases at higher C/N ratio due to the increased production of propionate, valerate and lactate.

Pre-treatments

Typically consortium of cultures has been utilized for production of H_2 from waste sources. However, hydrogenotrophic bacteria utilize hydrogen produced by *Clostridium* and *Enterobacter* [132]. FW may be utilized as a source of H_2-producing microflora and seed biomass is used to suppress hydrogen-consumers by pretreating with heat [133] (Table 10).

Kim et al. [134] have adopted various pre-treatments to choose microorganisms for hydrogen production. Among untreated FW, lactic acid bacteria species are the most abundant whereas, in the pre-treated FW, hydrogen-producing bacteria are dominant.

- Treatment of heat is effective for increasing hydrogen and butyrate production whereas it suppresses lactate production.

- Treatment of heat increases price in a large-scale process. Luo et al. [135] have studied various pre-treatment processes of inoculums and showed that pretreatment is not very important as it would only have cut-short the effects on the production of hydrogen [136].

Table 10. Production of Hydrogen from FW [64]

Type of waste	Type of vessel	Pretreatment	Microorganism	Duration (day)	Ref
Food waste	Fermenter (3.8 L)	-	As specified	7	[10]
FW with sludge	Fermenter/flask (200 ml)	-	As specified	3	[78]
FW	Fermenter/flask (500 ml)	-	As specified	6	[79]
FW	Fermenter (3 L)	-	As specified	5	[80]
FW	Fermenter (3 L)	-	As specified	60	[73]
FW	Fermenter (10 L)	-	As specified	150	[81]
FW	Fermenter/flask (500 ml)	-	As specified	2	[18]
FW	Fermenter (7.5 L)	heat pretreatment (90°C 20 min)	As specified	3	[26]
FW	Fermenter (4.5 L)	-	As specified	-	[82]
FW	Fermenter (1 L)	-	As specified	2	[83]

Hydrolysis of carbon sources plays a crucial role in the production of H_2 because the carbon and nutrient sources in fats and lipid are not bio-accessible that are present in wastewater and dairy waste for H_2 producing microorganisms [137,138]. Feng et al. [139] have reported bio-hydrogen production using anaerobic bacteria from apple pomace and river sludge. They have generated maximum bio-hydrogen along with other beneficial by-products such as ethanol, butyric acid and acetic acid. Carbohydrate-rich starch and/or cellulose are present in agricultural and food industry wastes which are easier to process for carbohydrate and hydrogen gas formation [140].

Anaerobic fermentation is an energy-saving and eco-friendly process for bioethanol production (Fig. 18). The FW through anaerobic acidification methods generates different organic acids, CO_2, H_2 and other intermediates. This process may be used to treat huge quantities of organic waste because the reactions involved in the production of hydrogen are fast in the absence of solar radiation. Biohydrogen is used as an alternate compressed gas to gasoline because it produces 2.8 times higher energy i.e. approximately 142.35 kJ/g and also 2.1 times more than methane. Carbohydrate-rich FW is suitable for biohydrogen production. Hydrogen may be produced using electrolysis of water only in areas where low-cost electricity is usually available.

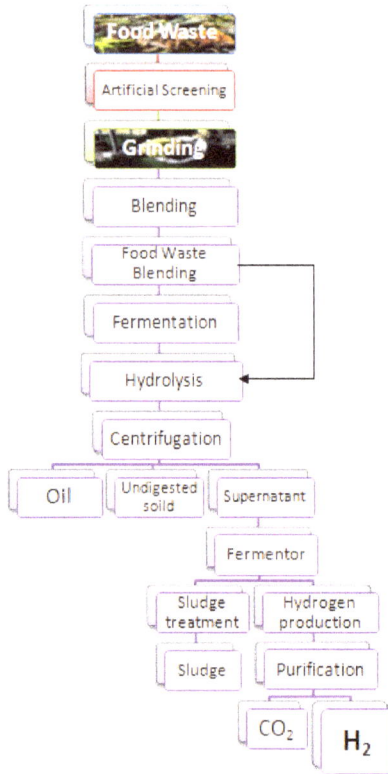

Figure 18. Flowsheet of biohydrogen production from FW

Hence, the utilization of bacteria has a greater impact on indirect conversion of renewable biomass and water into hydrogen. Hydrogen production using microorganisms is through: i) culturing bacteria under anaerobic light conditions i.e. photosynthetic bacteria, and ii) chemotrophic bacteria [141]. Shimizu et al. [142] studied hydrogen production from FW in the absence of microorganisms. Factors affecting hydrogen production are types of pre-treatment, materials and process configuration. The flowsheet of production of hydrogen from FW is shown in Fig. 18.

It has been found that waste from carbohydrate produces 20 times more hydrogen compared to protein-based or fat-based waste. Hence, the carbohydrate-based waste can be the ideal raw material for hydrogen production [143]. In the case of microorganisms, *Clostridium sporogenes* generate more hydrogen (1.61–2.36 mol/mol glucose) compared to other anaerobic bacteria, such as *Enterobacter* and *Bacillus* [144].

Many researchers have reported the use of consortium of microbial cultures to form hydrogen from food waste. However, hydrogenotrophic bacteria generated hydrogen from *Clostridium* and *Enterobacter* have been often used [145]. FW can also be an enriched source of hydrogen-producing microorganisms.

2.3.3 Biohythane

Hythaneis is the blend of biogases containing methane and 10-12% hydrogen by volume. Hythane produced from food processing waste through anaerobic digestion process has been recognized as a cost-effective biogas energy. The beneficial advantage over biohythane process includes high energy yield and low-cost input. Biohythane is usually produced from two-phase processes [146] i.e. first phase is hydrogenogenic phase, in which process parameters (pH of 5.5–6.5, thermophilic conditions) favouring the growth of hydrogenogenic bacteria are controlled and neutral pH (7.0–7.5) and mesophilic condition are employed in the second phase i.e. methanogenic stage [147]. The two-stage processes for bio-hythane production using palm oil effluent [148], fruit vegetable waste [149], and starch wastewater [150] have been found successful.

Conclusions

Food waste is a zero-value and non-consumable resource despite its high potential to be bioconverted into value-added products like biofuel. The strategic scientific management of huge waste generated globally by employing suitable technology will not only combat serious environmental issues but will also add to compensate for the ever-increasing need for energy. Thus, it is time to focus research on evolving various commercially viable inexpensive techniques for efficient recycling of wastes to different types of fuels *viz.* ethanol, hydrogen, methane, diesel etc.

References

[1]   L. Cadoche, G.D. Lopez, Assessment of size reduction as a preliminary step in the production of ethanol from lignocellulosic wastes. Biology of Wastes, 30 (1989) 153-157. https://doi.org/10.1016/0269-7483(89)90069-4

[2]   J. Sheehan, V. Cambreco, J. Duffield, M. Garboski, H. Shapouri, An overview of biodiesel and petroleum diesel life cycles, A report by US Department of Agriculture and Energy, Washington, D.C, 1998, pp. 1-35.

[3]   P. Awasthi, S. Shrivastava, A.C. Kharkwal, A. Varma, Biofuel from agricultural waste: a review. Int. J. Curr. Microbiol. App. Sci. 4 (2015) 470-477.

[4]   Digital trends. https://www.digitaltrends.com/

[5]   S.J.B. Duff, W.D. Murray, Bioconversion of forest products industry waste cellulosics to fuel ethanol: a review, Bioresource Technol. 55 (1996) 1-33. https://doi.org/10.1016/0960-8524(95)00122-0

[6]   M. Melikoglu, C.S.K. Lin, C. Webb, Analysing global food waste problem: pinpointing the facts and estimating the energy content, Cent. Eur. J. Eng, 3 (2013) 157-164. https://doi.org/10.2478/s13531-012-0058-5

[7]   FAO, FAOSTAT, 21.02.2013

[8]   U. Gustafsson, W. Wills, A. Draper, Food and public health: contemporary issues and future directions, Crit. Publ. Health 21 (2011) 385-393. https://doi.org/10.1080/09581596.2011.625759

[9]   Sharma N and Sharma N. 2016. Bioethanol production from alkaline hydrogen peroxide pretreated Populas deltoids wood, International Journal of Bioassays 5 (2) 4810-4816.

[10] The hindu business line. https://www.thehindubusinessline.com,https://www.eatthismuch.com,https://beverageindustrynews.com.ng

[11] L. Reijnders, Conditions for the sustainability of biomass based fuel use, Energy Pol. 34 (2006) 863-876. https://doi.org/10.1016/j.enpol.2004.09.001

[12] P. Vasudevan, S. Sharma, A. Kumar, Liquid fuel from biomass: an overview, J. Sci. Ind. Res, 64 (2005) 822-831.

[13] T.I.N. Ezejiofor, U.E. Enebaku, C. Ogueke, Waste to wealth-value recovery from agro-food processing wastes using biotechnology: A review, British Biotechnology Journal, 4 (2014) 2231-2927. https://doi.org/10.9734/BBJ/2014/7017

[14] IPCC Climate Change 2014: Mitigation of Climate Change (eds Edenhofer, O. et al.) Cambridge Univ. Press, 2014.

[15] V Dyk, E Chanchorn, M.W. Van Dyke, The Saccharomyces cerevisiae protein Stm1p facilitates ribosome preservation during quiescence, Biochem. Biophys. Res. Commun, 430 (2013) 745-50 https://doi.org/10.1016/j.bbrc.2012.11.078

[16] F Raganati, G Olivieri G, A Procentese, M.E. Salatino P, A Marzocchella A, Butanol production by bioconversion of cheese whey in a continuous packed bed reactor, Bioresour. Technol. 138 (2013) 259-265. https://doi.org/10.1016/j.biortech.2013.03.180

[17] WRAP. 2015. http://www.wrap.org.uk/sites/files/wrap/Household%20food%20waste%20restate d%20data%202007-2015.pdf

[18] B. Mahro, M. Timm, Potential of biowaste from the food industry as a biomass resource, Engineer. Life Sci. 7 (2007) 463-476. https://doi.org/10.1002/elsc.200620206

[19] T. Katami, A. Yasuhara, T. Shibamoto, Formation of dioxins from incineration of foods found in domestic garbage, Environmental Science and Technology, 38 (2004) 1062-1065. https://doi.org/10.1021/es030606y

[20] H. Ma, The utilization of acid-tolerant bacteria on ethanol production from kitchen garbage, Renew. Energy, 34 (2009) 1466-1470. https://doi.org/10.1016/j.renene.2008.10.020

[21] S.K. Han, H.S. Shin, Biohydrogen production by anaerobic fermentation of food waste, Int. J. Hydrogen Energ. 29 (2004) 569-577. https://doi.org/10.1016/j.ijhydene.2003.09.001

[22] Y. Ohkouchi, Y. Inoue, Impact of chemical components of organic wastes on l(+)-lactic acid production, Bioresource Technol. 98 (2007) 546-553. https://doi.org/10.1016/j.biortech.2006.02.005

[23] K. Sakai, Y. Ezaki, Open L-lactic acid fermentation of food refuse using thermophilic Bacillus coagulans and fluorescence in situ hybridization analysis of microflora, J. Biosci. Bioeng. 101 (2006) 457-463. https://doi.org/10.1263/jbb.101.457

[24] Q. Wang, Bioconversion of kitchen garbage to lactic acid by two wild strains of Lactobacillus species, Journal of Environmental Science and Health - Part A Toxic/Hazardous Substances and Environmental Engineering, 40 (2005) 1951-1962. https://doi.org/10.1080/10934520500184624

[25] S.Y. Yang, Lactic acid fermentation of food waste for swine feed, Bioresource Technol. 97 (2006) 1858-1864. https://doi.org/10.1016/j.biortech.2005.08.020

[26]   C. Zhang, The anaerobic co-digestion of food waste and cattle manure,
       Bioresource Technol. 129 (2013) 170-176.
       https://doi.org/10.1016/j.biortech.2012.10.138

[27]   Y. Koike, Production of fuel ethanol and methane from garbage by high-efficiency
       twostage fermentation process, J. Biosci. Bioeng. 108 (2009) 508-512.
       https://doi.org/10.1016/j.jbiosc.2009.06.007

[28]   Y. He, Recent advances in membrane technologies for biorefining and bioenergy
       production, Biotechnol. Adv. 30 (2012) 817-858.
       https://doi.org/10.1016/j.biotechadv.2012.01.015

[29]   J. Pan, Effect of food to microorganism ratio on biohydrogen production from
       food waste via anaerobic fermentation, Int. J. Hydrogen Energ. 33 (2008) 6968-
       6975. https://doi.org/10.1016/j.ijhydene.2008.07.130

[30]   M.S. Rao, S.P. Singh, Bioenergy conversion studies of organic fraction of MSW:
       Kinetic studies and gas yield–organic loading relationships for process
       optimization, Bioresource Technol. 95(2004) 173-185.
       https://doi.org/10.1016/j.biortech.2004.02.013

[31]   WRAP, Household food waste, 2015.
       http://www.wrap.org.uk/sites/files/wrap/Household%20food%20waste%20restate
       d%20data%202007-2015.pdf

[32]   A. Lundgren, T. Hjertberg, Ethylene from renewable resources, Surf. Renew.
       Resour. (2010) 109-126.

[33]   A.B. Thomsen, C. Medina, B.K. Ahring, Biotechnology in ethanol production,
       New Emerg Bioenergy Technol. 2 (2003) 40-44.

[34]   R.J. Kawa, C. Joanna, P. Elzbieta, Effect of raw material quality on fermentation
       activity of distillery yeast, Polish J. Food Nutri. Sci, 57 (2007) 275-279.

[35]   Joanna Kawa-Rygielska, Anna Czubaszek, Witold Pietrzak, Some aspects of
       baking industry wastes utilization in bioethanol production, Zeszyty Problemowe
       Postępów Nauk Rolniczych, 575 (2013) 71–77.

[36]   K. Ohgren, A. Rudolf, M. Galbe, G. Zacchi, Fuel ethanol production from steam-
       pretreated cornstover using SSF at higher dry matter content, Biomass Bioenerg.
       30 (2006) 863-869. https://doi.org/10.1016/j.biombioe.2006.02.002

[37]   H.S. Oberoi, P.V. Nadlani, L. Saida, S. Bansal, J.D. Hughes, Ethanol production
       from banana peels using statistically optimized simultaneous saccharification and
       fermentation process, Waste Manag. 31 (2011) 1576-1584.
       https://doi.org/10.1016/j.wasman.2011.02.007

[38]   D. Arapoglou, T. Varzakas, A. Vlyssides, C. Israilides, Ethanol production from potato peelwaste (PPW), Waste Manag. 30 (2010) 1898-1902. https://doi.org/10.1016/j.wasman.2010.04.017

[39]   J.W. Jensen, C. Felby, H. Jorgensen, G.O. Ronsh, N.D. Norholm, Enzymatic processing of municipal solid waste, Waste Manag. 34 (2010) 2497–2503. https://doi.org/10.1016/j.wasman.2010.07.009

[40]   J.H. Kim, J.C. Lee, D. Pak, Feasibility of producing ethanol from food waste, Waste Manag. 31 (2011) 2121-2125. https://doi.org/10.1016/j.wasman.2011.04.011

[41]   R.C. Agu, A.E. Amadife, C.M. Ude, A. Onyia, E.O. Ogu, M. Okafor, E. Ezejiofor, Combined heat treatment and acid hydrolysis of cassava grate waste (CGW) biomass for ethanol production, Waste Manag. 17 (1997) 91–96. https://doi.org/10.1016/S0956-053X(97)00027-5

[42]   K. Dewettinck, F. Van Bockstaee, B. Kühne, D. Van de Walle, T.M. Courtens, X. Gellynck, Nutritional value of bread: influence of processing, food interaction and consumer perception, J. Cereal Sci. 48 (2008) 243-257. https://doi.org/10.1016/j.jcs.2008.01.003

[43]   S.S. Joshi, R. Dhopeshwarkar, U. Jadav, R. Jadav, L. D'souza, D. Jayaprakash, Continuous ethanol production by fermentation of waste banana peels using flocculating yeast, Indian J. Chem. Technol. 8 (2001) 153-159.

[44]   S. Bhushan, K. Kalia, M. Sharma, B. Singh, P.S. Ahuja, Processing of apple pomace for bioactive molecules, Critical Review Biotechnology, 28 (2008) 285-296. https://doi.org/10.1080/07388550802368895

[45]   S. Pathania, N. Sharma, S. Handa, Utilization of horticultural waste (Apple Pomace) for multiple carbohydrase production from Rhizopus delemar F2 under solid state fermentation, Int. J. Genet. Eng. Biotechnol. 16 (2018) 181-189. https://doi.org/10.1016/j.jgeb.2017.10.013

[46]   J.W. Jensen, C. Felby, H. Jørgensen, Cellulase hydrolysis of unsorted MSW, Appl. Biochem. Biotechnol. 165 (2011) 1799-1811. https://doi.org/10.1007/s12010-011-9396-7

[47]   N. Sharma, T.C. Bhalla, A.K. Bhatt, H.O. Agrawal, Enhanced degradation of gamma irradiated forest biomass by a strain of Trichoderma viride isolated from forest soil, National Academy Science Letters, 16 (1993) 11-13.

[48]   D. Tandon, N. Sharma, R. Kaushal, Saccharification of pine needles by a potential celluloytic and hemicelluloytic strain of Penicillium notatum102 isolated from forest soil, Int. J. Biol. Pharm. Allied Sci. 1 (2012) 1344-1355.

[49]   Y.Q. Tang, Ethanol production from kitchen waste using the flocculating yeast
       Saccharomyces cerevisiae strain KF-7, Biomass and Bioenergy, 32 (2008) 1037-
       1045. https://doi.org/10.1016/j.biombioe.2008.01.027

[50]   J.V. Kumar, A. Shahbazi, R. Mathew, Bioconversion of solid food wastes to
       ethanol, Analyst 123 (1998) 497-502. https://doi.org/10.1039/a706088b

[51]   K.C. Kim, Saccharification of food wastes using cellulolytic and amylolytic
       enzymes from Trichoderma harzianum FJ1 and its kinetics, Biotechnol. Bioproc.
       Eng. 10 (2005) 52-59. https://doi.org/10.1007/BF02931183

[52]   Z. Ye, Use of starter culture of Lactobacillus plantarum BP04 in the preservation
       of dining hall food waste, World J. Microbiol. Biotechnol. 24 (2008) 2249-2256.
       https://doi.org/10.1007/s11274-008-9737-z

[53]   Q. Wang, Ethanol production from kitchen garbage using response surface
       methodology, Biochem. Eng. J. 39 (2008) 604-610.
       https://doi.org/10.1016/j.bej.2007.12.018

[54]   F. Tao, Ethanol fermentation by an acid-tolerant Zymomonas mobilis under non-
       sterilized condition, Process Biochem. 40 (2005) 183-187.
       https://doi.org/10.1016/j.procbio.2003.11.054

[55]   R.S. Tubb, Amylolytic yeasts for commercial applications, Trends Biotechnol. 4
       (1986) 98-104. https://doi.org/10.1016/0167-7799(86)90218-0

[56]   P. Tomasik, D. Horton, Enzymatic conversions of starch, Adv. Carbohydr. Chem.
       Biochem. 68 (2012) 59-436. https://doi.org/10.1016/B978-0-12-396523-3.00001-4

[57]   P. Ducroo, Improvements relating to the production of glucose syrups and purified
       starches from wheat and other cereal starches containing pentosans, In Chem.
       Abstr., E.P. Ep, Editor. 1987. p. 4704.

[58]   D. Cekmecelioglu, O.N. Uncu, Kinetic modeling of enzymatic hydrolysis of
       pretreated kitchen wastes for enhancing bioethanol production, Waste Manag. 33
       (2013) 735-739. https://doi.org/10.1016/j.wasman.2012.08.003

[59]   Y.S. Hong, H.H. Yoon, Ethanol production from food residues, Biomass
       Bioenergy 35 (2011) 3271-3275. https://doi.org/10.1016/j.biombioe.2011.04.030

[60]   N. Sharma, A. Sood, Biodegradation of agricultural residue by Bacillus sp. Strain
       CBS28 and CBS11 isolated from soil, Proceedings in Biotechnological Strategies
       in Agroprocessing, PCST Chandigarh (2000) 242-250.

[61]   R. Kaushal, N. Sharma, V. Dogra, Molecular characterisation of
       glycolslyhydrolysalses of Trichomerma harzanium WF5 a potential strain isolated
       from decaying wood and their application in bioconversion of popular wood to

ethanol under separate hydrolysis and fermentation, Biomass and Bioenergy 85 (2016) 243-251. https://doi.org/10.1016/j.biombioe.2015.12.010

[62]   N. Sharma, N. Sharma, Bioethanol production from NaOH+ H2O2 pretreated Populus deltoids wood using cocktail of in house and commercial enzymes under four different modes of separate hydrolysis and fermentation, World J. Pharmaceut. Res. 5 (2016) 661-676.

[63]   N. Sharma, T.C. Bhalla, A.K. Bhatt, H.O. Agrawal, Enhanced degradation of gamma irradiated forest biomass by a strain of Trichoderma viride isolated from forest soil, National Academy Science Letters, 16 (1993) 11-13.

[64]   E.U. Kiran, A.P. Trzcinski, W.J. Ng, Y. Liu, Bioconversion of food waste to energy: A review, Fuel, 134 (2014) 389-399. https://doi.org/10.1016/j.fuel.2014.05.074

[65]   K.D. Ma, Repeated-batch ethanol fermentation of kitchen refuse by acid tolerant flocculating yeast under the non-sterilized condition, Japan Journal of Food Engineering, 8 (2007) 275-279.

[66]   N. Sharma, Optimization of fermentation parameters for production of ethanol from kinnow waste and banana peels by simultaneous saccharification and fermentation, Indian J. Microbiol. 47(2007) 310-316. https://doi.org/10.1007/s12088-007-0057-z

[67]   K.D. Ma, Repeated-batch ethanol fermentation of kitchen refuse by acid tolerant flocculating yeast under the non-sterilized condition, Japan Journal of Food Engineering, 8 (2007) 275-279.

[68]   N. Sharma, N. Sharma, Evaluation of different pretreatments for enzymatic digestibility of forest residues and cellulase production by Bacillus stratosphericus N12 (M) under submerged fermentation, International Journal of Current Research, 9 (2017) 58430-58436.

[69]   N. Sharma, K.L. Bansal, B. Neopany, Enhanced biodegradation of forest waste under solid state fermentation by using a new modified technique, Indian Journal of Forestry, (2008) 112-117.

[70]   S. Yan, Fed batch enzymatic saccharification of food waste improves the sugar concentration in the hydrolysates and eventually the ethanol fermentation by Saccharomyces cerevisiae H058, Braz. Arch. Biol. Tech. 55 (2012) 183-192. https://doi.org/10.1590/S1516-89132012000200002

[71]   M. Ballesteros, Ethanol production from paper materials using a simultaneous saccharification and fermentation system in a fed-batch basis, World J. Microbiol. Biotechnol. 18 (2009) 559-561. https://doi.org/10.1023/A:1016378326762

[72] H. Krishna, T.J. Redd, G.V. Chowdary, Simultaneous saccharification and fermentation of lignocellulosic wastes to ethanol using a thermotolerant yeast, Bioresource Technol. 77 (2001) 193-196. https://doi.org/10.1016/S0960-8524(00)00151-6

[73] H. Ma, The utilization of acid-tolerant bacteria on ethanol production from kitchen garbage, Renewable Energy, 34 (2009) 1466–1470. https://doi.org/10.1016/j.renene.2008.10.020

[74] C. Xue, X. Zhao, C. Liu, L. Chen, F. Bai, Prospective and development of butanol as an advanced biofuel, Biotechnol. Adv., 31 (2013) 1575-1584. https://doi.org/10.1016/j.biotechadv.2013.08.004

[75] M. Shao, H. Chen, Feasibility of acetone–butanol–ethanol (ABE) fermentation from Amorphophallus konjac waste by Clostridium acetobutylicum ATCC 824, Process. Chem. 50 (2015) 1301-1307. https://doi.org/10.1016/j.procbio.2015.05.009

[76] P. Branduardi, V. Longo, N. Berterame, G. Rossi, D. Porro, A novel pathway to produce butanol and isobutanol in Saccharomyces cerevisiae, Biotechnol. Biofuels, 6 (2013) 68. https://doi.org/10.1186/1754-6834-6-68

[77] T. Si, Y. Luo, H. Xiao, H. Zhao, Utilizing an endogenous pathway for 1-butanol production in Saccharomyces cerevisiae, Metab. Eng. 22 (2014) 60-68. https://doi.org/10.1016/j.ymben.2014.01.002

[78] N. Ouellette, H. Rogner, D.S. Scott, Hydrogen-based industry from remote excess hydroelectricity, Int. J. Hydrogen Energ. 22 (1997) 397-403. https://doi.org/10.1016/S0360-3199(96)00098-5

[79] I.K. Muniraj, S.K. Uthandi, Z. Hu, L. Xiao, X. Zhan, Microbial lipid production from renewable and waste materials for second-generation biodiesel feedstock, Environ. Technol. Rev. 10 (2015) 10-18. https://doi.org/10.1080/21622515.2015.1018340

[80] A. Mondala, K. Liang, H. Toghian, R. Hernandez, T. French, Biodiesel production by in situ transesterification of municipal primary and secondary sludges, Bioresour Technol. 100 (2009) 1203-1210. https://doi.org/10.1016/j.biortech.2008.08.020

[81] H.J. Berchmans, S. Hirata, Biodiesel production from crude Jatropha curcas L. seed oil with a high content of free fatty acids, Bioresour Technol. 99 (2008) 1716-1721. https://doi.org/10.1016/j.biortech.2007.03.051

[82]   D.Y.C. Leung, X. Wu, M.K.H. Leung, A review on biodiesel production using catalyzed transesterification, Appl. Energ. 87 (2010), 1083-1095. https://doi.org/10.1016/j.apenergy.2009.10.006

[83]   H.N. Bhatti, M.A. Hanif, M. Qasim, A.U. Rehman, Biodiesel production from waste tallow, Fuel, 87 (2008) 2961-2966. https://doi.org/10.1016/j.fuel.2008.04.016

[84]   Z. Helwani, M.R. Othman, N. Aziz, W.J.N. Fernando, J. Kim, Technologies for production of biodiesel focusing on green catalytic techniques: A review, Fuel Process Tech. 90 (2009) 1502-1514. https://doi.org/10.1016/j.fuproc.2009.07.016

[85]   I.M. Atadashi, M.K. Aroua, A.A. Aziz, High quality biodiesel and its diesel engine application: a review, Renew. Sustain. Energ. Rev. 14 (2010) 1999-2008. https://doi.org/10.1016/j.rser.2010.03.020

[86]   M.W. Formo, Ester reactions of fatty materials, J. Am. Oil Chem. Soc. 31 (1954) 548-559. https://doi.org/10.1007/BF02638571

[87]   S.C. Low, G.K. Gan, K.T. Cheong, Separation of methyl ester from water in a wet neutralization process, J. Sustain. Energy Environ. 2 (2011) 15-19.

[88]   J.A. Siles, M.A. Martín, A.F. Chica, A. Martín, Anaerobic co-digestion of glycerol and waste waster derived from biodiesel manufacturing, Bioresourc. Technol. 101 (2010) 6315-6321. https://doi.org/10.1016/j.biortech.2010.03.042

[89]   J.M.N. Van Kasteren, A.P. Nisworo, A process model to estimate the cost of industrial scale biodiesel production from waste cooking oil by supercritical transesterification, Resour. Conservat. Recycl. 50 (2007) 442-458. https://doi.org/10.1016/j.resconrec.2006.07.005

[90]   W. Zhang, L. Zhang, A. Li, Enhanced anaerobic digestion of food waste by trace metal elements supplementation and reduced metals dosage by green chelating agent [S, S]-EDDS via improving metals bioavailability, Water Res. 84 (2015) 266-277. https://doi.org/10.1016/j.watres.2015.07.010

[91]   C.S. Zhang, H.J. Su, J. Baeyens, T.W. Tan, Reviewing the anaerobic digestion of food waste for biogas production, Renew. Sustain. Energy. Rev. 38 (2014) 383-392. https://doi.org/10.1016/j.rser.2014.05.038

[92]   N.N.A.N. Yusuf, S.K. Kamarudin, Z. Yaakob, Overview on the current trends in biodiesel production, Energ. Convers. Manag. 52 (2011) 2741-2751. https://doi.org/10.1016/j.enconman.2010.12.004

[93]   A.N. Phan, T.M. Phan, Biodiesel production from waste cooking oils, Fuel, 87 (2008) 3490-3496. https://doi.org/10.1016/j.fuel.2008.07.008

[94] W.N.N. Omar, S.N.A. Amin, Optimization of heterogeneous biodiesel production from waste cooking palm oil via response surface methodology, Biomass and Bioenergy, 35 (2011) 1329-1338. https://doi.org/10.1016/j.biombioe.2010.12.049

[95] K.T. Tan, K.T. Lee, A.R. Mohamed, Potential of waste palm cooking oil for catalyst- free biodiesel production, Energy, 36 (2011) 2085-2088. https://doi.org/10.1016/j.energy.2010.05.003

[96] A. Banerjee, R. Chakraborty, Parametric sensitivity in trans esterification of waste cooking oil for biodiesel production - A review, Resour. Conservat. Recycl.53 (2009) 490-497. https://doi.org/10.1016/j.resconrec.2009.04.003

[97] B.K. Barnwal, M.P. Sharma, Prospects of biodiesel production from vegetable oils in India, Renew. Sustain. Energy Rev. 9 (2005) 363-378. https://doi.org/10.1016/j.rser.2004.05.007

[98] D. Wang, S. Czernik, E. Chornet, Production of hydrogen from biomass by catalytic steam reforming of fast pyrolysis oils, Energy Fuels, 12 (1998) 19-24. https://doi.org/10.1021/ef970102j

[99] D. Wang, S. Czernik, D. Montané, M. Mann, E. Chornet, Biomass to hydrogen via pyrolysis and catalytic steam reforming of the pyrolysis oil and its fractions, Ind. Eng. Chem. Res. 36 (1997) 1507-1518. https://doi.org/10.1021/ie960396g

[100] A. Demirbas, Yields of hydrogen-rich gaseous products via pyrolysis from selected biomass samples, Fuel, 80 (2001) 1885-1891. https://doi.org/10.1016/S0016-2361(01)00070-9

[101] A. Demirbas, D. Gulu, Acetic acid, methanol and acetone from lignocellulosics by pyrpolysis, Energy Edu. Sci. Technol. 1 (1998) 111-115.

[102] N. Takezawa, M. Shimokawabe, H. Hiramatsu, H. Sugiura, T. Asakawa, H. Kobayashi, Steam reforming of methanol over Cu/ZrO2. Role of ZrO2 support, React. Kinet. Catal. Lett. 33 (1987) 191-196. https://doi.org/10.1007/BF02066722

[103] R.G. Phillips, I.J.H. Roberts, P.W. Ingham, J.R.S. Whittle, The Drosophila segment polarity gene patched is involved in a position-signalling mechanism in imaginal discs. Develop. 110 (1990) 105-114.

[104] H.P. Brown, A.J. Panshin, C,C, Forsaith, Textbook of Wood Technology, Vol. II, New York, McGraw-Hill, 1952

[105] H.A. Sorensen, Energy conversion systems, New York, Wiley, 1983

[106] G. Grassi, Modern bioenergy in the European Union, Renew. Energy, 16 (1999) 985-990. https://doi.org/10.1016/S0960-1481(98)00347-4

[107] D.O. Hall, C.F. Rosillo, R.H. Williams, J. Woods, Biomass for energy: supply prospects. In: renewable energy-sources for fuels and electricity, T.B. Johansson,

H. Kelly, A.K.N. Reddy, R.H. Williams, (Eds.), Washington, DC, Island Press, 1993

[108] C. Azar, K. Lindgren, B.A. Andersson, Global energy scenarios meeting stringent CO2 constraints-cost-effective fuel choices in the transportation sector, Energy Pol. 31 (2003) 961-976.

[109] A. Cadenas, S. Cabezudo, Biofuels as sustainable technologies: perspectives for less developed countries, Technol. Forecasting Social Change, 58 (1998) 83-103. https://doi.org/10.1016/S0040-1625(97)00083-8

[110] C. Difiglio, Using advanced technologies to reduce motor vehicle greenhouse gas emissions, Energy Pol. 25 (1997) 1173-1178. https://doi.org/10.1016/S0301-4215(97)00109-2

[111] M. Morita, K. Sasaki, Factor influencing the degradation of garbage in methanogenic bioreactors and impacts on biogas formation, Appl. Microbiol. Biotechnol. 94 (2102) 575-582. https://doi.org/10.1007/s00253-012-3953-z

[112] M. Chartrain, M, L. Katz, C. Marcin, M. Thien, S. Smith, F. Fisher, K. Goklen, P. Salmon, T. Brix, K. Price, R. Greasham, Purification and characterization of a novel bioconverting lipase from Pseudomonas aeruginosa MB 5001, Enzyme Microb. Technol. 15 (1993) 575-580. https://doi.org/10.1016/0141-0229(93)90019-X

[113] J.P. Lee, J.S. Lee, S.C. Park, Two-phase methanization of food wastes in pilot scale. Appl. Biochem. Biotechnol. 77 (1999) 585-593. https://doi.org/10.1385/ABAB:79:1-3:585

[114] Anonymous, Research P: Worldwide power generation capacity from biogas will double by 2022, Boulder, USA, 2012.

[115] N. Rasit, A. Idris, R. Harun, W.A.K.G. Azilina, Effects of lipid inhibition on biogas production of anaerobic digestion from oily effluents and sludges: an overview, Renew. Sustain. Energy. Rev. 45 (2015) 351-358. https://doi.org/10.1016/j.rser.2015.01.066

[116] C. Mao, Y. Feng, X. Wang, G. Ren, Review on research achievements of biogas from anaerobic digestion, Renew. Sustain. Energy. Rev. 45 (2015) 540-555. https://doi.org/10.1016/j.rser.2015.02.032

[117] T.P.T. Pham, R. Kaushik, G.K. Parshetti, R. Mahmood, Food waste-to-energy conversion technologies: Current status and future directions, Waste Manag. 38 (2015) 399-408. https://doi.org/10.1016/j.wasman.2014.12.004

[118] P. Aichinger, Synergistic co-digestion of solid-organic-waste and municipal-sewage-sludge: 1 plus 1 equals more than 2 in terms of biogas production and

solids reduction, Water Res. 87 (2015) 416-423.
https://doi.org/10.1016/j.watres.2015.07.033

[119] E. Mara-ón, Co-digestion of cattle manure with food waste and sludge to increase biogas production, Waste Manag. 32 (2012) 1821-1825
https://doi.org/10.1016/j.wasman.2012.05.033

[120] L. Zhang, Y.W. Lee, D. Jahng, Anaerobic co-digestion of food waste and piggery wastewater: Focusing on the role of trace elements, Bioresour. Technol. 102 (2011) 5048-5059. https://doi.org/10.1016/j.biortech.2011.01.082

[121] C.J. Banks, Y. Zhang, Y. Jiang, S. Heaven, Trace element requirements for stable food waste digestion at elevated ammonia concentrations, Bioresour. Technol. 104 (2012) 127-135 https://doi.org/10.1016/j.biortech.2011.10.068

[122] W. Edelmann, H. Engeli, M. Gradenecker, Co-digestion of organic solid waste and sludge from sewage treatment, Water Sci. Technol. 41 (2000) 213-221.
https://doi.org/10.2166/wst.2000.0074

[123] R.A. Labatut, L.T. Angenent, N.R. Scott, Conventional mesophilic vs. thermophilic anaerobic digestion: a trade-of between performance and stability? Water Res. 53 (2014) 249-258. https://doi.org/10.1016/j.watres.2014.01.035

[124] J.H. Ebner, R.A. Labatut, J.S. Lodge, A.A. Williamson, T.A. Trabold, Anaerobic co-digestion of commercial food waste and dairy manure: Characterizing biochemical parameters and synergistic effects, Waste Manag. 52 (2016) 286-294.
https://doi.org/10.1016/j.wasman.2016.03.046

[125] G. Knothe, Fuel properties of highly polyunsaturated fatty acid methyl esters. Prediction of fuel properties of algal biodiesel, Ener. Fuel, 26 (2012) 5265-5273.
https://doi.org/10.1021/ef300700v

[126] V.K. Vijay, E-newsletter of biogas forum of India, BiGFIN, 1 (2010) 1-29.

[127] J.X.W. Hay, T.Y. Wu, J.C. Juan, J.M. Jahim, Biohydrogen production through photo fermentation or dark fermentation using waste as a substrate: overview, economics, and future prospects of hydrogen usage, Biofuels Bioprod. Biorefin. Biofpr. 7 (2013) 334-352. https://doi.org/10.1002/bbb.1403

[128] A. Ghimire, L. Frunzo, F. Pirozzi, E. Trably, R. Escudie, P.N.L. Lens, G. Esposito, A review on dark fermentative biohydrogen production from organic biomass: process parameters and use of by-products, Appl. Energy, 144 (2015) 73-95.
https://doi.org/10.1016/j.apenergy.2015.01.045

[129] M.Y. Azwar, M.A. Hussain, A.K.W. Abdul, Development of biohydrogen production by photobiological, fermentation and electrochemical processes: A

review, Renew. Sustain. Energy. Rev. 31 (2014) 158-173.
https://doi.org/10.1016/j.rser.2013.11.022

[130] K.Y. Show, Biohydrogen production: Current perspectives and the way forward, Int. J. Hydrogen Energ. 37 (2012) 15616-15631.
https://doi.org/10.1016/j.ijhydene.2012.04.109

[131] D.H. Kim, Experience of a pilot-scale hydrogen-producing anaerobic sequencing batch reactor (ASBR) treating food waste, Int. J. Hydrogen Energ. 35 (2010) 1590-1594. https://doi.org/10.1016/j.ijhydene.2009.12.041

[132] C.L. Li, H.H.P. Fang, Fermentative hydrogen production from wastewater and solid wastes by mixed cultures, Crit. Rev. Environ. Sci. Technol. 37 (2007) 1-39.
https://doi.org/10.1080/10643380600729071

[133] E. Elbeshbishy, Single and combined effect of various pretreatment methods for biohydrogen production from food waste, Int. J. Hydrogen Energ. 36 (2011) 11379-11387. https://doi.org/10.1016/j.ijhydene.2011.02.067

[134] D.H. Kim, S.H. Kim, H.S. Shin, Hydrogen fermentation of food waste without inoculum addition, Enzym. Microb. Tech. 45 (2008) 181-187.
https://doi.org/10.1016/j.enzmictec.2009.06.013

[135] G. Luo, Evaluation of pretreatment methods on mixed inoculum for both batch and continuous thermophilic biohydrogen production from cassava stillage, Bioresource Technol. 101 (2010) 959-964.
https://doi.org/10.1016/j.biortech.2009.08.090

[136] X. Wang, Y.C. Zhao, A bench scale study of fermentative hydrogen and methane production from food waste in integrated two-stage process, Int. J. Hydrogen Energ. 34 (2009) 245-254. https://doi.org/10.1016/j.ijhydene.2008.09.100

[137] G. Cai, B. Jin, P. Monis, C. Saint, Metabolic flux network and analysis of fermentative hydrogen production, Biotechnol. Adv. 29 (2011) 375-387.
https://doi.org/10.1016/j.biotechadv.2011.02.001

[138] A. Ghimire, L. Frunzo, L. Pontoni, G. Antonio, P.N.L. Lens, G. Esposito, F. Pirozzi, Dark fermentation of complex waste biomass for biohydrogen production by pretreated thermophilic anaerobic digestate, J. Environ. Manag. 152 (2015) 43-48. https://doi.org/10.1016/j.jenvman.2014.12.049

[139] X. Feng, H. Wang, Y. Wang, X. Wang, J. Huang, Biohydrogen production from apple pomace by anaerobic fermentation with river sludge, Int. J. Hydrogen Energ. 30 (2009) 1-7.

[140] I.K. Kapdan, F. Kargi, Bio-hydrogen production from waste materials, Enzyme Microb. Technol. 38 (2006) 569-582. https://doi.org/10.1016/j.enzmictec.2005.09.015

[141] C.T. Gray, H. Gest, Biological formation of molecular hydrogen, Science, 148 (1965) 186-192. https://doi.org/10.1126/science.148.3667.186

[142] S. Shimizu, A. Fujisawa, O. Mizuno, T. Kameda, T. Yoshioka, Fermentative hydrogen production from food waste without inocula, In: 5th international workshop on water dynamics, AIP Conf Proc, 987 (2008) 171-174. https://doi.org/10.1063/1.2896968

[143] F. Vendruscolo, Biohydrogen production from starch residues, International Journal of Chemical, Molecular, Nuclear, Materials and Metallurgical Engineering, 8 (2014) 1400-1406.

[144] L. A. Hawkes, A.C. Broderick, M.H. Godfrey, R.J. Godley, Investigating the potential impacts of climate change on a marine turtle population. Glob. Change. Bio. 13 (2007 b). 923-932. https://doi.org/10.1111/j.1365-2486.2007.01320.x

[145] C.L. Li, H.H.P. Fang, Fermentative hydrogen production from wastewater and solid wastes by mixed cultures, Crit. Rev. Environ. Sci. Technol. 37 (2007) 1-39. https://doi.org/10.1080/10643380600729071

[146] S.B. Pasupuleti, S.V. Mohan, Single-stage fermentation process for high-value biohythane production with the treatment of distillery spent-wash, Bioresour. Technol. 189 (2015) 177-185. https://doi.org/10.1016/j.biortech.2015.03.128

[147] Z. Liu, C. Zhang, Y. Lu, X. Wu, L. Wang, B. Han, X. Xing, States and challenges for high-value biohythane production from waste biomass by dark fermentation technology, Bioresour. Technol. 135 (2013) 292-303. https://doi.org/10.1016/j.biortech.2012.10.027

[148] C. Mamimin, A. Singkhala, P. Kongjan, B. Suraraksa, P. Prasertsan, T. Imai, S. Thong, Two-stage thermophilic fermentation and mesophilic methanogen process for biohythane production from palm oil mill effluent, Int. J. Hydrogen Energ. 40 (2015) 6319-6328. https://doi.org/10.1016/j.ijhydene.2015.03.068

[149] X. Jia, M. Li, B. Xi, C. Zhu, Y. Yang, T. Xia, C. Song, H. Pan, Integration of fermentative biohydrogen with methanogenesis from fruit-vegetable waste using different pre-treatments, Energy Convers. Manag. 88 (2014) 1219-1227. https://doi.org/10.1016/j.enconman.2014.02.015

[150] S. Roy, D. Banerjee, M. Dutta, D. Das, Metabolically redirected biohydrogen pathway integrated with biomethanation for improved gaseous energy recovery, Fuel, 158 (2015) 471-478. https://doi.org/10.1016/j.fuel.2015.05.060

Microbial Fuel Cells: Materials and Applications

Materials Research Foundations **46** (2019) 221-248

Materials Research Forum LLC

doi: http://dx.doi.org/10.21741/9781644900116-9

Chapter 9

Microbial Desalination Cell: An Integrated Technology for Desalination, Wastewater Treatment and Renewable Energy Generation

V.R.V. Ashwaniy[1], M. Perumalsamy[1]*

[1]Department of Chemical Engineering, National Institute of Technology, Tiruchirappalli-620015, India

* mpsamy@nitt.edu

Abstract

Conversion of seawater to potable water is necessary to meet the demands of future generations. The energy-intensive and un-sustainable desalination technologies led to the development of bio-electrochemical systems called microbial desalination cells (MDCs). MDC configuration has been developed for a decade to overcome the challenges faced in operation and other phenomena. This chapter is aimed to discuss MDCs based on different types of cathodes chemical, air, and bio-cathode and their applications. The technical challenges such as pH imbalance in the anode, membrane fouling, electrode material selection and ohmic resistance are also included.

Keywords

Microbial Desalination Cell, Chemical Cathode, Air Cathode, Bio-Cathode

List of abbreviations

MDC-Microbial desalination cell

RO-reverse osmosis

GHG-green house gas

CEM-cation exchange membrane

AEM-anion exchange membrane

rMDC-recirculation microbial desalination cell

c-SMDC-S-separator coupled stacked circulation microbial desalination cell

M-MDC-multistage microbial desalination cell

PMDC-photosynthetic microbial desalination cell

CoTMPP-cobalt tetramethoxyphenylporphyrin

HRT-hydraulic retention time

PS-photosystems

NADP-nicotinamine adenine dinucleotide phosphate

MEDC-microbial electrolysis and desalination cell

MEDCC-microbial electrolysis desalination and chemical production cell

SMDC-submersible microbial desalination cell

CSTR- continuous stirred tank reactor

C-SMDC-non-separator coupled circulation stacked microbial desalination cell

UMDC-up flow microbial desalination cell

Contents

1.   Introduction

Clean water represents 3% of the total water resources on earth of which only 1% is accessible and the rest 2% is locked under glaciers and icecaps. However, urbanization, industrial and population growth resulted in water deterioration and shortage. Hence there is a need to develop new sources of water to meet the increase in water demand. The water desalination technology is identified as a potential technology to overcome the water shortage problem, i.e., conversion of sea water into potable water, yet the present traditional techniques such as thermal desalination and reverse osmosis are energy consuming [1, 2]. For example, the pressure required for desalination of seawater by reverse osmosis process ranges from 55 to 82 bars and the energy consumption for the typical size of seawater reverse osmosis (RO) unit of 24000 m^3/day ranges from 4 to 6 kWh/m^3. More than 60% of the world's desalination plants utilise fossil fuels as the source for multi-stage flash distillation [3]. Thus, the present desalination processes contribute to the degradation of the environment by the emission of excessive heat, greenhouse gas (GHG) and high concentration of brine. Hence the challenges in the traditional process led to the development of a cheap, efficient and environmentally friendly method called microbial desalination cell technology for desalination of seawater.

1.1   Microbial desalination cell and its evolution

The new type of bio-electrochemical systems (BES) called microbial desalination cell (MDC) was developed by the integration of microbial fuel cell (MFC) and electro-dialysis cell for desalination and wastewater treatment in a single reactor [4, 5]. The MFC consists of the anode for treating domestic and industrial wastewaters and the cathode separated by membranes such as cation exchange membranes (CEMs), anion exchange membrane (AEMs), bipolar membranes, and ultrafiltration. Insertion of the desalination chamber separated by AEM and CEM in the MFC led to the development of MDC. Cao et al. proposed a first small-scale setup of microbial desalination reactor with a salt water chamber volume of 3 mL [4]. It was further increased to 250 L to use at large scale by

Zuo et al. [6]. Effective selection of cathode electron acceptor plays the vital role in the desalination process and bioelectricity generation. Many researchers utilized a chemical cathode called potassium ferricyanide during initial studies of microbial desalination cell. To overcome the toxic, expensive and non-sustainability nature of a chemical cathode, Mehanna et al., [1] used an air cathode and obtained substantial (43–67%) desalination of water using equal volumes of anode and salt solution. Expensive catalyst usage and high energy requirement to maintain dissolved oxygen led to the development of using aerobic microbes called bio-cathodes. Wen et al. suggested the use of an aerobic bio-cathode consisting of carbon felt and bacterial catalyst during their study that produced 609 mV of bioelectricity with bio-cathode which was 136 mV higher than that of the air cathode. The sustainable and effective desalination performance of the bio-cathode proved to be a promising technology when compared with the air cathode without catalyst and ferricyanide [7]. Another study reported utilizing algae, *Chlorella vulgaris* as algae bio-cathode. The passive bio-cathodes performed better than the air cathode in desalination, COD removal and produced high-value products from biomass. [8]. Thereafter many studies concentrated on using bio-cathodes because of their sustainability and cost-effectiveness.

1.2   Microbial desalination cell construction and general principle

The basic configuration of MDC comprised of three compartments (anode, desalination chamber, and cathode) separated by using AEM and CEM and clamped by gaskets to inhibit the water flow between compartments. Fig 1 depicts the schematic representation of a MDC. The compartments are usually made with materials such as polymethyl methacrylate, borosil glass, plexi glass, polycarbonate, acrylic plates, etc. An AEM made of polystyrene cross-linked with divinylbenzene with quaternary ammonia as functional group separates the anode and middle desalination chamber. The CEM made of polystyrene gel cross-linked with divinylbenzene with sulphonic acid separates desalination chamber and cathode. The basic materials such as carbon felt, carbon cloth, carbon brush, graphite granule, graphite fiber granules, graphite rod, carbon fiber felt are generally used as electrodes in a MDC. The reactor can be operated in open circuit and also as a closed circuit by connecting an external resistance.

In the anode compartment, the multiplication of exoelectrogens form bio-films on the electrode material and breaks the organic matter into protons, electrons, and other by-products. The electrons produced by the oxidation of organic matter at the anode are transferred to the solid electrode and migrate to the cathode compartment through the external circuit. In the cathode chamber, the electrons and protons combine with the oxygen (terminal electron acceptor) and form clean water. The difference in potential

created across the electrodes of cathode and anode chambers generates electricity. The continuous oxidation of organic matter by exoelectrogens in the anaerobic environment leads to wastewater treatment. The desalination process is achieved by movement of ions such as anions (Cl^-, NO_3^-, and SO_4^{2-}) and cations (Na^+, Ca^{2+}, Mg^{2+}) across the AEM and CEM respectively. The transfer occurs to maintain electro-neutrality condition due to the release of electrons by the bioelectrochemical oxidation process at the anode that causes the transfer of anions. Similarly, the reduction reaction occurs at the cathode since oxygen accepts electrons and initiates the movement of cations into the cathode chamber. [4]. Thus the bio-electrochemical effect on MDC leads to simultaneous wastewater treatment, desalination, and electricity generation. The rate of desalination is the key factor in MDC function and depends on the initial salt concentration of the sample to be desalinated. The reaction taking place at anode and cathode chamber is given below:
At the anode:

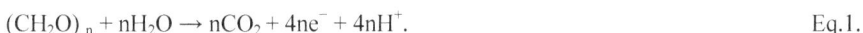

$$(CH_2O)_n + nH_2O \rightarrow nCO_2 + 4ne^- + 4nH^+.$$ Eq.1.

At the cathode:

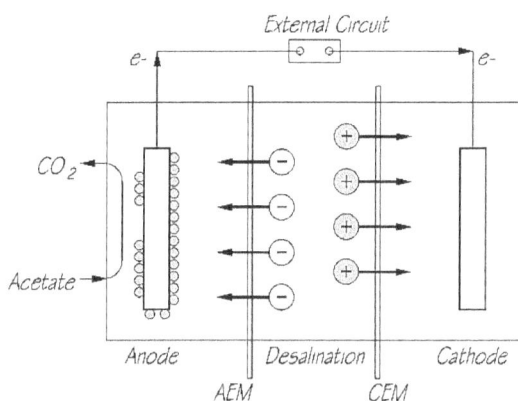

$$O_2 + 4ne^- + 4nH^+ \rightarrow 2H_2O.$$ Eq.2.

Figure 1. Schematic representation of Microbial Desalination Cell

Continuous research on MDC led to the development of different kinds of bio-electrochemical systems for integrated desalination and wastewater treatment. Table 1 represents the various design configurations with their advantages, special features and challenges to be tackled for a significant impact on the performance and efficiency (adapted from references 2, 9, 10)

Table1. Different configurations of MDC (adapted from 2, 9, 10)

MDC configuration	Advantages	Challenges	Special features	Ref
Air cathode MDC	High reduction potential in the cathode chamber	Rise in anolyte pH with time decreasing	Cathode exposed to O_2 as a terminal electron acceptor	[1] [11] [12]
Bio-cathode MDC	Self-generating and sustainable. Enhanced desalination with reduced start-up time.	A small reduction of cell potential during the batch mode of operation.	Enhanced reduction reactions with the help of microbes	[6] [7]
Stack structure MDC	Improved charge transfer efficiency, Increased energy recovery and improved desalination rate.	Increase in number of desalination chambers resulted in declined current and total desalination rate	Alternating IEMs (AEMs and CEMs)	[11] [13] [14]
Recirculation MDC (rMDC)	Increased separation of ion pairs from salt water and power efficiency. Reduction in the pH imbalance.	–	Sequential recirculation of electrolyte i.e. anolyte and catholyte across the chambers and low concentration of buffer used.	[5] [15]
Capacitive adsorption capability MDC (cMDC)	Reduction in the pH imbalance. No ion migration problem. Increased desalination rate.	The effect of increased ion concentration on anode biofilm activity and community on the electrode was not studied	Incorporating capacitive deionization, double layer capacitor on the surface of the electrode, acid-producing chamber and bipolar membrane.	[16] [17]

Upflow MDC	Increased desalination efficiency and power density. Efficient fluid mixing with the chamber. Easier to scale up.	Decrease in pH at a higher TDS removal rate	The tubular reactor containing two compartments and separated by IEMs. Continuous operation with improved performance in desalination and current production.	[18]
Osmotic MDC	Improved desalination efficiency and performance. Enhanced substrate removal from wastewater.	In-depth research is important to overcome challenges such as system scale-up, FO membrane fouling, reduced energy consumption	Forward Osmosis membrane replaced AEM, potassium ferricyanide used as a catalyst, increase water flux to dilute saltwater	[13] [19] [20]
Bipolar membrane MDC	High perm-selectivity. . Enhanced desalination efficiency and long time duration of bipolar membranes. Low water splitting voltage and electric resistance.	Membrane requires the additional voltage since that produced by the cell is not sufficient to the potentials needed to operate the membrane.	Four-chambered Bipolar MDC is formed by placing the bipolar membrane placed next to Anode Chamber; t anion and cation selective layers laminated together to make a Bipolar membrane.	[1] [13]
Decoupled MDC	Easy to scale up Easy to control liquid volume ratios Easy to repair and replace damaged parts	–	Electrodes are made from stainless steel mesh wrapped with carbon cloth, anode and cathode units placed directly in salt solution.	[21]
Separator coupled stacked circulation MDC (c-SMDC-S)	Check-in pH imbalance and improved Coulombic efficiency. No biofouling and smooth operation of the system for a longer period.	-	Glass fibre attached to water-facing one side of the cathode and acts as a separator	[22]

Ion-exchange resin coupled MDC	Stabilised ohmic resistance Reduced energy consumption Enhances charge transfer efficiency and desalination rate.	–	Desalination chamber packed with mixed anions and cations exchange resins.	[22]
Microbial electrolysis desalination and chemical production	Higher desalination rate and low pH fluctuation	–	Involved an acid production chamber and a bipolar membrane	[14] [23] [24]
Submerged microbial desalination-denitrification cell	No bacterial leakage into groundwater. No additional energy treatment for nitrate	The whole process ended up with the depletion of nitrate and ionic strength of groundwater. The nitrogen species in the cathodic effluent required further treatment prior to discharge. Adapted external nitrification of anodic effluent was beneficial to the current generation and nitrate removal rate but total nitrogen removal not possible.	MDC integrated with denitrification system forms submerged microbial desalination-denitrification cell. Removal of nitrate ions from groundwater are for electricity generation	[25]

| Multi-stage microbial desalination cell (M-MDC) | Simultaneous enhanced treatment and self-driven desalination of real domestic wastewater. Enhanced nitrogen removal due to coefficient biological nitrification/denitrification and electrical migration. Increased organics removal due to multi-stage anaerobic/ oxic conditions of the anode and cathode. | Partial desalination (43.4–75.7%). and relatively low current and desalination efficiency. Aeration of biocathode increases the cost and energy. | Two alternating anode and biocathode chambers, with AEM and CEM at opposite positions as in the conventional MDC. A concentrate chamber with deionized water with mixed ion exchange resins. | [7] |
| Photosynthetic microbial desalination cells (PMDCs) | Beneficial use of algae as a passive biocathode by *in- situ* oxygen generation and COD removal. Enhanced desalination and electricity production compared to chemical cathode | – | Microbial solar desalination cells supported by a photosynthetic microorganism, i.e., microalgae (*Chlorella Vulgaris* sp.), as a biocatalyst in the biocathode. | [8] [26] |

2.   Different types of MDC based on cathode electron acceptor

The desalination performance and bioelectricity generation of MDC greatly depend on the cathode and its potential to carry out reduction reaction enabling product formation. The rate of the reaction depends on the desired product formation at the cathode chamber by accepting electrons from the oxidation process at the anode via an external circuit. The different types of cathodes include chemical, air and bio-cathodes.

2.1   MDC with the chemical cathode

The chemical cathode such as ferricyanide is the commonly used catholyte as the electron acceptor in MDC studies. Fig 2 represents the schematic working mechanism of the chemical cathode. Ferricyanide reduces to ferrocyanide by accepting electrons from the anode. The reaction mechanism is as follows:

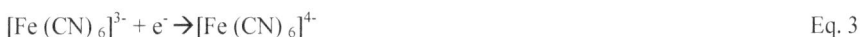

$$[Fe(CN)_6]^{3-} + e^- \rightarrow [Fe(CN)_6]^{4-} \qquad \text{Eq. 3}$$

Microbial Fuel Cells: Materials and Applications   Materials Research Forum LLC
Materials Research Foundations **46** (2019) 221-248   doi: http://dx.doi.org/10.21741/9781644900116-9

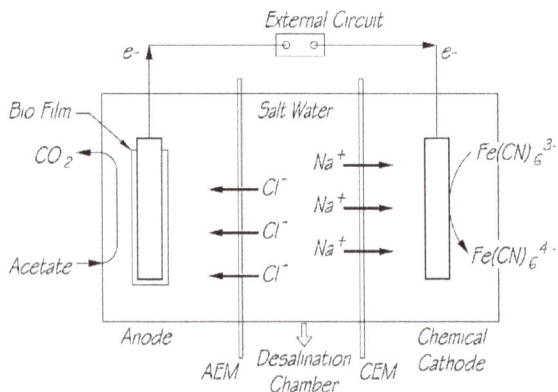

Figure 2. Schematic representation of MDC with the chemical cathode

Cao et al. [4] used ferricyanide as the catholyte because of its high cathodic potential and faster reduction kinetics. However, the significant disadvantage of using chemical catholyte is that it requires continuous replacement, expensive and toxic to the environment. Moreover, it is not sustainable for large-scale operation. Ferricyanides can be used for water softening process in MDC as tested by Brastad et al. [27]. In a study, MDC with $KFe(CN)_6$ catholyte produced $1,532 \pm 14$ mW/m^3 for the first hour and decreased to 379 mW/m^3 after 10 h, while MDC with marine algae catholyte generated 384 ± 5 mW/m^3 during the first hour of operation. However, the stable voltage produced in algae catholyte was greater than for the chemical catholyte. Efficient desalination occurred by using algae catholyte when compared with chemical catholyte with lower power generation [28].

2.2   Microbial desalination cell with the Air cathode

The shortcomings of the chemical cathode led to the development of air cathodes. The air cathode can be defined as a cathode that utilizes oxygen as terminal electron acceptor. Pt, cobalt tetramethoxyphenylporphyrin (CoTMPP) and activated carbon are different types of air cathodes [29, 30]. Fig 3 illustrates the schematic diagram of MDC working with an air cathode. The oxygen combines with protons in the cathode chamber, and electrons generated by oxidation process from the anode chamber leads to the formation of clean water as shown in eq.2.

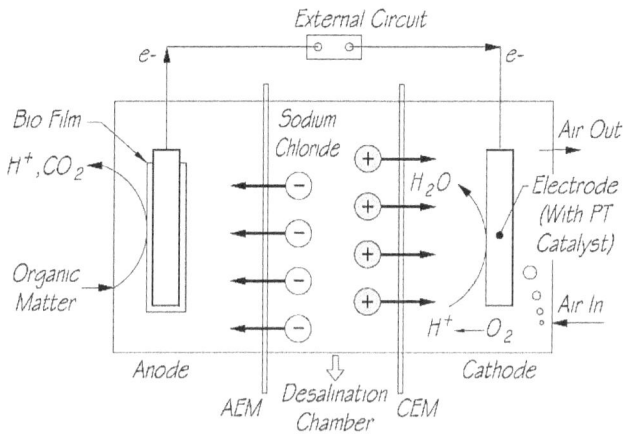

Figure 3. Schematic representation of MDC with air cathode

Oxygen is primarily used because of its high availability and high reduction potential. [31]. It is not toxic when compared with chemical catholytes. The significant disadvantage is that under ambient conditions, air cathodes undergo slow redox kinetics. To minimize the activation overpotential linked with oxygen reduction, expensive catalysts s (e.g. platinum) required. Another disadvantage is the utilization of mechanical equipment to maintain the necessary optimal dissolved oxygen concentration [16]. Thus, the high energy requirement increases the capital cost of the reactor. In another perspective, the reduction of oxygen to water requires four electrons which are not achieved. The decrease in oxygen at times forms hydrogen peroxide thus rendering only a two electron transfer reaction. Hydrogen peroxide is considered a strong oxidizer and results in electron or membrane degradation. It also acts as a disinfectant and inhibits the biofilm formation on the electrode.

Mehanna et al. [1] used air cathode and anolyte containing 2 g/L acetate in cubical shaped chambers, achieved desalination efficiency up to 63% and found superior to using ferricyanide as catholyte. Upflow microbial desalination cell with an air cathode containing platinum wire operated in a continuous mode reported to remove salt efficiently. The results showed that at a hydraulic retention time (HRT) of 4 days, more than 99% of NaCl removal from the initial salt concentration of 30 g TDS/L and current production of approximately 62 mA can be achieved. At 1 or 4 days HRT, the charge transfer efficiencies were 98.6% or 81% [18]. Zuo et al. fabricated a modularized filtration air cathode MDC (FMDC) using nitrogen-doped carbon nanotubes and Pt-carbon as filtration material and cathode. Salt removal efficiency and COD removal efficiencies were 93.6% and 97.3% respectively when real wastewater was circulated from anode to cathode and finally to the middle membrane stack. [32].

2.3 Microbial desalination cell with bio-cathode

The new research paved the way for the exploitation of microorganisms and their usage in the cathode chamber. Bio-cathodes called biological cathodes can be defined as microbial population present over the surface of the cheap electrode material or in the electrolyte that catalyzes the cathodic reaction. Microalgae and cyanobacteria are considered as potential bio-cathodes. Algal MDC is a new approach for environmentally sustainable desalination and bioelectricity generation method. Fig 4 portrays a schematic working model of MDC with a bio-cathode. Fig 5 denotes the fabricated image of a bio-cathode MDC [35]. The Oxygenic photosynthesis is carried out in algae that eliminate the mechanical aeration usage. Oxygenic photosynthesis requires three membrane-bound proteins such as Photosystems PSI, PSII, and cytochrome b6F complex for in electron transport from water to nicotinamide adenine dinucleotide phosphate (NADP). Plastoquinones and plastocyanins are small mobile molecules that help in the transportation of electrons between protein complexes thus leading to the photosynthetic energy conversion. [33].The advantages of bio-cathode include 1) enhanced desalination of water, 2) sustainable and self-generating by the production of oxygen in the cathode chamber and 3) reduction in start-up time [11]. Various by-products can also be obtained from the bio-cathode such as biofuels from algal biomass.

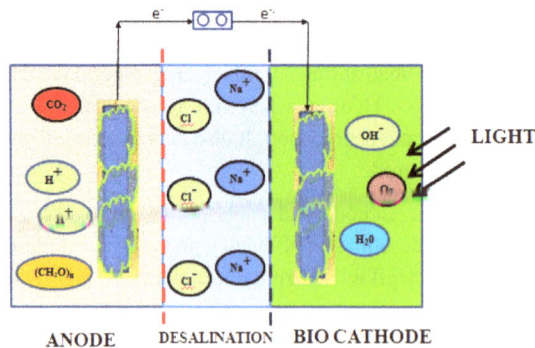

Figure 4. Schematic representation of MDC with bio- cathode

Figure 5. Fabricated image of Bio-cathode Microbial Desalination Cell [36]

The algae in the presence of light consume carbon dioxide and release oxygen and other by-products that help in the formation and growth of new cells. The oxygen produced combines with protons in the cathode chamber and with electrons from the oxidation process at the anode forms the clean water. The general mechanism using algae as bio-cathode is as follows:

Carbon dioxide + water + light energy → Carbohydrates + oxygen + new cells

The reaction mechanism at the cathode is as shown in e.q.2.

Wen et al. [7] first proposed the use of the bacterial catalyst in the cathode chamber of microbial desalination cells. The study produced maximum voltage, Columbic efficiency, desalination performance for the bio-cathode when compared with the air cathode. During the aerobic metabolic process, more hydrogen ions and electrons were consumed and resulted in higher removal of salt and substrate. The photosynthetic microbial desalination cell led to the utilization of microalgae *Chlorella vulgaris* as passive bio-cathode. The study produced the power density of 151 mW/m^3 and 40% desalination performance with 0.9:1:0.5 volumetric ratios of the anode, desalination chamber, and cathode. Bio-cathodes catalyzed by microbes produced steady output voltage when compared with the chemical catholytes [8]. Zamanpour et al. also reported simultaneous desalination and the power generation using microalgae *Chlorella vulgaris* as bio-cathode and dairy wastewater in the anode chamber. The results showed that MDC with a salinity of 35 g/L had a removal rate of 0.341 g/L/day which was 1.5 times greater than MDC with the saline concentration of 15 g/L. Higher algal growth (38%) in 35 g/L saline water proved that higher saline concentration positively increases the desalination performance. [34]. Microalgae *Scenedesmus abundans* also had the positive effect of desalination coupled with petroleum wastewater treatment and bioelectricity generation. [35].

Although MDCs with the chemical and air cathodes have their respective advantages; MDCs with bio-cathodes are found to be sustainable and efficient in enhancing the desalination process. Table 2 summarises the different types of cathodes used in MDCs along with their advantages and disadvantages.

Table2. Summary of different types of MDC cathodes

MDC Configurations	Materials used	Advantages	Disadvantages
Chemical cathode MDC	• Ferricyanide • Phosphate buffer solution	• High cathodic potential • Faster reduction kinetics	• The need for continuous replacement • Expensive • Toxic to the environment
Air cathode MDC	• Oxygen with Platinum catalyst • Cobalt tetramethoxyphenylp orphyrin (CoTMPP) • Activated carbon	• Easy availability • High reduction potential • Not Toxic	• Utilization of expensive catalyst, • High energy requirement • Formation of hydrogen peroxide
Bio-cathode MDC	• Microalgae such as *Chlorella vulgaris, Scenedesmus abundans* etc • Cyanobacteria	• Enhanced desalination of water • Sustainable and self-generating • Reduction in start-up time	• Maintenance of essential growth conditions

3.   Applications of microbial desalination cell

The important application of MDC is desalination. Further, it carries out the following diversified applications instead of desalination.

3.1   Water softening

The available groundwater is probably hard due to the presence of positively charged multivalent ions such as calcium (Ca^{2+}) and magnesium (Mg^{2+}). [36]. The most commonly used method for water softening is ion exchange. The major disadvantage of the process is a generation of large amount of concentrated solutions of acids and salts that cause a problem during the regeneration process. [37][38].The other frequently used softening process accepted by US EPA is chemical precipitation, i.e., lime softening. [39] Other softening methods include carbon nanotubes, nanofiltration, capacitive deionization, electrodialysis and reverse osmosis which involve energy and are cost intensive. [40]

Kristein et al. [27] reported the first bench scale laboratory experiment to study the effectiveness of using MDC in the water softening process. It was found that MDC removed more than 90% of hardness in tested samples and also 89% of the arsenic, 97% of the copper, 99% of the mercury, and 95% of the nickel at the testing concentrations in a synthetic solution. Since it is the first study on water softening, it has some drawbacks to be considered such as membrane fouling, scaling, retention period and economics of the study. The enzymatic mode of MDC was developed for water softening process. [41]. Enzymatic oxidation of glucose by glucose dehydrogenase generates bioelectricity and removes 46% of hardness. The removal efficiency was decreased by the composition of hard waters. The performance of the enzymatic MDC can be enhanced by improving the current generation and bio-anode stability. Hemalatha et al. also proposed MDC as an alternative technique for ground and surface water treatment. [42].

3.2   Production of chemicals and gases

Integrated BES system called microbial electrolysis and desalination cell (MEDC) helps in the production of hydrogen, desalination of sea water and wastewater treatment in a single reactor. Lab scale study achieved 1.6 ml/h of hydrogen production from the cathode chamber with the application of 0.8V potential difference and concurrently desalinated 10 g/L NaCl in the middle chamber with removal efficiency of 98.8% [43]. A new set up called microbial electrolysis desalination and chemical production cell (MEDCC) was developed from MEDC by the insertion of acid production chamber and a bipolar membrane. The MEDCC simultaneously produced HCl and NaOH in the cell along with desalination rates of 46-86% within 18 h using 10 g NaCl/L in the middle chamber with 1.0 V applied voltage [24]. Stack construction of MEDCC enhanced the desalination rates and chemical production when compared with MEDCC. Desalination rate of 0.58 ± 0.02 mmol/h was achieved in the four desalination chamber MEDCC with the AEM and CEM stack structure, which was 43% higher than MEDCC. Maximum acid

Materials Research Foundations **46** (2019) 221-248   doi: http://dx.doi.org/10.21741/9781644900116-9

and alkali production rates were 46 and 8% higher than MEDCC respectively [23]. Although system scaling up poses an important challenge, MEDCC concept effectively recovers high-value chemicals.

3.3   Remediation of contaminated water and nutrients recovery

In general, groundwater remediation employs active or passive methods to remove contaminants. Groundwater possess a high concentration of nitrates due to improper discharge of untreated wastewater with increased use of fertilizers [44]

Zhang et al. developed a novel technique called microbial desalination denitrification cell (SMDCC) to *in-situ* removal of nitrate from groundwater and simultaneous energy production and wastewater treatment. The SMDCC with 12 h wastewater retention time and 10 Ω external resistances produced 3.4 A/m^2 of current density and removed 90.5% nitrate from groundwater [25]. Another study concentrated on the development of an innovative method to remove ammonia in a continuous stirred tank reactor (CSTR) by submersible desalination cell (SMDC) to recover ammonia. In the batch experiment, the average ammonia recovery rate of 80 g-$N/m^2/d$ and the maximum power density of 0.71 ± 0.5 W/m^2 generated at 2.85 A/m^2 during 30 days of operating time was achieved. The mechanisms responsible for the ammonia transportation in this study were due to current driven NH_4^+ migration and free NH_3 diffusion. [45]. A microbial nutrient recovery cell has been developed to purify wastewater and the recovery of nutrient ions by taking the advantage of energy present in the wastewater. The removal efficiencies were > 82% for COD, >96% for NH_4^+-N, and >64% for PO_4^{3-}-P in all the operational cycles. [46]. Thus, the above novel and cost-efficient techniques were beneficial in nutrients recovery and renewable energy production.

4.   Parameters affecting the performance of microbial desalination cell in desalination

Effective performance of a process or technology can be made only by means of controlling certain factors. The performance of MDC reactor is essentially influenced by several factors such as reactor configurations, mode of operation, operational conditions, volumetric ratios of chambers, substrate concentration, bacteriological media, initial salt concentration, conductivity, charge transfer efficiency, internal and external resistances, electrode materials, membrane and intermembrane distance, resin packing, hydraulic retention time, pH imbalance and electrolyte recirculation rate. The important parameters are discussed below in detail.

4.1   Mode of microbial desalination cell operation

Different modes of MDC operation such as batch, cyclic batch and continuous mode affect MDC performance and stability. For example, four microbial desalination cells operated under continuous flow conditions showed total NaCl removal efficiency of 97 ± 1% at an HRT for 2 days. The continuous flow study eliminated substantial pH change and hence increased the desalination rate when compared with the single reactor. [47]. In UMDC continuous mode of operation, uniform substrate distribution enhanced current distribution thus increasing desalination efficiency [48] and produced the maximum power density of 30.8 W/m^3 in the study carried out by Jacobson et al. [18]. In batch mode, the increase in internal resistance due to low electrolyte conductivity decreased MDC performance [49]. Atieh Ebrahimi et al. compared batch and continuously operated MDC in terms of power generation and salt removal efficiency. The batch MDC generated the maximum power density of 13.9 W/m^3 and salt removal rate of 68.1% whereas continuously operated MDC produced the maximum power density of 15.9 W/m^3 and average salt removal of 80%. [12].

4.2   Substrate concentration

The substrate as the fuel of microbial desalination cell affects the performance of MDCs. Anolyte and catholyte are the sources of the nutrient, organic matter and reservoir for ion species. [8] Different substrates such as acetate [1, 4], phosphate buffer solution [5], cellulose degrading rumen microbial consortium [26], synthetic wastewater containing acetate [18], wastewater [12, 51], and dewatered sludge [52] have been used by many researchers in their studies to enhance the exoelectrogenic bacteria activity in the anode chamber. Meanwhile, ferricyanide [4], PBS with NaCl [53], bio-cathode inoculation from the topsoil with the solution [52], *Chlorella vulgaris* as biocathode [8], *Scenedesmus abundance* as biocathode [35] and many more have been used as catholyte substrates. These studies have revealed that concentration losses due to the nutrient gradient in the substrate which significantly influenced the MDC performance and efficiency. In a study, MDCs were acclimated to different acetate concentrations (1-2 g/L) in a fed-batch cycle. In lower substrate concentration, 43% reduction in NaCl conductivity was observed. Current density with 1 g/L of acetate was high (2.80 Am^2) when compared with 2 g/L (less than 1.00 Am^2). Higher substrate concentration reported inhibiting the anode performance [1]. Luo et al. [15] observed that the power output from the MDC was four times higher than the control MFC without desalination function when wastewater was used as a sole substrate. Increase in conductivity by 2.5 times and stability in anolyte pH maintained the microbial activity in the anode.

4.3 Salt concentration

Generally, desalination efficiency depends on the effect of initial salt concentration. The desalination performance is basically measured as conductivity change in the desalination chamber. The desalination efficiency increases with increase in salinity of saline solution i.e., a higher concentration between the desalination chamber and electrode containing chamber. Due to increased internal resistance and lower electrical conductivity, low salt concentration leads to lower desalination efficiency. Yang et al. [54] compared the results of MDCs having different initial salt concentrations and MFC with no desalination chamber. When the initial NaCl concentration was increased from 5 to 30 g/L, the current and power density generation increased from 2.82 mA and 158.2 mW/m^2 to 3.17 mA and 204.5 mW/m^2 respectively in MDC whereas the internal resistances decreased from 2432.0 to 2328.4Ω. The junctional potential i.e., the passage of ions across AEM and CEM increased in MDC with the increase in initial NaCl concentration. In a study conducted by Mehanna et al., [1], the desalination performance of osmotic MDC and control MDC for three different initial saline concentrations (2, 10 and 20 g/L) were investigated. Control MDC showed lower removal efficiency when compared with osmotic MDC at all the variable salinities. The changes occurred due to the difference in osmotic pressure that contributed to the desalination of water.

4.4 pH imbalance

In MDC operation, pH variation is considered as a universal phenomenon. The reaction between protons released by the oxidation of bio-pollutants and the anions in the desalination chamber results in a pH drop (acidic) over a period. Meanwhile, an oxygen reduction reaction in the cathode chamber leads to accumulation of hydroxide ions thus increasing pH within the cathode chamber. [4, 18, 55]. Researchers tackled the pH imbalance based on different perspectives such as adding acids or bases [4, 18, 14] or increasing the anolyte volume [4]. Alternatively, the recirculation of the solutions between anode and cathode to avoid the inhibition of bacterial metabolism was found to be a promising method. Maximum power density was 931 ± 29 mW/m^2 with a 50 mM phosphate buffer solution (PBS) and 776 ± 30 mW/m^2 with 25 mM PBS with recirculation process. The power densities obtained were higher than those achieved without recirculation. [5]. The above recirculation technique was found to eliminate the pH imbalances effectively. Youpeng Qu et al. [47] reported that hydraulic flow through the desalination cells, i.e., anode solution from the first reactor flowed into the cathode and then to the anode of the next reactor, connected can avoid pH fluctuations. But, the transfer of organic matter and microbes during recirculation stimulated the growth of biofilm on the cathode thus reducing the catalytic activity. The new concept called

Materials Research Forum LLC

doi: http://dx.doi.org/10.21741/9781644900116-9

separator coupled circulation stacked microbial desalination cell (c-SMDC-S) to avoid the bio-film growth in the cathode and increase desalination efficiency. The performance of c-SMDC-S was compared with other configurations such as non-separator coupled circulation stacked microbial desalination cell (c-SMDC) and the regular SMDC. It was found that the operation lasted for two months with mild pH variation of 6.8 to 7.9 with desalination efficiency of 65-37% [24] [23][21]. Another technology to overcome pH fluctuation problem was addressed by microbial capacitive desalination cell (MCDC). Insertion of specially designed membrane assembly consisting of cation exchange membrane and layers of activated carbon cloth (ACC) in the set up effectively increased the desalination efficiency, 7 to 25 times than the traditional capacitive deionization process [16]. In another study, pH variation was decreased by 54% by dividing the desalination chamber into upper and lower compartments. The upper part was used for desalination, and lower compartment acted as a medium for proton transfer [56].

4.5   External resistance

External resistance is one of the essential factors that affect the performance of microbial desalination cells. The circuit is usually closed through external resistance. The external resistance controls the flow of electrons from the anode to the cathode affecting the cell potential and current outputs of the microbial desalination cell according to Ohm's Law ($V=I\ R_{ext}$). The approximate total internal resistance of the cell can be obtained by the polarisation curve with variable external resistances. [57]. Sevda et al. [55] reported that COD removal and desalination efficiency were directly dependent on external resistance. At very high resistance (5000-10000 Ω), the flow of electrons was less, and hence upflow microbial desalination cell (UMDC) behaved near to open circuit thus leading to the reduction in COD removal. Higher COD removal was obtained when the reactor was operated at the high resistance (100-1000 Ω) and higher desalination efficiency obtained under lower resistance (0.1 to 1 Ω).   Chen et al. [14] proposed that low external resistance (<10 Ω) could cause the unstable behaviour of exoelectrogenic bacteria and a decrease in production of electrons. Maximum power production was recorded at 10 Ω in this study. With the above results, conclusions can be drawn that important factors such as desalination rate, current generation, power density, COD removal depend on external resistance.

5.   Challenges and future prospects

Despite many advantages such as lower energy or chemical consumption and the added benefit of wastewater treatment, bioelectricity production, and production of high-value by-products, MDC suffers few limitations. It is crucial to overcoming the barriers of

MDC to use as a pre-treatment process for the energy-intensive technologies. The important restriction is the constructive reactor design. The innovation in reactor configuration, selection of prominent materials, and thorough knowledge about biochemical mechanisms ensure the high level of mutual benefits such as power amplification and reduction in capital cost of desalination. The MDCs power densities can be enhanced by reducing the electrochemical losses such as ohmic losses, activation overpotential, and mass transfer limitations. The losses at the cathode can also be minimized by low electrode spacing, large anode surface area (m^2/m^3), and usage of the efficient chemical catalyst. Anolyte and catholyte pH should be taken care of, as the pH fluctuations inhibit the biofilm formation and growth. Chloride ions alter crossing AEM combine with protons in the anode to form hydrochloric acid and thus make the medium acidic. Meanwhile, sodium ions combine with hydroxide ions forming sodium hydroxide in the cathode chamber thus increasing the alkalinity in the cathode chamber. Hence extensive buffering is essential to avoid the pH changes.

Another drawback is the increase in ohmic resistance which occurs due to a decrease in conductivity and salinity, limiting the desalination performance and bioelectricity production. The above scenario is particularly seen in low salinity solutions. The sustainability of MDC operation is affected majorly by another parameter called membrane fouling/biofouling. Deposition of soluble cations such as Mg^{2+} and Ca^{2+} cause scaling phenomenon resulting in amplification of MDC's external resistance. Additionally, the ions react with natural organic matter in water and form scales on the membrane. [58, 59]. Hence, modification in the ion exchange membrane is essential to overcome the scaling problems.

Another issue to be sorted out is scale-up of MDC. The liquid volume ranges from hundreds to thousands. A large-scale MDC holding 105 L liquid volume was tested and found to produce bio-electricity of 2000 mA and the removal of 9.2 $kg/m^3/d$ salt. [60]. The system efficiency was mainly affected by fluctuations in environmental conditions such as temperature, pH, membrane fouling, etc. and also the scale-up faced difficulty due to the usage of a significant amount of pH buffers in the reactor.

MDC can be used as a pre-treatment for the RO process, or it can be used as a stand-alone process for treatment of brackish water. MDC technology can also be used as a water softening process for sustainable treatment of wastewater. Use of MDC as a pre-treatment to RO can decrease the impact of membrane-related issues and other operational costs. Hence significant research is necessary to understand the loopholes of technology to improve the overall system's efficiency.

Conclusion

The general concepts about various aspects of MDC have been discussed in this chapter. In-depth researches on MDCs have led to the development of various reactor configurations and utilization of different forms of cathodes. Finally, the photosynthetic MDC is declared as an efficient technology utilizing biological cathodes such as microalgae. Usage of bio-cathodes increased because of their sustainable nature, increased desalination efficiency, reduction in start-up time and production of valuable biomass. The ability of bio-cathodes can be improved with the recent molecular biology and bioinformatics techniques. Identification of better bacterial and algal species can improve the overall efficiency of bio-cathode microbial desalination cells.

Other than desalination process, MDC plays a vital role in other valuable applications such as water softening process, production of chemicals, acids and bases, hydrogen gas production, remediation of contaminated water and also helps in recovery of nutrients.

A process is considered efficient only after controlling defined operational parameters. The continuous mode of MDC operation, selection of effective anode and cathode substrates, pH balance, initial salt concentration, usage of lower external resistance are some of the important changes that can be done to enhance the working of MDC process.

MDC technology came into existence less than a decade and had multiple outcomes from a single reactor. Further development in reactor configurations, selection of better electrode materials, and optimization of other operational parameters can extensively increase the integrated process of desalination performance, wastewater treatment, and electricity generation.

Hence MDC is found to be a new, green, environmentally friendly, efficient and sustainable technology.

References

[1] M. Mehanna, T. Saito, J. Yan, M. Hickner, X. Cao, X. Huang, B.E. Logan, Using microbial desalination cells to reduce water salin¬ity prior to reverse osmosis, Ener. Environ. Sci. 3 (2010) 1114. https://doi.org/10.1039/c002307h

[2] A. Carmalin Sophia, V.M. Bhalambaal, E.C. Lima, M. Thirunavoukkarasu, Microbial desalination cell technology: Contribution to sustainable waste water treatment process, current status and future applications, J. Environ. Chem. Eng. 4 (2016) 3468–3478. https://doi.org/10.1016/j.jece.2016.07.024

[3] V.G. Gude, N. Nirmalakhandan, S. Deng, Renewable and sustainable approaches for desalination. Renew. Sustain. Energy Rev. 14 (2010) 2641-2654. https://doi.org/10.1016/j.rser.2010.06.008

[4] X. Cao, X. Huang, P. Liang, K. Xiao, Y. Zhou, X. Zhang, B.E. Logan, A new method for water desalination using microbial desalination cells, Environ. Sci. Technol. 43 (2009) 7148–7152. https://doi.org/10.1021/es901950j

[5] Y. Qu, Y. Feng, X. Wang, J. Liu, J. Lv, W. He, B.E. Logan, Simultaneous water desalination and electricity generation in a microbial desalination cell with electrolyte recirculation for pH control, Bioresour. Technol. 106 (2012) 89–94. https://doi.org/10.1016/j.biortech.2011.11.045

[6] K. Zuo, F. Liu, S. Ren, X. Zhang, P. Liang, X. Huang, A novel multi-stage microbial desalination cell for simultaneous desalination and enhanced organics and nitrogen removal from domestic wastewater, Environ. Sci. Water Res. Technol. 2 (2016) 832–837. https://doi.org/10.1039/C6EW00196C

[7] Q. Wen, H. Zhang, Z. Chen, Y. Li, J. Nan, Y. Feng, Using bacterial catalyst in the cathode of microbial desalination cell to improve wastewater treatment and desalination. Bioresour. Technol. 125 (2012) 108–113 https://doi.org/10.1016/j.biortech.2012.08.140

[8] B. Kokabian, V.G. Gude, Photosynthetic microbial desalination cells (PMDCs) for clean energy, water and biomass production. Environ. Sci.: Processes Impacts, 15 (2013) 2178. https://doi.org/10.1039/c3em00415e

[9] H.M. Saeed, G.A. Husseini, S. Yousef, J. Saif, S. Al-Asheh, A. Abu Fara, S. Azzam, R. Khawaga, A. Aidan, Microbial desali¬nation cell technology: A review and a case study, Desalina¬tion. 359 (2015) 1–13.

[10] H. Jingyu, D. Ewusi-Mensah, EyramNorgbey, Microbial desalination cells technology: a review of the factors affecting the process, performance and efficiency, Desal. Water Treat. 87 (2017) 140–159.

[11] G. Gude, B. Kokabian, V. Gadhamshetty, Beneficial bioelectrochemical systems for energy, water, and biomass production, Microb. Biochem. Technol. 6 (2013) 1–14.

[12] X. Zhang, W. He, L. Ren, J. Stager, P.J. Evans, B.E. Logan, COD removal characteristics in air-cathode microbial fuel cells, Bioresour Technol. 176 (2015) 23–31 https://doi.org/10.1016/j.biortech.2014.11.001

[13] Y. Kim, B.E. Logan, Microbial desalination cells for energy production and desalination, Desalination, 308 (2013) 122–130. https://doi.org/10.1016/j.desal.2012.07.022

[14] X. Chen, X. Xia, P. Liang, X. Cao, H. Sun, X. Huang, Stacked microbial desalination cells to enhance water desalination effi¬ciency, Environ. Sci. Technol. 45 (2011) 2465–2470. https://doi.org/10.1021/es103406m

[15] H. Luo, P. Xu, T.M. Roane, P.E. Jenkins, Z. Ren, Microbial desalination cells for improved performance in wastewater treatment electricity production, and desalination, Bioresour. Technol. 105 (2012) 60–66. https://doi.org/10.1016/j.biortech.2011.11.098

[16] C. Forrestal, P. Xu , Z. Ren, Microbial desalination cell with capacitive adsorption for ion migration control, Bioresour. Technol. 120 (2012) 332–336. https://doi.org/10.1016/j.biortech.2012.06.044

[17] L. Yuan, X. Yang, P. Liang, L. Wang, Z.H. Huang, J. Wei, X. Huang, Capacitive deionization coupled with microbial fuel cells to desalinate low-concentration salt water, Bioresour. Technol. 110 (2012) 735–738. https://doi.org/10.1016/j.biortech.2012.01.137

[18] K.S. Jacobson, D.M. Drew, Z. He, Efficient salt removal in a continuously operated upflow microbial desalination cell with an air cathode, Bioresour. Technol. 102 (2011) 376–380. https://doi.org/10.1016/j.biortech.2010.06.030

[19] C. Huang, T. Xu, Electrodialysis with bipolar membranes for sus¬tainable development, Environ. Sci. Technol. 40 (2006) 5233–5243. https://doi.org/10.1021/es060039p

[20]   B. Zhang, Z. He, Improving water desalination by hydrauli¬cally coupling an osmotic microbial fuel cell with a microbial desalination cell, J. Membr. Sci. 441 (2013) 18–24. https://doi.org/10.1016/j.memsci.2013.04.005

[21]   X. Chen, P. Liang, Z. Wei, X. Zhang, X. Huang, Sustainable water desalination and electricity generation in a separator coupled stacked microbial desalination cell with buffer free electrolyte circulation, Bioresour. Technol. 119 (2012) 88-93. https://doi.org/10.1016/j.biortech.2012.05.135

[22]   A. Morel, K. Zuo, X. Xia, J. Wei, X. Luo, P. Liang, X. Huang, Microbial desalination cells packed with ion-exchange resin to enhance water desalination rate, Bioresour. Technol. 118 (2012) 43–48. https://doi.org/10.1016/j.biortech.2012.04.093

[23]   S. Chen, G. Liu, R. Zhang, B. Qin, Y. Luo, Y. Hou, Improved performance of the microbial electrolysis desalination and chemical-production cell using the stack structure, Bioresour. Technol. 116 (2012b) 507-511. https://doi.org/10.1016/j.biortech.2012.03.073

[24]   S. Chen, G. Liu, R. Zhang, B. Qin, Y. Luo, Development of the microbial electrodialysis and chemical-production cell for desalination as well as acid and alkali productions, Environ. Sci. Technol. 46 (2012) 2467–2472. https://doi.org/10.1021/es203332g

[25]   Y. Zhang, I. Angelidaki, A new method for in situ nitrate removal from groundwater using submerged microbial desalination edenitrification cell (SMDDC), Water Res. 47 (2013)1827-1836. https://doi.org/10.1016/j.watres.2013.01.005

[26]   G.M. Girme, Algae powered microbial desalination cells, 58. MSc Thesis, Graduate School of the Ohio State Uni¬versity, Ohio. (2014)

[27]   S. Kristen Brastad, Zhen He, Water softening using microbial desalination cell technology, Desalination. 309 (2013) 32–37 https://doi.org/10.1016/j.desal.2012.09.015

[28]   D. BeenishSaba, A. Christy, Z. Yu, C.A. Co, A. Park, Simultaneous power generation and desalination of microbial desalination cells using nannochloropsissalina (marine algae) versus potassium ferricyanide as catholytes, Environ. Eng. Sci. 34 (2017).

[29]   S. Cheng and B. E. Logan, Evaluation of catalysts and membranes for high yield biohydrogen production via electrohydrogenesis in microbial electrolysis cells

(MECs), Water Sci. Technol. 58 (2008) 853–857.
https://doi.org/10.2166/wst.2008.617

[30]   F. Zhang, S. Cheng, D. Pant, G. V. Bogaert and B. E. Logan, Power generation using an activated carbon and metal mesh cathode in a microbial fuel cell Electrochem. Commun. 11 (2009) 2177–2179.
https://doi.org/10.1016/j.elecom.2009.09.024

[31]   S.M. Rismani-Yazdi, A.D. Carver, O.H. Christy, Tuovinen, Cathodic limitations in microbial fuel cell: An overview, J. Power Sour. 180 (2008) 683-694.
https://doi.org/10.1016/j.jpowsour.2008.02.074

[32]   K. Zuo, Z. Wang, X. Chen, X. Zhang, J. Zuo, P. Liang, X. Huang, Self-driven desalination and advanced treatment of wastewater in a modularized filtration air cathode microbial desalination cell, Environ. Sci. Technol. 50 (2016) 7254–7262.
https://doi.org/10.1021/acs.est.6b00520

[33]   L. Taiz, E. Zeiger, Plant Physiology, Sinauer Associate (2010).

[34]   M.K. Zamanpour, H.R. Kariminia, M. Vosoughi, Electricity generation, desalination and microalgae cultivation in a biocathode-microbial desalination cell, J. Environ. Chem. Eng. 5 (2017) 843–848.
https://doi.org/10.1016/j.jece.2016.12.045

[35]   V.R.V. Ashwaniy, M. Perumalsamy, Reduction of organic compounds in petro-chemical industry effluent and desalination using Scenedesmus abundans algal microbial desalination cell, J. Environ Chem. Eng. 5 (2017) 5961-5967.
https://doi.org/10.1016/j.jece.2017.11.017

[36]   Water Quality Research Council, What makes water hard & how can it be improved, Water Rev. 5 (1990) 1–2

[37]   M.A. Burris, in: Soft Water, Hard Choice? Government Engineering, (2004) 20–21.

[38]   S.R. Maguin, P.C. Martyn, in: Notification of the Continued Prohibition on Brine Discharges from Self-Regenerating Water Softeners and the Imposition of New Chloride Discharge Requirements at Santa Clarita Valley Businesses, 2 (2010).

[39]   R.A. Bergman, Membrane softening versus lime softening in Florida: a cost comparison update, Desalination 102 (1995) 11–24. https://doi.org/10.1016/0011-9164(95)00036-2

[40] M.A. Tofighy, T. Mohammadi, Permanent hard water softening using carbon nanotube sheets, Desalination 268 (2011) 208–213. https://doi.org/10.1016/j.desal.2010.10.028

[41] M.A. Arugula, K.S. Brastad, S.D. Minteer, Z. He, Enzyme catalyzed electricitydriven water softening system. Enzyme Microb. Technol. 51 (2012) 396-401. https://doi.org/10.1016/j.enzmictec.2012.08.009

[42] M. Hemalatha, S.K. Butti, G. Velvizhi, S. Venkata Mohan, Microbial mediated desalination for ground water softening with simultaneous power generation. Bioresour. Technol. 242 (2017) 28-35. https://doi.org/10.1016/j.biortech.2017.05.020

[43] H. Luo, P.E. Jenkins, Z. Ren, Concurrent desalination and hydrogen generation using microbial electrolysis and desalination cells. Environ. Sci. Technol. 45 (2010) 340-344. https://doi.org/10.1021/es1022202

[44] J.N. Galloway, J.D. Aber, J.W. Erisman, S.P. Seitzinger, R.W. Howarth, E.B. Cowling, B.J. Cosby, The nitrogen cascade, Bioscience 53 (2003) 341–356 https://doi.org/10.1641/0006-3568(2003)053[0341:TNC]2.0.CO;2

[45] Y. Zhang, I. Angelidaki, I, Submersible microbial desalination cell for simultaneousammonia recovery and electricity production from anaerobic reactorscontaining high levels of ammonia. Bioresour. Technol. 177 (2015) 233-239. https://doi.org/10.1016/j.biortech.2014.11.079

[46] X. Chen, D. Sun, X. Zhang, P. Liang, X. Huang, Novel self-driven microbial nutrient recovery cell with simultaneous wastewater purification, Scientific Reports 5 (2015) 15744. https://doi.org/10.1038/srep15744

[47] Y. Qu, Y. Feng, J. Liu, W. He, X. Shi, Q. Yang, J. Lv, B.E. Logan, Salt removal using multiple microbial desalination cells under continuous flow conditions, Desalination 317 (2013) 17–22. https://doi.org/10.1016/j.desal.2013.02.016

[48] K.S. Jacobson, D.M. Drew, Z. He, Use of a liter-scale microbial desalination cell as a platform to study bioelectrochemical desalination with salt solution or artificial seawater, Environ. Sci. Technol. 45 (2011) 4652–4657. https://doi.org/10.1021/es200127p

[49] Q. Ping, B. Cohen, C. Dosoretz, Z. He, Long-term investigation of fouling of cation and anion exchange membranes in microbial desalination cells, Desalination 325 (2013) 48–55. https://doi.org/10.1016/j.desal.2013.06.025

[50]   D.A. Ebrahimi, G.N. Kebria, D. Youse, Effect of batch vs. continuous mode of operation on microbial desalination cell performance treating municipal wastewater, Iranian Journal of Hydrogen & Fuel Cell 4 (2016) 281-290.

[51]   A. Aidan, G.A. Husseini, H. Yemendzhiev, V. Nenov, A. Rash¬eed, H. Chekkath, Y. Al-Assaf, Microbial desalination cell (MDC) in the presence of activated carbon, Adv. Sci, Eng. Med¬. 6 (2014) 1100–1104.

[52]   F. Meng, J. Jiang, Q. Zhao, K. Wang, G. Zhang, Q. Fan, L. Wei, J. Ding, Z. Zheng, Bioelectrochemical desalination and electricity generation in microbial desalination cell with dewatered sludge as fuel, Bioresour Technol. 157 (2014) 120–126 https://doi.org/10.1016/j.biortech.2014.01.056

[53]   G.C. Gil, I.S. Chang, B.H. Kim, M. Kim, J.K. Jang, H.S. Park, H.J. Kim, Operational parameters affecting the performance of a mediator-less microbial fuel cell., Biosens. Bioelectr.,18 (2003) 327–334. https://doi.org/10.1016/S0956-5663(02)00110-0

[54]   1Euntae Yang, Mi-Jin Choi, Kyoung-Yeol Kim, Kyu-Jung Chae, In S. Kim (2014): Effect of initial salt concentrations on cell performance and distribution of internal resistance in microbial desalination cells, Environmental Technology 36 (2015) 852-860. https://doi.org/10.1080/09593330.2014.964333

[55]   S. Sevda, H. Yuan, Z. He, I.M. Abu-Reesh, Microbial desalina¬tion cells as a versatile technology: Functions, optimization and prospective, Desalination. 371 (2015) 9–17. https://doi.org/10.1016/j.desal.2015.05.021

[56]   E. Yang, M.J. Choi, K.Y. Kim, I.S. Kim, Improvement of biohydrogen generation and seawater desalination in a microbial electrodialysis cell by installing the direct proton transfer pathway between the anode and cathode chambers, Desalination Water Treat. 51 (2013) 6362-6369. https://doi.org/10.1080/19443994.2013.780997

[57]   T.A. Bower, A.D. Christy, O. Tuovinen, L. Zhao, Voltage Self-Amplification and Signal Conditioning for Enhanced Microbial Fuel Cell Performance, Ohio, 2013.

[58]   L. Bazinet, M. Araya-Farias, Effect of calcium and carbonate concentrations on cationic membrane fouling during electrodialysis, J. Colloid Interface Sci. 281 (2005) 188–96. https://doi.org/10.1016/j.jcis.2004.08.040

[59]   C. Casademont, G. Pourcelly, L. Bazinet, Effect of magnesium/calcium ratio in solutions subjected to electrodialysis: Characterization of cation-exchange membrane fouling, J. Colloid Interface Sci. 315 (2007) 544–54. https://doi.org/10.1016/j.jcis.2007.06.056

Microbial Fuel Cells: Materials and Applications   Materials Research Forum LLC
Materials Research Foundations **46** (2019) 221-248   doi: http://dx.doi.org/10.21741/9781644900116-9

[60] F. Zhang, Z. He, Scaling up microbial desalination cell system with a post-aerobic process for simultaneous wastewater treatment and seawater desalination, Desalination 360 (2015) 28–34 https://doi.org/10.1016/j.desal.2015.01.009

Microbial Fuel Cells: Materials and Applications
Materials Research Foundations **46** (2019) 249-288

Materials Research Forum LLC
doi: http://dx.doi.org/10.21741/9781644900116-10

Chapter 10

Biofuels from Food Processing Wastes

Rouf Ahmad Dar*[1], Mudasir Yaqoob[2], Manisha Parmar[1], Urmila Gupta Phutela[3]

[1]Department of Microbiology, Punjab Agricultural University Ludhiana-141004, Punjab, India

[2]Department of Food Science and Technology, Punjab Agricultural University, Ludhiana-141004, Punjab, India

[3]Department of Renewable Energy Engineering, Punjab Agricultural University, Ludhiana-141004, Punjab, India

roufdar-mb@pau.edu; roufulramzan086@gmail.com

Abstract

Food processing wastes (FPWs) are produced in large quantity. The issue of waste is increasing unabatingly and is posing a threat to the environment. The management of these wastes is needed. FPWs are managed through various approaches; however, the biofuel production is the most feasible solution and alluring the researchers across the globe as it furnishes a substantial alternative to conventional fuels, thus reducing the greenhouse emissions significantly. It is a worthwhile means of utilizing food processing wastes compared to other valorisation processes. This chapter reviews the status of FPWs across the globe, biofuel production from food processing wastes, factors affecting and strategies for enhancing biofuel production from food processing wastes.

Keywords

Food Processing Wastes, Biofuel, Conventional Fuels, Greenhouse Emissions, Environment

Contents

1.   Background

Food processing waste (FPW) is obtained from different food processing industries as the last product which has not been recycled and is abandoned as waste. The food processing industries are one of the principal industries of the present times. These supply a large number of products ranging from industrial to human use. The food wastage occurs significantly almost at every step in the processing industries. However, the wastage at consumer and supply chain levels, the degradability of crop and food products, are responsible for the production of substantial quantities of wastes from food processing industries [1]. It not only involves the monetary loss [2] but also has an immense influence on climate change. This wastage costs approximately 2600 billion USD annually, which is almost comparable to the gross domestic product (GDP) of France or the UK [3]. The wastage in the USA at the consumer level is 10 times more food per capita compared to that in Southeast Asia. On the other hand, approximately 0.925 billion people endure hunger. This raises the questions related to food security and imposes a big challenge on food security, which will be increasing with the passage of time. As per the McKinsey consulting report, if the wastage at consumer is curbed by 30%, it would save approximately 0.100 billion acres of cropland by 2030 [4].

The main priorities of food waste management are to prevent food waste and recovery of food to feed needy people. However, about 95% of the food waste is dumped at landfill sites [5]. The incessant increase in population demands the increased production of food. This increase in population had increased the price of oil which in turn has increased the food production cost. These things have made it imperative to look for the advanced approaches to food processing waste management [6]. FPWs contain carbohydrates, oils, fats, proteins and organic acids [5,7]. FPWs are the potential substrates for the production of different types of value-added biochemicals, biofuels, feed and enzymes etc. [8]. Production of biofuels from wastes is of more significance than other valorization processes [5] as it significantly can influence the food prices by alleviating the dependency on conventional fuels. Besides this, biofuels being renewable, biodegradable in nature and emitting acceptable quality gases make them more demanding as compared to other value-added products. Thus, this chapter is focused on various aspects of FPWs like status of FPWs across globe, characterization, management, energy generation technologies and valorisation to various kinds of biofuels. Apart from this, we focused on various factors affecting biofuel generation and possible strategies for its enhancement.

2.   Status of production of food processing wastes across the globe

2.1   Food processing waste

Waste is presently a major concern worldwide, more importantly in the developing countries (China, India, etc.) and in Europe. Wastes can be classified into industrial, sanitary, agricultural and solid urban residues based on their origin. Wastes generated by food processing companies are a good example of a pre-consumer type of waste, produced on a large scale globally [5].

FPWs are defined as residuals which remain after processing a primary product. Food industries generate large proportions of solid and liquid wastes, resulting from the preparation, production and consumption of food; due to losses during processing; non-appropriate transport systems; contamination problems during storage and inappropriate packaging [9]. FPWs include vegetable and fruit peels, pits, seeds, cheese whey, blood, bone, process water, tofu whey, wastewater treatment sludge and so on. These wastes consisting of worthwhile nutrients represent valuable biomass. Food effluents are rich in biodegradable components with high chemical oxygen demand (COD) and biological oxygen demand (BOD) contents. If FPWs are left untreated and unmanaged, their uncontrolled decomposition will pollute the environment due to the release of toxic materials and methane [10].

2.2   Sources

The sources of food residuals are determined by the products being processed. The wastes produced generally include liquid and solid constituents. The product-specific waste from the food industry is characterized by its high proportion of organic material. This can range from tea leaves, potato and orange peels to old bread [11]. The various sources generating food processing wastes are as follows:

2.2.1   Meat industry

The bulk of the waste in the meat industry is generated during slaughtering [11]. The liquid and solid wastes generated include grease, oils and fats, cooking waste, hair and feathers from butchering or slaughtering sites [12]. Depending on the type of the animal being slaughtered, 40-50% of the animal weight remains as by-products that are often undervalued as these are not utilized in food due to their chemical or physical characteristics. These by-products can be categorized into edible or non-edible as hides and bones belong to the non-edible by-products and blood to the edible category [13].

2.2.2   Seafood wastes

Seafood production is a diverse and large industry. Disposal of seafood waste depends on species, processing method and plant location. The solid by-products produced during fish processing consist of scarps, heads, rejected fish, offal, skin, bones and tail [14]. The amount varies between 30-65% in case of fish canning and 50-75% in the case of fish filleting, smoking and salting.

2.2.3 Dairy sources

As such the dairy industry is not associated with severe environmental problems; it must still continually consider its environmental interaction, as dairy pollutants are mainly of organic in nature. Manufacturing of dairy products produces different types of solid and liquid wastes as well as some by-products. These may contain off-specification goods, damaged or out-dated products, solids, cheese, whey, curd, and milk sludge including lactose, proteins, fats etc. making their exploitation necessary [15].

2.2.4 Fruit and vegetable processing

During fruit processing, 30-50% of the by-products are produced depending upon the type of fruit being processed [14]. These by-products can be distinguished as (a) pre-processing by-products including stems, stalks and rotten fruits from sorting processes, and (b) by-products from processing such as pulp, seeds, peels and pomace. The main constituents of the wastewater of fruit and vegetable processing plants contain sugars, starches, pectin, vitamins and other components of the cell wall.

2.2.5 Pulses and oil seed industry

A number of waste streams are generated during the processing of oil seeds. Emission in the air also takes place like grain dust, hexane solvent loss, odour during the meal drying and deodorization. The wastes from this industry are generated in the major processes like milling, extraction, deodorizing, caustic refining, acidulating, packaging, tank car washing, margarine production, bleaching winterizing, mayonnaise production and salad dressing [16]. Grain dust is generated from a range of sources during milling and handling of oilseeds.

Separation and dehulling of cotyledons during milling operation of pulses involve the use of strong abrasive forces. This creates `broken grains and powder products. The commercial pulse milling process generates almost 75% of finished products and 25% of by-products like powder (up to 12%); husk (up to 14%); broken (up to 13%), unprocessed and shriveled seeds.

2.2.6 Miscellaneous sources

The other miscellaneous sources of food processing waste may include wastewater from juice, soda or fruit bottling; breweries, bakeries, distilleries and sugar processing units.

2.3 Production across the globe

Food processing and agriculture are the two vital components for the well being of society. The production of fruits, vegetables, cattle, grains, fish etc; the transportation and storage of the farm products to processing plants; and the production of food in ready-to-use forms improve and sustain the quality of human life. This intern increases the incidence of waste produced during each method of processing.

According to the Food and Agricultural Organization (FAO) of the United Nations [17], generally $1/3^{rd}$ of the food produced across the globe for human consumption is wasted. It amounts roughly 310 billion dollars for the developing countries and US$ 680 billion for the industrialized countries. It has been estimated that almost 30% of the cereals; 40-50% root crops, fruits and vegetables; 20% of the oil seeds, meat and dairy products and 35% of the fish get wasted every year. Wastage worldwide is shown in Fig.1.

In 2015, the price of the food wasted from U.S. homes, manufacturers and consumer-facing businesses was estimated at approximately 218 billion U.S. dollars. There has been almost a loss of 8.8% of cereals, 38.8% of roots and tubers, 16.8% of oilseeds and pulses, 24.7% of fruits and vegetables, 9.2 % of meat, 17.7% of fish and seafoods and 5.2 % of dairy products during their production and processing in North America and Oceania [18]. There is almost a loss of 6% of the food during processing in Latin America and the Caribbean in the year 2015

According to the data from Stenmark et al. [19], the European Union contributes most of the food waste from households (47 million tonnes) and the processing sector (17 million tonnes). These two sectors contribute about 72% of EU food waste, even though there is large uncertainty around the estimate for the processing sector. Of the remaining 28 per cent of food waste, 11 million tonnes (12%) comes from food service, 9 million tonnes (10%) comes from production and 5 million tonnes (5%) comes from wholesale and retail sectors.

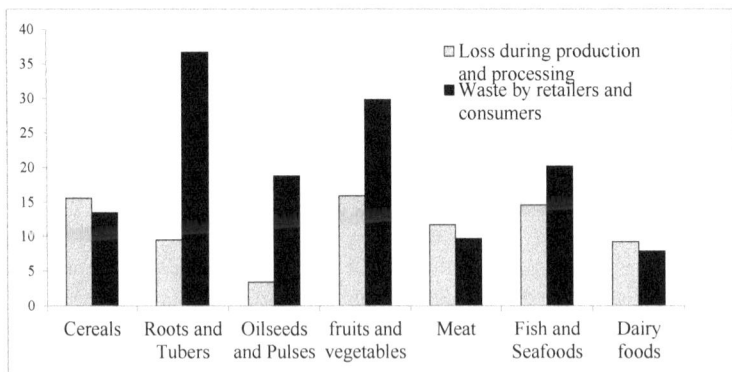

Fig .1. The share of food loss and waste worldwide (the Year 2015), Adapted from [20]

Microbial Fuel Cells: Materials and Applications   Materials Research Forum LLC
Materials Research Foundations **46** (2019) 249-288   doi: http://dx.doi.org/10.21741/9781644900116-10

In Switzerland, there is almost 61% of food waste generated in the agricultural sector, 22% in the processing industry, 13% in the catering industry and 4% in the large supermarket chains. Potato-processing industry has the highest rate at around 190,000 tonnes of losses. The lowest proportion of losses is in the cereals and baking industry (21,000 tonnes) [21].

India processes only 2% of its produce although it being the second largest producer of cereals and fruits, third in marine production and also one of the leading countries in terms of livestock in the world. According to the United Nations development programme, up to 40% of the food produced in India is wasted. According to the agriculture ministry, Govt. of India, Rs. 50,000 crore worth of food produced is wasted every year in the country. India loses more than $9 billion in food waste. Much of that is led by fruit and vegetable produce, oil seeds and fisheries [22] as shown in the Figure. 2.

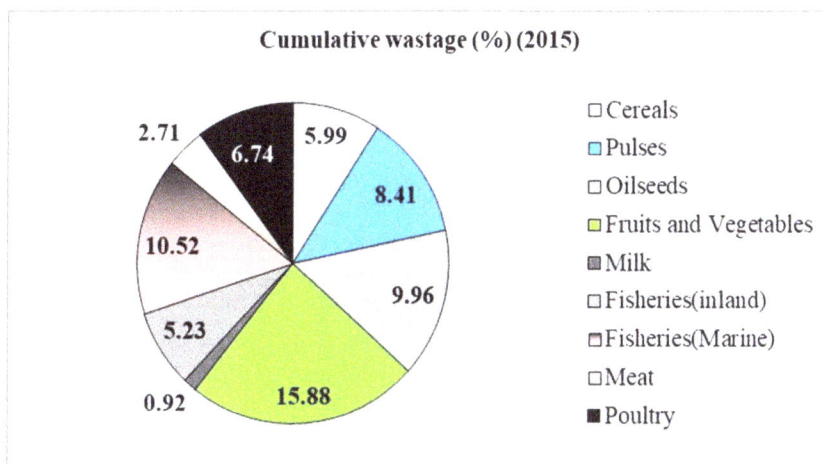

Fig. 2. Data: Ministry of food processing Industries, 2015 [22]

Food losses in industrialized countries are as high as in developing countries, but in developing countries, more than 40% of the food losses occur at post-harvest and processing levels, while in industrialized countries, more than 40% of the food losses occur at retail and consumer levels. Knowing of waste volumes produced is important, as substrate availability is a major constraint in waste conversion processes. However, the amount of food processing waste generated is increasing every year as food production increases to support the growing population [23].

3.   Characterization and composition of food processing wastes

The FPWs, in general, are characterized by high chemical oxygen demand (COD) and biological oxygen demand (BOD) along with many other recoverable nutrients like potassium, nitrogen and phosphorus [24]. Macronutrients include proteins, carbohydrates and lipids. Micronutrients consist of minerals like sodium, calcium, magnesium, zinc, iron, manganese and sulfur. In addition, certain chemical and physical properties like total solids, pH, and ash content are also present in the solid and liquid streams. The wastewater from the food production industries is considered nontoxic because it contains little hazardous and non-biodegradable compounds [23]

The different types of wastes generated by food manufacturing operations reflect diverse types of processes and ingredients being carried out; for example washing of root vegetables, like sugar beet, give rise to elevated levels of total soluble solids (TSS) in the effluent. Additional processing of vegetables, involving dicing and/or peeling can give rise to the elevated dissolved solids (e.g. sugar in fruit processing). Joshi et al. [25] reported that wastes from vegetable industries like tomatoes, peas and carrots contain elevated BOD levels and are rich sources of many nutrients like fibers, minerals, vitamins, etc.

Brewing and cereal processing create carbohydrate-rich effluent. The processing of legumes creates an effluent rich in proteins. Effluents from oilseed processing contain fats as suspended matter. Milk processing creates liquid wastes with a varying quantity of dissolved protein, lactose and suspended fat. The poultry and meat processing industries generate effluents rich in both fat and proteins. In most of the examples, there are also some particulate wastes, (1 mm in size), in addition to the fine suspended matter.

3.1   Proximate properties

The proximate properties involve the set of methods related to the nutritional value of food or feed. The proximate analysis estimates the percentage of the main components of food waste, like carbohydrates, fats, proteins, moisture, etc.

Commonly, fruit plus vegetable wastes contain water (80–90%) and little amount of fat and proteins [26]. All the byproducts generated during the processing retain the same chemical properties as that of the raw material. The different compositions of some food wastes are listed in Table 1.

Potato peel contains 64.5% carbohydrates, 3.4% sugars 11.2% moisture, proteins 13.5% and 7.6% ash [27]. However, the composition may vary from different geographical locations, growing conditions, potato varieties and additional parameters. Currently, its maximum part is used to feed livestock [28]. Cassiva pomace contains a maximum of water and non-soluble dietary fiber while as cassava peels contain high crude fiber content (10–30% dry matter basis) and a low protein content of about less than 6% on dry matter basis. They also contain natural anti-nutritional factors, decreasing the availability of phosphorous in the non-ruminants [29]. Pulses are considered as the chief sources of proteins. These also contain dietary fibers, carbohydrates, minerals and vitamins [30]. Husks are important sources of tannins and insoluble dietary fibers (75–87%) [31].

Table. 1. The composition of different vegetable wastes

S no.	Food waste	Carbohydrates %	Protein %	Fats %	Moisture %	Fiber %	Ref.
1	Orange peel	-	8.72	1.57	4.15	63.45	[32]
2	Carrot pomace	-	10.06	1.75	4.61	69.85	[32]
3	Green pea peels	-	13.27	1.34	4.28	71.25	[32]
4	Potato peel	64.5%	13.5	-	11.2 dw	-	[27]
5	Cassava pomace	-	1-4	-	75-85	35	[33]
6	Poultry processing	-	27.1	55.3	-	4.1%	[34]
7	Rice bran	49.04	9.28	2.90	13.32	20.63	[35]
8	Palm kernel	8.01	6.30	5.62	15.35	61.02	[35]
9	Shrimp residue	-	49	4.9	-	18.2	[36]
10	Funnel	61.77	16.07		9.85	61.77	[37]
11	Husk tomato	70.94	13.23	-	80.50	9.85	[37]

In the poultry processing plant, the effluents normal percentage of fat, protein, ash, and fiber on a dry-weight basis are 55.3, 27.1, 6.1, and 4.1% respectively [34]. Fat is the major component, averaging 55.3% of the particulate matter in wastewater streams. Proteins are the second most predominant constituent. Major sources of proteins in poultry processing waste include blood and muscle tissues. Crude proteins mostly constitute more than 75% of feathers [38]. There are generally two important sources of inorganic materials which result in ash. The first is the dirt and manure detached from birds within the feather and scalder picking operations [34], while the second is the wastewater produced from the cleaning of the live haul area. Fiber is the minimum in the dry matter. Fiber content within particulates is mostly from manure, ingesta, and extra plant-containing materials detached from birds throughout the slaughtering process. The main fiber sources in the feed are remnants of plant cells that are resistant to hydrolysis by enzymes in the alimentary tract. These include hemicelluloses, cellulose, lignin and pectin [39].

3.2 Ultimate properties

The ultimate properties of food processing waste involve the percentage of carbon, hydrogen, oxygen, nitrogen and sulphur. The presences of halogens are also included due to the concern over the production of chlorinated compounds during incineration. The ultimate properties are usually used to distinguish the chemical composition of the waste and to get the proper mix of the processing waste to attain a proper C/N ratio for biological decomposition processes. The carbon/nitrogen ratios represent the accessibility of organic substrate in relation to the availability of the main nutrients. For the metabolism by microorganisms, these require the micro, macro, and trace elements as the primary nutrients. Phosphorus compounds are present in wastewater in the form of organophosphorus derivatives, water-soluble phosphates and soluble orthophosphate.

Table 2. Characteristics of some food wastes

Food Waste	TN (mg/L)	pH	SS (g/L)	BOD (g O_2/L)	COD (g O_2/L)	Ref.
Olive oil waste	1.2-1.5	3-5	65.0	43.0	100	
Apples	-	5.9	0.45	9.6	18.70	[41]
Carrots	-	8.7	4.12	1.35	2.30	
Tomatoes	-	7.9	0.95	1.02	1.50	
Cattle	-	6.7-9.3	-	0.90-4.62	3.0-12.87	
Hog	14.3	7.3	-	1.95	3.015	[42]
Cheese industry whey	-	5.2	0.188-2.330	0.19-6.22	0.377-2.214	
Milk industry	-	7.1-8.1	0.36-0.92	0.71-1.41	-	
Wheat	-	3.35	-	-	90.75	[43]
Potato	-	3.69	-	-	48.95	[44]

TN=Total nitrogen; SS= Suspended solids; BOD= Biological oxygen demand and COD=Chemical oxygen demand

Fruits residues have elevated levels of carbon/nitrogen (C/N) ratio in comparison to vegetable wastes. Vegetables (V) and fruits (F) show uniqueness in terms of sulphur and carbon/nitrogen contents. All the F&V samples don't show optimal sulphur and nitrogen content as feedstock for anaerobic digestion. The optimal ratio for C/N should be less than 25 and the optimal ratio of nitrogen/sulphur (N/S) must be 15-20 [40]. Nutrients such as potassium, magnesium, sodium, calcium, and the major micro and trace elements must be present in adequate amounts for aerobic/anaerobic digestion.

Potatoes represent a separate category that reported 21.8% of total solids, 17.4% of volatile solids on a wet basis, and a C/N ratio of 23. The seafood and meat processing waste streams are rich sources of nitrogen. Also, the detergents used for the cleaning operations provide phosphorous to the waste streams. The characteristics of various food wastes are summarized in Table 2.

4.   Management and value-added products from food processing wastes

Food industries generate large proportions of wastes, both liquid and solids, which are produced during the preparation, production and consumption of food. These wastes create rising pollution and disposal problems and thereof a loss of precious nutrients and biomass [45]. All industrial processings give rise to residue materials which cannot be generally used at the point of source and are usually released into the environment. The environmental legislation has considerably contributed to the beginning of sustainable waste management practices globally. In view of the challenges in certain areas of the food processing industry, efforts are being made to optimize to deal out with technologies that can help in the reduction of waste. FPWs can be converted into useful products of high value as a by-product, for use as feed or food after biological treatment and even as raw materials for other industries,

The instability of the waste due to its high water content enables the pathogenic microorganisms to grow on it. The high-fat content in case of the meat industry is prone to oxidation and therefore escalates the spoilage due to the enzymatic activity [11]. The conventional valorization methods like composting; anaerobic digestion; incineration and utilizing food processing residues for animal feed and fertilizer are in use since decades. Over some years, new management practices have evolved that invoke the utilization and recovery of valuable components from the food wastes.

The vegetable and fruit processing wastes have remained underutilized in spite of being potential resources of carbohydrates and phenolics. Extraction of polysaccharides like cellulose, arabinans and pectins, processing of dietary fibre, production of sugars, and phenolics are presently the chief ways to upgrade these residues.

Microbial Fuel Cells: Materials and Applications   Materials Research Forum LLC
Materials Research Foundations **46** (2019) 249-288   doi: http://dx.doi.org/10.21741/9781644900116-10

The cheese industry involves the utilization of whey, lactose and the different types of proteins present in the waste streams. Seafoods is rich sources of minerals and omega-3 fatty acids. Some of the food processing wastes and their utilization are presented in Table 3.

Table. 3.   Food wastes, their byproducts and utilization.

Vegetable	Part used	Resources Utilized	Ref.
Tomato	crushed and Dried seeds and skins of the fruit	Lycopene, beta-caretone, hydroxycinnamic acid derivatives, flavonols (quercetin derivatives), flavanones and naringenin chalcone.	[46,47]
Carrot	Pomace	Beta-carotene, Coumarins and hydroxycinnamates and	[48]
Red beet	Pomace and peel	Betelians, betacyanins, betaxanthins, coumaric acid, cyclodopa glucoside derivatives and ferulic acid	[47]
Potato	Peel	chlorogenic, gallic, protocatechuic and caffeic acids. Antioxidents	[49]
Onion	Scale tissues and External membranes	Flavonoids (quercetin) and organosulphur compounds	[50,51]
Lettuce	Low-quality lettuce heads, stems and External leaves	Caffeoylquinic acid; caffeoyl tartaric acid derivatives; flavones and flavonols; chlorogenic acid and chicoric acid	[52]
Brassicaceae	low-quality florets stem, Leaves,	hydroxycinnamates (mainly sinapic acid derivatives), Isothiocyanates and glycosylated flavonoids,	[53]
Apple juice processing	Apple pomace, apple press cake	Flavonoids, chlorogenic acid, glycosides, pectins, natural sweeteners, antioxidants, essential oils and fibers	[54]
Grape pomace	Skin seeds pulp and stalks	PUFA, polyphenols, dietary fibre, flavonoids like catechin, proanthiocyaninds and epicatechins	[55–57]
Fish processing waste	Skin, scraps, bones, head and tail	Gelatin, collagen omega-3 fatty acids, and essential oils	[58]
Sugar beet processing	Sugar beet pulp	Ethanol, starch, oils, proteins, fibers, bioplastics, polyamides, poly lactic acids	[59,60]

5. Energy generation technologies

The energy from the food processing wastes can be generated by followings modes:

- Thermochemical processes
- Biochemical/Biological processes

5.1 Thermochemical Processes:

Thermochemical conversion is a reasonable route to produce energy and other chemical products from low-value biomass resources [61] which include incineration, gasification, pyrolysis, liquefaction and hydrothermal carbonization (Fig. 3).

Incineration

Incineration has been used extensively to produce energy from waste materials, and thus it decreases the amount of waste generation significantly [62]. It is a well-established technology involving the combustion and conversion of wastes into heat and energy [63]. Incineration decreases the volume of wastes by 80% and hence proven to be fruitful for the countries having insufficient territory for landfills.

The heat produced by incinerating food wastes can be utilized to run steam turbines for generation of electricity or could be used to operate machines involved in food processing. Food wastes usually have lower oxygen content but higher nitrogen, ash, and energy contents than wood [64]. However, the high moisture content and the emission of dioxins make some of the food processing wastes unsuitable for incineration [65].

Gasification

It is defined as the process of conversion of biomass at moderately high temperatures into syngas (combustible gas mixtures) and tar by the simultaneous oxidation and pyrolysis under inadequate oxygen supply [66,67]. Syngas consists of carbon monoxide (CO), hydrogen (H_2), carbon dioxide (CO_2), methane (CH_4), nitrogen (N_2) and water vapours (H_2O). Syngas can be used to produce heat by its combustion [68].

Pyrolysis

Pyrolysis refers to the conversion of biomass into syngas at higher temperatures (around 1000°C) in an inert atmosphere. It can also be defined as the thermal decomposition of the organic matrix to produce solid, liquid and gaseous products [69,70]. The most favourable temperature range for the formation of pyrolysis products is between 625 and 775K. Yield maximization of various pyrolysis products depends upon different conditions like liquid products need low temperature, short gas residence time period and low heating rate whereas charcoal needs low temperature as well as low heating rate process. However, fuel gas needs high temperature, low heating rate and long gas residence time period [70,71].

Liquefaction

Liquefaction is defined as a thermochemical process operating at high pressure and low temperature in the presence of a catalyst. The liquefaction produces a liquid product. The process is not in demand because of the complicated fuel feeding systems and the requirement of more expensive reactors compared to pyrolysis processes [72].

Hydrothermal carbonization

Hydrothermal carbonization (HTC) is the thermal conversion technology employed to convert food processing wastes into a valuable, energy-rich resource. It is alluring the attention of researchers to generate energy resource from wastes having high moisture content (80–90%). It can be also defined as a process of conversion of food wastes into a valuable, energy resource under endogenous pressures and moderately low temperature (180–350 °C) [73–75]. Hydrothermal carbonization of various substrates including lignocellulosic biomass has been studied recently [73,76,77]. It was confirmed that food waste could be favourably treated by HTC ensuing in the production of hydrochar (high energy product). Lin et al. [5] described the positive energy balances of hydrothermal carbonization of food waste from local restaurants. In comparison to other conversion methods, HTC has various advantages [78]. Furthermore, the operation of the process at higher temperatures helps in the elimination of pathogens and inactivation of other crucial organic contaminants. Thus, this process produces a sterilized, clean, easy to store and transport energy-rich resource. The other advantages of this process are the recovery of some useful chemicals and nutrients from the HTC process water for use/reuse and the liquids containing nitrogen species for use as fertilizer [73,75,78].

5.2   Biochemical/Biological conversion processes:

The processes categorized under this category are anaerobic digestion, alcohol fermentation and biodiesel formation (Fig. 3).

Anaerobic digestion

Anaerobic digestion (AD) is a process where microorganisms break down biodegradable materials, such as food wastes, manure, and sewage sludge, in the absence of oxygen, to produce biogas [79,80]. Biogas is primarily made of methane and carbon dioxide and can be utilized as the source of energy like natural gas. The nutritious liquor and a solid residue are the other products obtained from anaerobic digestion. This solid residue can be used as a soil conditioner in order to minimize the need for chemical fertilizers and to

Microbial Fuel Cells: Materials and Applications   Materials Research Forum LLC
Materials Research Foundations **46** (2019) 249-288   doi: http://dx.doi.org/10.21741/9781644900116-10

alleviate erosion and nutrient run-off problems. The solid residue also helps in retention of water by soil and enhances growth and yield of plants [81,82]. AD involves hydrolysis, fermentation, acetogenesis, and finally methanogenesis. This process is being carried out in digesters at the waste treatment sites like in case of dairy waste and distillery effluent digestion [79]. It has been utilized to valorise fruit and vegetable processing wastes [83], food processing wastewater [84], and slaughterhouse wastes [85]. This technology seems very suitable for the food processing wastes. The process is beneficial as it generates renewable energy, reduces greenhouse gas emissions, and involves management and rerouting of waste [75,86]. Disposing of food and other organic materials in landfills results in methane production, a potent greenhouse gas. Methane escapes directly into the atmosphere, so diversion of food wastes from these landfills to digesters decreases methane emissions from these sites. Codigestion is another process related to anaerobic digestion. It refers to the digestion of more than one substrate together to enhance performance and stability of the process. In this process, the energy-rich organic waste materials can be added to a dairy or wastewater digester with excess capacity to increase its digestion process. The anaerobic digestion can be performed at both mesophilic (25-45°C) and thermophilic (55-70°C) temperatures. However higher temperature enhances the extent as well as the rate of biogas production but it demands greater control and sophisticated instrumentation. Operation of the process at psychrophilic range (5-15°C) is a simple and low-cost option but it results in low performance [80]. The implementation of the AD by food processing plants is fascinating as it partly or fully meets the processing energy demands, reduces the common problems of smell and vermin efficiently by building digesters on the sites where wastes are generated and managed effectively. Besides, the advances in anaerobic digestion technologies have made them more feasible for the processing of various food wastes [79,87]. The production of methane by anaerobic digestion is an appropriate solution for food processing waste management [88].

Alcohol fermentation

Fermentation is the age-old process familiar to humans. It is used to produce foods, feeds and other well-known compounds by the utilization of microorganisms. The food processing wastes can be used to produce bioethanol and biobutanol (the liquid biofuels). These biofuels are the potential alternative sources of energy to replace petrol and diesel (conventional liquid fuels). Alcoholic fermentation producing liquid biofuels is advantageous as it can significantly reduce the greenhouse gas emissions [89].

Biodiesel formation

Biodiesel, which is a potential alternative biofuel to conventional fuels can be generated from food processing wastes either through direct transesterification (by using acid or alkaline catalysts) or transesterification of microbial oils by different oleaginous microorganisms. Commercially, it is generated by transesterification using an alcohol and a catalyst. It consists of the conversion of triglycerides (oil) to methyl esters (biodiesel) and a by-product (glycerol) [90]. Most of the biodiesel produced today use a base catalysis reaction [91,92] which provides advantages of high conversion yields at low temperatures and pressures and the minimal side reactions as well as reaction times [93,94].

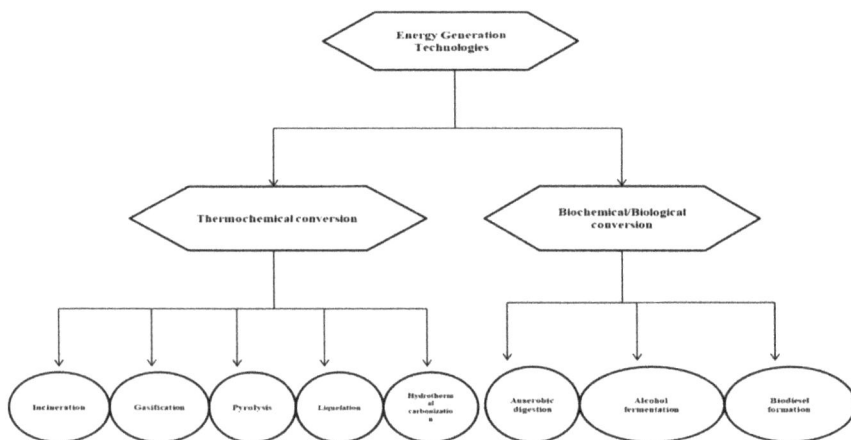

Fig. 3. Energy generation technologies

6.   Food processing wastes as a bioenergy source

6.1   Bioethanol

Food processing wastes are potential sources of bioethanol production. The bioethanol is produced through simultaneous saccharification and fermentation (SSF) or hydrolysis and fermentation (SHF) processes from starchy wastes by *Saccharomyces cerevisiae* [95]. Ethanol production from food processing wastes involves a different strategy compared to other conversion methods. The hydrolysis of starch to sugars (glucose) is enhanced by enzymes like α-amylase and glucoamylase. The food wastes from fruit, vegetable, and grain processing sectors and from cafeterias are rich in carbohydrates. These can be utilized as a substrate for the production of bioethanol [79]. Sugar juice obtained from food wastes like fruit pomace can be directly utilized to produce ethanol [96]. However, the lignocellulosic and starchy food wastes need to be hydrolyzed to fermentable sugars before fermentation. After hydrolysis, fermentation is carried out either by yeast or bacteria (*Zymomonas mobilis* and *Clostridium thermocellum*). The FPWs can be pretreated by various methods using acid, alkali, thermal and enzymatic processes to breakdown the complex lignocellulosic framework [97,98]. These pretreatments though assist and expedite the bioethanol production but the degradation of soluble sugars (especially at harsh conditions) can result in the formation of various inhibitors viz., furfural [99]. The bioethanol has been produced from different food wastes such as banana peel [100], grape pomace [101], lactose-rich dairy waste [102,103], whey permeate [104], potato peel waste [105,106], citrus waste [107,108], cafeteria food waste [109], and kitchen waste [99]. The other substantial sources for ethanol generation are wastewater from fruit, vegetable, and starch processing. Ultrafiltration is an efficient approach to recover the solid residues in wastewater [110].

6.2   Biobutanol

Biobutanol fuel is one of the second generation alcoholic fuels. It can be utilized as a transportation fuel. It is the higher member of straight chain alcohols, has four carbon atoms rather than two as in ethanol [89]. The usage of biobutanol as a transportation fuel is gaining interest because of its lower volatility and poor hydroscopic nature, better corrosion resistance and higher energy density compared to ethanol [111,112]. Unlike ethanol, it can be blended with gasoline [113]. It can be easily produced from food processing wastes and cellulosic substrates. The conventional method for its production is acetone–butanol–ethanol (ABE) fermentation process (an anaerobic process) by *Clostridia* spp [114].

The food industry wastes and any carbohydrate-containing substrates especially having high water content (such as whey) are suitable sources of biofuel production (biobutanol) [115]. The starchy wastes from processing industries like inedible dough, bread and butter liquid are suitable feedstocks for biobutanol (0.3 g butanol per g of food waste) [116]. Huang et al [117] recommended that food processing waste as a better substrate for butanol production using *Clostridium beijerinckii*. In comparison to expensive glucose, the utilization of wastes as a substrate for butanol production is advantageous because of their cost-effectiveness, more yield, and lesser residual sugars. Apart from these merits, the process does not involve the use of hydrolytic enzymes. All the above-mentioned merits make FPWs potential feedstocks for n-butanol production at commercial scale.

6.3   Biogas

Biogas from food processing wastes is produced by the anaerobic digestion process. Biogas is a concoction of CH_4 (50-60%), CO_2 (30-40%), H_2 (1-5%), N_2 (0.5%), CO, H_2S and water vapors (traces). Biogas is considered to be a propitious alternate source of energy with many applications: heat or electricity generation; its liquefaction into methanol and chemical feedstocks; similar to that of compressed natural gas (CNG) [72]. Anaerobic digestion has extensively been used for biogas (methane) production from food processing wastes like wastewater from processing industries producing beer, fruit jams and potatoes etc. [118]. The rate-limiting step in biogas formation is the hydrolysis of the substrate, as some substrates like food processing wastes consisting of animal fats, protein and lignocellulosic biomass are complex in nature [89,119]. These sort of wastes demand a pretreatment step to accelerate the hydrolysis process. This results in the production of simpler units (sugars, fatty acids, amino acids) from complex organic materials. These are then converted to sufficient volatile fatty acids which are further converted to acetate which is the basic material for the methanogenic bacteria to produce methane [65]. The quality of biogas is predominantly regulated by C/N ratio of raw material. Both higher and lower C/N ratios affect the methane production rate as nitrogen uptake by methanogens increases with higher C/N ratio whilst lower C/N ratio results in accretion of ammonia. Thus, selection of substance should be done carefully as substrates with abundant degradable organic matter or volatile solids produce considerable biogas as compared to other raw materials [120]. The other parameters influencing biogas formation are temperature, inhibitors, pH, volatile solids, organic loading rate (OLR), total solids, hydraulic retention time (HRT), internal pressure, particle size, and mixing.

6.4   Biohydrogen

The high calorific value, easy availability and environment-friendly nature make biohydrogen a prospective biofuel [121,122]. It is carbon neutral and has a tendency to replace the conventional fuels thus saving the depletion of oil reserves [123]. There are various methods through which biohydrogen is produced. Fermentative H_2 generation via light dependent and dark fermentation processes is advantageous. Light-dependent processes like photolysis and photo-fermentation are aerobic in nature while as the dark fermentation occurs under anaerobic conditions [124]. The combined photo-dark fermentations for food processing wastes have been economical [125]. The dark fermentation among all these processes is comparatively favourable due to higher yield and lesser production cost [126,127].

The suitable materials for biohydrogen production should be rich in carbohydrate but be nitrogen deficient. Among the various substrates assessed, food processing wastes being comparatively cheaper and have been an ideal biodegradable organic matter for biohydrogen production through dark fermentation [7,128]. The type of cultures whether pure, co-culture or mixed also influence the biohydrogen yield [121]. Dong et al. [129] and Logan et al. [130] obtained enhanced biohydrogen yield from 15 to 55 % from wastes produced from potato processing industries. The various other wastes utilized for biohydrogen generation are carrot, cabbage, lettuce and sugar beet wastes. De Gioannis et al. [131] studied the H_2 production from various types of cheese under different pH values (6.5-7.5) and reported the highest yield of 170 ml H_2/kg of total organic content (TOC).

6.5   Biodiesel

Biodiesel according to American Society for Testing and Materials Standard (ASTM) is defined as the monoalkyl esters of long chain fatty acids that are obtained from animal fat, waste cooking oils and vegetable oils through a chemical reaction known as transesterification. It has properties similar to diesel [132]. It can be easily blended with diesel and engines do not need any modifications to utilize it. The major advantage of biodiesel is that it has zero carbon emission [89]. Food-grade vegetable oils like soyabean oil and rapeseed are currently used for the production of biodiesel in the U.S. and Europe. The biodiesel production from such substrates is not economically viable as it is costly. The substantial substrates for biodiesel production are animal fats, waste cooking oils, and restaurant grease as these are also produced in higher quantity by food processing industries [133]. The quantities of fatty residues generated by the catering industry, food processing industry, wastewater treatment plants and autonomous sanitation are about 32, 29, 23, and 16% respectively [134].

7.   Factors affecting biofuel production from food processing wastes

7.1   Factors affecting biodiesel production

Biodiesel is regarded as a potential alternative to petrodiesel fuel. In recent times, biodiesel fuel production has gained interest due to its environmental benefits [135]. Vegetable oil is considered to be the best material to produce biodiesel [136]. But higher viscosity of vegetable oil limits its direct application as fuel for diesel engines [137]. However, the viscosity of vegetable oils can be minimized by using different processes namely blending, pyrolysis, micro-emulsification and transesterification. Of these methods, transesterification is extensively employed for reducing viscosity and improving the fuel property [137]. There are various factors that influence this process including alcohol to oil molar ratio, reaction time, reaction temperature, nature and the amount of catalyst as well as the raw material composition [138].

The molar ratio of alcohol: One of the most influential factors affecting the conversion efficiency, biodiesel yield as well as the production cost is the molar ratio of alcohol. Transesterification is an equilibrium reaction, so in order to circumvent reversible reactions and to cause the reaction to proceed forward, a surfeit of alcohol is required. The stoichiometric molar ratio of alcohol to oil for transesterification is 3:1; higher molar ratios enhance the miscibility and increase the contact between triglyceride and alcohol molecules [139]. The increase in the concentration of alcohol favours the formation of esters from fats in a short period. Hence, there is a direct relation of the yield of biodiesel with the concentration of alcohol but only up to a specific concentration level, beyond which the only cost of downstream processing increased [140]. The range of molar ratios of 6:1-30:1 has been accepted universally. Among various alcohols used for transesterification, methanol, due to its low cost, has been commonly used. The volumetric ratio of methanol and ethanol to oil has a great effect on biodiesel yield. Maximum biodiesel yield of 99.5% at 1:6 oil/methanol ratio was observed. In comparison, biodiesel yield using methanol continuously increases with the increase in methanol molar ratio [141].

Temperature: Temperature affects the yield of biodiesel by increasing the rate of reaction and shortening the reaction time. However, at higher temperatures saponification of triglycerides gets accelerated [142], which vaporizes methanol and reduces the biodiesel yield [143]. Depending on the oils or fats used, the optimum temperature may vary from 50 to 60°C [144]. Eevera [145] concluded that the highest production of biodiesel can be obtained at 50°C. The yield started to decrease when the temperature goes beyond 55 °C, which may be ascribed to the fact that methanol, with boiling point 64.7°C, vaporizes at a higher temperature and remains in this form leading to less conversion. Abbah [146] found that conversion of biodiesel was less at 30°C but as temperature increases, there was an increase in the conversion rate, but it decreases if the temperature increased beyond 60°C. The maximum yield was obtained at 55°C. Hence, temperature determines the rate of reaction and the ethyl esters production while keeping other factors (reaction time, stirring speed, alcohol oil ratio, etc.) constant.

Reaction Time: Reaction time has a direct effect on fatty acid esters conversion, with maximum ester conversion within less than 90 minutes. Prolonged reaction time results in a reduction of biodiesel production as it favours reversible reaction and can result in depletion of esters and soap production [144,147]. Naik et al. [148] obtained a higher biodiesel yield from longer mixing time and concluded that the reaction time of 120 minutes gave better results than other reaction times used. However, no accretion in conversion was observed with excessively higher reaction time as it favours hydrolysis of esters. However, Banihani [149] conducted an experiment at different reaction times: 30, 60, 90 and 120 minutes while keeping other parameters constant and found that the best results were obtained from 90-minute reaction duration and the yields were lower when the reaction time was fixed at 120 min. This was due to the fact that when the longer reaction time was used, the rate of soap formation increases.

Effect of Catalyst Concentration: Sodium hydroxide or potassium hydroxide is the most frequently used catalyst for biodiesel production. The characteristics of raw material and procedure applied for the transesterification have been the decisive factors for selecting the kind and quantity of catalyst [144]. Hossain and Mazen [150] concluded that different catalysts have a distinct effect on the production of biodiesel. In an experiment, where sodium hydroxide and potassium hydroxide were used as catalysts at a concentration level of 0.5% NaOH gave better yield (71.2%) of biodiesel while KOH resulted in 68.9% of biodiesel yield. It has been generally considered that with the increase in the quantity of catalyst, the production of fatty acid alkyl esters also increased, as more active sites are available by the addition of a substantial amount of catalyst [151]. On the other hand, a limited quantity of catalyst resulted in the partial conversion of triglycerides into esters. A study carried out by Dorado et al. [152] revealed that higher amount of catalyst in the process results in the formation of soap which enhances the viscosity of the reacting substances and lowers the rate of biodiesel production.

7.2   Factors affecting biohydrogen production

Generally, factors like hydrolysis and pretreatments methods are considered to affect biohydrogen production from food waste.

Hydrolysis rate: Biohydrogen production is limited by the hydrolysis rate which in turn is affected by the composition of food waste material. Bacteria ferment carbohydrates, protein and lipids to volatile fatty acids which further get hydrolyzed by acetogenic bacteria to acetate, carbon dioxide and hydrogen [153]. The hydrolysis of carbohydrate occurs at a faster rate as compared to proteins and lipids. Carbohydrate-rich substrates produce twenty times more biohydrogen than protein and fat-rich substrates [154]. Lipids and fats result in flotation, clogging and mass transfer problems in an anaerobic fermentor. High temperature counteracts clogging problems during anaerobic degradation of lipid [155].

pH: The activity of enzymes and the rate of biohydrogen production are influenced by pH. pH range of 6.3-7.8 is considered to be optimum for hydrogen-consuming bacteria [156]. Yasin et al. [157] carried out the fermentation of food waste at 55 °C to examine the effects of different pH (5.0, 5.5 and 6.0) values on biohydrogen production. It was observed that maximum biohydrogen production was found at pH values of 5.5 and 6.0, which showed that biohydrogen production from food waste was pH-dependent with hydrogen yields of 79, 76 and 23 mmol H_2/L-media/d for pH 5.5, 6.0 and 5.0, respectively.

7.3 Factors affecting bioethanol production

Effect of Temperature: Temperature greatly affects the fermentation. The optimum temperature for fermentation using yeast is 30°C. Similar results were obtained with maximum bioethanol yield of 52% at 30°C. Temperature above optimum range can cause thermal stress and reduces the reproduction rates of the yeast [158].

Effect of pH: The pH is an important factor that influences ethanol production. The pH range of 4.0–5.0 is considered to be optimum for maximum bioethanol production. During a study, the maximum ethanol production of 410 g kg^{-1} h^{-1} was at pH 5.0, with a conversion efficiency of 61.93% while ethanol production rate at pH4.0 was 310 g kg^{-1} h^{-1} [159].

Effect of immobilization: Cell immobilization improves the stability of the cells, the ability to ferment and stress tolerance. It has been reported for improving ethanol yields. At 30°C and pH 6.0, free cells of *Saccharomyces cerevisiae* produce 86.90 g L^{-1} ethanol after 48 h as compared to the immobilized cells produces 102.70 g L^{-1} ethanol under similar conditions [160]. Kashid and Ghosalkar [161] observed that immobilization positively affects the fermentative metabolism of *Pichia*, improving the ethanol yield from 0.40 to 0.43 g/g and productivity from 0.31 to 0.51 g/L/h for acid treated corn cob hydrolysate.

8. Strategies for enhancing food processing waste-based biofuel production

Common technologies employed to convert food-waste-to-energy includes biological e.g. anaerobic digestion and fermentation, thermal and thermochemical technologies (e.g. pyrolysis, gasification, incineration etc.) [75]. In order to improve biofuels production from food waste, various other strategies have been explored.

Pretreatment of initial substrate: Pretreatment of the substrate is a common strategy used to enhance biofuels production. O'Leary [162], used pre hydrolysed whey with β-galactosidase for ethanol production of *Kluyveromyces fragilis* and *Saccharomyces cerevisia.*

Use of strains with high ethanol tolerance: An important feature of ethanol-producing yeast is ethanol tolerance. The concentration of ethanol above 8% (v/v) causes hyperpolarization of phospholipid present in organelles and lipid bilayer of cell membranes. Hyperpolarization increases the membrane fluidity and decreases the integrity of the membrane [163]. Therefore, developing strains with high ethanol tolerance is of great economic value to the industries. Thammasittirong et al. [163] have reported increased tolerance of *Saccharomyces cerevisiae* NR1 random UV-C mutagenesis. UVNR56 mutant exhibited improved tolerance to ethanol and higher viability, even when 15% (v/v) ethanol was present and showed a considerably higher viability throughout the fermentation from molasses. A new method for isolating ethanol-tolerant mutants of *Escherichia coli* KO11 uses a combination of solid and liquid mediums. By repetitive subculturing and successive transfers in liquid and solid media, it was possible to select a mutant that produces 60 g ethanol per litre from xylose in 72 hours [164].

High cell concentration: Higher cell density results in faster fermentation as compared to those fermentations where lower biocatalyst concentrations are used [165]. Adding more cells at the initial stages of fermentation is the most common method to achieve high cell density; these cells can be further reused and recycled in the bioreactor. Flocculation forms large complexes of yeast cells and is a usual way of immobilizing yeast cells. In a study, flocculating yeast strain KF-7, while growing on molasses in a two-stage continuous fermentation process resulted in a high ethanol concentration of 80 g/L [166].

Genetic engineering strategies: Use of genetic engineering strategies to produce biodiesel from yeast, plant and algae is one of the latest approaches. The oil content of the plant can be increased by genetically modifying the lipid biosynthetic pathway [167]. It has been observed that an increase in the activity of acetyl-CoA carboxylase increases the lipid synthesis by enhancing the utilization of malonyl-CoA in the biosynthetic pathway. Also, changing the composition of the fatty acid of plant seed oil improves the biodiesel production. In an attempt to enhance the biodiesel production from *Arabidopsis*, it was observed that overexpression of WRINKLED1 (WRI1) mRNA is directly associated with the increase in seed oil. WRI1 regulates the seed storage metabolism and its expression under the CMV 35S promoter increases the seed oil content and also accumulate triacylglycerols in developing seedlings [168].

Various metabolic engineering methods have been used for constructing lactose-consuming *Saccharomyces cerevisiae* strains, using the lactose genes of the yeast *Kluyveromyces lactis*, *Escherichia coli* and *Aspergillus niger [169]*.

Hybrid processes for improving biofuels production: In the past few years, studies have been carried out to produce hydrogen by integrating dark fermentation process (first stage) and photo fermentative process (second stage) [170]. The volatile fatty acids (major deterrent for the hydrogen production) formed in the first stage of the fermentation can be used as a substrate in the second stage.

A two-stage process for improved hydrogen production was developed where glucoamylse was first used to hydrolyse the food waste. The hydrolysate was then further utilized as a substrate for batch fermentation and continuous fermentation processes. The highest cumulative production of hydrogen was observed with a yield of 245.7 mL H_2/g glucose in the batch system [171].

Similarly, studies have been carried out to produce hydrogen and methane from food waste in a two-stage anaerobic digester. Both the fermentations were carried out in continuously stirred tank fermentors and the effect of continuous circulation on the efficiency and stability of processes was also investigated. Maximum H_2 production achieved was 3 L hydrogen per litre per day and maximum methane was 2.9 L methane per litre per day and about 70% degradation of volatile fatty acids was also observed [172].

Conclusion

Food processing wastes are produced at an alarming rate. The management of these wastes is of utmost importance. Biofuel production from these wastes is the most feasible option as it helps in alleviating the dependence on conventional fuels and also reduces greenhouse emissions. The various approaches of biofuel production (like anaerobic digestion, alcoholic fermentation and thermochemical conversion methods) from FPWs are promising and well-consolidated methods. There is still the need to explore and research the various aspects of valorisation of FPWs, the dissemination of knowledge to common masses regarding the efficient utilization of food processing wastes. The government and industries should join hands to give research wings from laboratories to commercial scale.

References

[1]   S. Li, X. Yang, Biofuel production from food wastes, in: R.Luque, C.S.K. Lin, K. Wilson, J. Clark (Eds.), Handbook of Biofuels Production, Woodhead Publishing, Elsevier, Duxford CB22 4QH, UK, 2016, pp. 617–653. https://doi.org/10.1016/B978-0-08-100455-5.00020-5

[2]   S. Zorya, N. Morgan, L.D. Rios, R. Hodges, B. Bennet, Missing food : The case of postharvest grain losses in sub-saharan Africa. documents.worldbank.org/curated/en/358461468194348132/Missing-food-the-case of-postharvest-grain-losses-in-Sub-Saharan-Africa, 2011 (11 October 2018).

[3]   FAO, Food wastage footprint: full cost-accounting.
        www.fao.org/publications/card/en/c/5e7c4154-2b97-4ea5-83a7-be9604925a24/,
        2014 (7 October 2018).

[4]   R. Dobbs, J. Oppenheim, F. Thompson, M. Brinkman, M. Zomes, Resource
        Revolution : Meeting the world's energy, materials, food, and water needs.
        McKinsey Global Institue. https://www.mckinsey.com/business-
        functions/sustainability-and-resource-productivity/our-insights/resource
        revolution, 2011 (3 September 2018).

[5]   C.S.K. Lin, L.A. Pfaltzgraff, L. Herrero-Davila, E.B. Mubofu, S. Abderrahim, J.H.
        Clark, A.A. Koutinas, N. Kopsahelis, K. Stamatelatou, F. Dickson, S.
        Thankappan, Z. Mohamed, R. Brocklesby, R. Luque, Food waste as a valuable
        resource for the production of chemicals, materials and fuels. Current situation and
        global perspective, Energy Environ. Sci. 6 (2013) 426–464.
        https://doi.org/10.1039/c2ee23440h

[6]   H. Ma, Q. Wang, D. Qian, L. Gong, W. Zhang, The utilization of acid-tolerant
        bacteria on ethanol production from kitchen garbage, Renew. Energy. 34 (2009)
        1466–1470. https://doi.org/10.1016/j.renene.2008.10.020

[7]   E. Uçkun Kiran, A.P. Trzcinski, W.J. Ng, Y. Liu, Bioconversion of food waste to
        energy: A review, Fuel. 134 (2014) 389–399.
        https://doi.org/10.1016/j.fuel.2014.05.074

[8]   C. Zhang, G. Xiao, L. Peng, H. Su, T. Tan, The anaerobic co-digestion of food
        waste and cattle manure, Bioresour. Technol. 129 (2013) 170–176.
        https://doi.org/10.1016/j.biortech.2012.10.138

[9]   F. Girotto, L. Alibardi, R. Cossu, Food waste generation and industrial uses: A
        review, Waste Manag. 45 (2015) 32–41.
        https://doi.org/10.1016/j.wasman.2015.06.008

[10]   K.Waldron, Handbook of Waste Management and Co-Product Recovery in Food
        Processing, 1st ed., Woodhead Publishing Limited: Cambridge, UK, 2007.

[11]   W. Russ, R. Meyer-Pittroff, Utilizing Waste Products from the Food Production
        and Processing Industries, Crit. Rev. Food Sci. Nutr. 44 (2004) 57–62.
        https://doi.org/10.1080/10408690490263783

[12]   USEPA, Industrial food processing waste analyses.
        https://www.epa.gov/sites/production/files/201601/documents/msw_task9_industri
        alfoodprocessingwasteanalyses_508_fnl_2.pdf. 2011(accessed 13 September
        2018).

[13]   J.K. Sahu, Introduction to Advanced Food Processing Engineering, ist ed., CRC Press, Taylor & Francis Group, 2016.

[14]   N. Pap, E. Pongrácz, L. Myllykoski, R. Keiski, Waste minimization and utilization in the food industry, in J.K. Sahu Introduction to Advanced Food Processing Engineering, ist ed., CRC Press, Taylor & Francis Group, 2016. pp. 595–630. https://doi.org/10.1201/b16696-23

[15]   A.R. Prazeres, F. Carvalho, J. Rivas, Cheese whey management: A review, J. Environ. Manage. 110 (2012) 48–68. https://doi.org/10.1016/j.jenvman.2012.05.018

[16]   P. B. Helkar, A. Sahoo, N. Patil, Review: food industry by-products used as a functional food ingredients, Int. J. Waste Resour. 6 (2016) 1–6. https://doi.org/10.4172/2252-5211.1000248

[17]   FAO, Key facts on food loss and waste you should know! save food: global initiative on food loss and waste reduction, Food and Agriculture Organization of the United Nations. http://www.fao.org/save-food/resources/keyfindings/en/, 2012 (accessed 30 August 2018).

[18]   Statista, The Statistics Portal, Share of food loss/waste in North America and Oceania in 2015, by type. https://www.statista.com/statistics/525520/share-of-food-waste-and-loss-by-food-category-north-america-oceania/, 2015. (accessed August 29, 2018).

[19]   Å. Stenmark, C. Jensen, T. Quested, G. Moates, Estimates of European food waste levels., 2016. https://doi.org/10.13140/RG.2.1.4658.4721

[20]   Statista, The Statistics Portal, Share of food loss/waste globally in 2015, by type. https://www.statista.com/statistics/525535/share-of-food-waste-and-loss-by-category-globally/, 2015 (accessed 30 August 2018).

[21]   Federal Office for the Environment (FOEN), Food waste. https://www.bafu.admin.ch/bafu/en/home/topics/waste/guide-to-waste-az/biodegradable-waste/types-of-waste/lebensmittelabfaelle.html,2018 (accessed 30 August 2018).

[22]   Ministry of Food Processing Industries (MOFPI), Annual Report 2016-17 http://mofpi.nic.in/documents/reports/annual-report, 2015 (accessed 24 August 2018).

[23]   S. Hegde, J.S. Lodge, T.A. Trabold, Characteristics of food processing wastes and their use in sustainable alcohol production, Renew. Sustain. Energy Rev. 81 (2018) 510–523. https://doi.org/10.1016/j.rser.2017.07.012

[24]   W. Qasim, A. V. Mane, Characterization and treatment of selected food industrial effluents by coagulation and adsorption techniques, Water Resour. Ind. 4 (2013) 1–12. https://doi.org/10.1016/j.wri.2013.09.005

[25]   V.K. Joshi, A. Pandey, D.K. Sandhu, Fermentation technology for food industry waste utilization. In: V.K. Joshi, P Ashok (Eds.), Biotechnology: food fermentation, microbiology, biochemistry and technology, vol II., Educational Publishers and Distributors, New Delhi, 1999, pp. 1291-1348.

[26]   N. Mirabella, V. Castellani, S. Sala, Current options for the valorization of food manufacturing waste: a review, J. Clean. Prod. 65 (2014) 28–41. https://doi.org/10.1016/j.jclepro.2013.10.051

[27]   M.E.M. Mabrouk, A.M.D. El-Ahwany, Production of β-mannanase by Bacillus amylolequifaciens 10A1 cultured on potato peels, Afr. J. Biotechnol. 7 (2008) 1123–1128. https://doi.org/10.5897/AJB08.047

[28]   A. Al-Weshahy, V.A. Rao, Potato peel as a source of important phytochemical antioxidant nutraceuticals and their role in human health - A review, in: V. Rao (Ed.), Phytochemicals as Nutraceuticals - Global Approaches to Their Role in Nutrition and Health InTech, 2012: pp. 207–224. https://doi.org/10.5772/30459.

[29]   A.M. Mullen, C. Álvarez, M. Pojić, T.D. Hadnadev, M. Papageorgiou, Classification and target compounds, in: C. Galanakis (Ed.), Food Waste Recovery Academic Press, Elsevier, 2015, pp. 25–57. https://doi.org/10.1016/B978-0-12-800351-0.00002-X

[30]   B.K. Tiwari, A. Gowen, B.M. McKenna, Pulse foods : processing, quality and nutraceutical applications, first ed., Academic Press, Elsevier, 2011.

[31]   T.K. Girish, V.M. Pratape, U.J.S. Prasada Rao, Nutrient distribution, phenolic acid composition, antioxidant and alpha-glucosidase inhibitory potentials of black gram (*Vigna mungo* L.) and its milled by-products, Food Res. Int. 46 (2012) 370–377. https://doi.org/10.1016/j.foodres.2011.12.026

[32]   A.M. Sharoba, M.A. Farrag, A. El-Salam, Utilization of some fruits and vegetables waste as a source of dietary fiber and its effect on the cake making and its quality attributes, J. Agroaliment. Process. Technol. 19 (2013) 429–444.

[33]   W. Kosoom, N. Charoenwattanasakun, Y. Ruangpanit, S. Rattanatabtimtong, S. Attamangkune, Physical, chemical and biological properties of cassava pulp, in: 47th Kasetsart Univ. Annu. Conf. Kasetsart, Kasetsart, 2015, pp. 117–124.

[34]   B.H. Kiepper, W.C. Merka, D.L. Fletcher, Proximate composition of poultry processing wastewater particulate matter from broiler slaughter plants, Poult. Sci. 87 (2008) 1633–1636. https://doi.org/10.3382/ps.2007-00331

[35]   B.J. Akinyele, O.O. Olaniyi, D.J. Arotupin, Bioconversion of selected agricultural wastes and associated enzymes by volvariella volvacea: an edible mushroom, Res. J. Microbiol. 6 (2011) 63–70. https://doi.org/10.3923/jm.2011.63.70

[36]   A.P. Sánchez-Camargo, M.Â. Almeida Meireles, B.L.F. Lopes, F.A. Cabral, Proximate composition and extraction of carotenoids and lipids from Brazilian redspotted shrimp waste (*Farfantepenaeus paulensis*), J. Food Eng. 102 (2011) 87–93. https://doi.org/10.1016/j.jfoodeng.2010.08.008

[37]   H.A. Hashem, M.M. Abul-Fadl, M.T.M. Assous, M.S.M.A. Abo-Zaid, Improvement the nutritional value of especial biscuits (children school meal) by using some fruits and vegetables, J. Appl. Sci. Res. 9 (2013) 5679–5691.

[38]   P.G. Dalev, Utilisation of waste feathers from poultry slaughter for production of a protein concentrate, Bioresour. Technol. 48 (1994) 265–267. https://doi.org/10.1016/0960-8524(94)90156-2

[39]   S. Damodaran, K.L. Parkin, O.R. Fennema, Fennema's food chemistry, fourth ed., CRC Press, New York, 2007.

[40]   D. Deublein, A. Steinhauser, Biogas from Waste and Renewable resources: An Introduction, Wiley VCH, Weinheim, 2008.

[41]   P. Thassitou, I. Arvanitoyannis, Bioremediation: a novel approach to food waste management, Trends Food Sci. Technol. 12 (2001) 185–196. https://doi.org/10.1016/S0924-2244(01)00081-4

[42]   M.R. Kosseva, Sources, characterization, and composition of food industry wastes, in: M.R. Kosseva, C. Webb (Eds.), Food Industry Wastes, Elsevier, London UK, 2013, pp. 37–60. https://doi.org/10.1016/B978-0-12-391921-2.00003-2

[43]   M. Hutnan, M. Hornak, I. Bodík, V. Hlavacka, Anaerobic treatment of wheat stillage, Chem. Biochem. Eng. 17 (2003) 233–241.

[44]   E. Cibis, C.A. Kent, M. Krzywonos, Z. Garncarek, B. Garncarek, T. Miśkiewicz, Biodegradation of potato slops from a rural distillery by thermophilic aerobic bacteria, Bioresour. Technol. 85 (2002) 57–61. https://doi.org/10.1016/S0960-8524(02)00069-X

[45]   H. Panda, The Complete Book on Managing Food Processing Industry Waste, first edition., National Institute of Industrial Research, New Delhi, 2011.

[46]   T. Baysal, S. Ersus, D.A.J. Starmans, Supercritical CO_2 extraction of β-carotene and lycopene from tomato paste waste, J. Agric. Food Chem. 48 (2000) 5507–5511. https://doi.org/10.1021/jf000311t

[47]   A. Schieber, F. Stintzing, R. Carle, By-products of plant food processing as a source of functional compounds—recent developments, Trends Food Sci. Technol. 12 (2001) 401–413. https://doi.org/10.1016/S0924-2244(02)00012-2

[48]   İ. Çinar, Effects of cellulase and pectinase concentrations on the colour yield of enzyme extracted plant carotenoids, Process Biochem. 40 (2005) 945–949. https://doi.org/10.1016/j.procbio.2004.02.022

[49]   N.S. Salim, A. Singh, V. Raghavan, Potential utilization of fruit and vegetable wastes for food through drying or extraction techniques, Nov Tech Nutr. Food Sci. 1 (2017) 1–12.

[50]   I. Erlund, Review of the flavonoids quercetin, hesperetin, and naringenin. Dietary sources, bioactivities, bioavailability, and epidemiology, Nutr. Res. 24 (2004) 851–874. https://doi.org/10.1016/j.nutres.2004.07.005

[51]   H. Tapiero, D.M. Townsend, K.D. Tew, Organosulfur compounds from alliaceae in the prevention of human pathologies, Biomed. Pharmacother. 58 (2004) 183–193. https://doi.org/10.1016/j.biopha.2004.01.004

[52]   R. Llorach, F.A. Tomás-Barberán, F. Ferreres, Lettuce and chicory byproducts as a source of antioxidant phenolic extracts, J. Agric. Food Chem. 52 (2004) 5109–5116. https://doi.org/10.1021/jf040055a

[53]   F. Vallejo, A. Gil-Izquierdo, A. Pérez-Vicente, C. García-Viguera, In vitro gastrointestinal digestion study of broccoli inflorescence phenolic compounds, glucosinolates, and vitamin C, J. Agric. Food Chem. 52 (2004) 135–138. https://doi.org/10.1021/jf0305128

[54]   S. Bhushan, K. Kalia, M. Sharma, B. Singh, P.S. Ahuja, Processing of apple pomace for bioactive molecules, Crit. Rev. Biotechnol. 28 (2008) 285–296. https://doi.org/10.1080/07388550802368895

[55]   B. Aliakbarian, A. Fathi, P. Perego, F. Dehghani, Extraction of antioxidants from winery wastes using subcritical water, J. Supercrit. Fluids. 65 (2012) 18–24. https://doi.org/10.1016/j.supflu.2012.02.022

[56]   D. Kammerer, A. Claus, A. Schieber, R. Carle, A novel process for the recovery of polyphenols from grape (*Vitis vinifera* L.) Pomace, J. Food Sci. 70 (2005) C157–C163. https://doi.org/10.1111/j.1365-2621.2005.tb07077.x

[57] Y. Lu, L. Yeap Foo, The polyphenol constituents of grape pomace, Food Chem. 65 (1999) 1–8. https://doi.org/10.1016/S0308-8146(98)00245-3

[58] Norland products, Fish Gelatin Products. https://www.norlandprod.com/Fishdefault.html, 2013 (accessed 30 August 2018).

[59] N. Jabeen, I. Majid, G.A. Nayik, Bioplastics and food packaging: A review, Cogent Food Agric. 1 (2015) 1–6. https://doi.org/10.1080/23311932.2015.1117749

[60] V.M. Vučurović, R.N. Razmovski, Sugar beet pulp as support for Saccharomyces cerivisiae immobilization in bioethanol production, Ind. Crops Prod. 39 (2012) 128–134. https://doi.org/10.1016/j.indcrop.2012.02.002.

[61] Y. Wang, L. Yan, CFD studies on biomass thermochemical conversion, Int. J. Mol. Sci. 9 (2008) 1108–1130. https://doi.org/10.3390/ijms9061108

[62] H.H. Khoo, T.Z. Lim, R.B.H. Tan, Food waste conversion options in Singapore: Environmental impacts based on an LCA perspective, Sci. Total Environ. 408 (2010) 1367–1373. https://doi.org/10.1016/j.scitotenv.2009.10.072

[63] F.R. McDougall, J.P. Hruska, Report: the use of life cycle inventory tools to support an integrated approach to solid waste management, Waste Manag. Res. 18 (2000) 590–594. https://doi.org/10.1177/0734242X0001800610

[64] P.A. Caton, M.A. Carr, S.S. Kim, M.J. Beautyman, Energy recovery from waste food by combustion or gasification with the potential for regenerative dehydration: A case study, Energy Convers. Manag. 51 (2010) 1157–1169. https://doi.org/10.1016/j.enconman.2009.12.025

[65] B. Digman, D.S. Kim, Review: Alternative energy from food processing wastes, Environ. Prog. 27 (2008) 524–537. https://doi.org/10.1002/ep.10312

[66] H.B. Goyal, D. Seal, R.C. Saxena, Bio-fuels from thermochemical conversion of renewable resources: A review, Renew. Sustain. Energy Rev. 12 (2008) 504–517. https://doi.org/10.1016/j.rser.2006.07.014

[67] N.S. Barman, S. Ghosh, S. De, Gasification of biomass in a fixed bed downdraft gasifier - A realistic model including tar, Bioresour. Technol. 107 (2012) 505–511. https://doi.org/10.1016/j.biortech.2011.12.124

[68] R.C. Brown, T.R. Brown, Biorenewable Resources Engineering: New Products from Agriculture, second ed., Wiley-Blackwell, 2003.

[69] S. Yaman, Pyrolysis of biomass to produce fuels and chemical feedstocks, Energy Convers. Manag. 45 (2004) 651–671. https://doi.org/10.1016/S0196-8904(03)00177-8

[70] N. Canabarro, J.F. Soares, C.G. Anchieta, C.S. Kelling, M.A. Mazutti, Thermochemical processes for biofuels production from biomass, Sustain. Chem. Process. 1 (2013) 1-10. https://doi.org/10.1186/2043-7129-1-22

[71] M. Balat, M. Balat, E. Kirtay, H. Balat, Main routes for the thermo-conversion of biomass into fuels and chemicals. Part 1: Pyrolysis systems, Energy Convers. Manag. 50 (2009) 3147–3157. https://doi.org/10.1016/j.enconman.2009.08.014

[72] A. Demirbas, Waste management, waste resource facilities and waste conversion processes, Energy Convers. Manag. 52 (2011) 1280–1287. https://doi.org/10.1016/j.enconman.2010.09.025

[73] N.D. Berge, K.S. Ro, J. Mao, J.R. V. Flora, M.A. Chappell, S. Bae, Hydrothermal carbonization of municipal waste streams, Environ. Sci. Technol. 45 (2011) 5696–5703. https://doi.org/10.1021/es2004528

[74] S.K. Hoekman, A. Broch, C. Robbins, Hydrothermal carbonization (HTC) of lignocellulosic biomass, Energy and Fuels. 25 (2011) 1802–1810. https://doi.org/10.1021/ef101745n

[75] T.P.T. Pham, R. Kaushik, G.K. Parshetti, R. Mahmood, R. Balasubramanian, Food waste-to-energy conversion technologies: Current status and future directions, Waste Manag. 38 (2015) 399–408. https://doi.org/10.1016/j.wasman.2014.12.004

[76] Z. Liu, A. Quek, S. Kent Hoekman, R. Balasubramanian, Production of solid biochar fuel from waste biomass by hydrothermal carbonization, Fuel. 103 (2013) 943–949. https://doi.org/10.1016/j.fuel.2012.07.069

[77] G.K. Parshetti, S. Kent Hoekman, R. Balasubramanian, Chemical, structural and combustion characteristics of carbonaceous products obtained by hydrothermal carbonization of palm empty fruit bunches, Bioresour. Technol. 135 (2013) 683–689. https://doi.org/10.1016/j.biortech.2012.09.042

[78] L. Li, R. Diederick, J.R.V. Flora, N.D. Berge, Hydrothermal carbonization of food waste and associated packaging materials for energy source generation, Waste Manag. 33 (2013) 2478–2492. https://doi.org/10.1016/j.wasman.2013.05.025

[79] L.J. Wang, Production of bioenergy and bioproducts from food processing wastes: a review, Am. Soc.Agr. Biol. Eng. 56 (2013) 217–229.

[80] G.M. Hall, J. Howe, Energy from waste and the food processing industry, Process Saf. Environ. Prot. 90 (2012) 203–212. https://doi.org/10.1016/j.psep.2011.09.005

[81] H.N. Chanakya, T. V. Ramachandra, M. Vijayachamundeeswari, Resource recovery potential from secondary components of segregated municipal solid

wastes, Environ. Monit. Assess. 135 (2007) 119–127.
https://doi.org/10.1007/s10661-007-9712-4

[82] N. Guermoud, F. Ouadjnia, F. Abdelmalek, F. Taleb, A. Addou, Municipal solid waste in Mostaganem city (Western Algeria), Waste Manag. 29 (2009) 896–902. https://doi.org/10.1016/j.wasman.2008.03.027.

[83] V.C. Kalia, V. Sonakya, N. Raizada, Anaerobic digestion of banana stem waste, Bioresour. Technol. 73 (2000) 191–193. https://doi.org/10.1016/S0960-8524(99)00172-8

[84] S. Di Berardino, S. Costa, A. Converti, Semi-continuous anaerobic digestion of a food industry wastewater in an anaerobic filter, Bioresour. Technol. 71 (2000) 261–266.

[85] E. Salminen, J. Rintala, Anaerobic digestion of organic solid poultry slaughterhouse waste: A review., Bioresour. Technol. 83 (2002) 13–26. https://doi.org/10.1016/S0960-8524(01)00199-7

[86] M.R. Kosseva, Management and Processing of Food Wastes, in M. Moo-Young (Ed.), Comprehensive Biotechnology, Elsevier, B.V., 2011, pp. 557-593. https://doi.org/10.1016/B978-0-08-088504-9.00393-7

[87] R. Gebauer, Mesophilic anaerobic treatment of sludge from saline fish farm effluents with biogas production, Bioresour. Technol. 93 (2004) 155–167. https://doi.org/10.1016/j.biortech.2003.10.024

[88] I. Muhammad Nasir, T.I. Mohd Ghazi, R. Omar, Production of biogas from solid organic wastes through anaerobic digestion: a review, Appl. Microbiol. Biotechnol. 95 (2012) 321–329. https://doi.org/10.1007/s00253-012-4152-7

[89] Z. Zhang, I.M. O'Hara, S. Mundree, B. Gao, A.S. Ball, N. Zhu, Z. Bai, B. Jin, Biofuels from food processing wastes, Curr. Opin. Biotechnol. 38 (2016) 97–105. https://doi.org/10.1016/j.copbio.2016.01.010

[90] V.H. Perez, E.G. Silveira Junior, D.C. Cubides, G.F. David, O.R. Justo, M.P.P. Castro, M.S. Sthel, H.F. De-Castro, Trends in Biodiesel Production: Present Status and Future Directions, in: S.S. da Silva, A.K. Chandel (Eds.), Biofuels in Brazil, Springer International Publishing, New York, 2014, pp. 281–302. https://doi.org/10.1007/978-3-319-05020-1_13

[91] B. Amigun, F. Müller-Langer, H. von Blottnitz, Predicting the costs of biodiesel production in Africa: learning from Germany, Energy Sustain. Dev. 12 (2008) 5–21. https://doi.org/10.1016/S0973-0826(08)60415-9

[92] E.T. Altikriti, A.B. Fadhil, M.M. Dheyab, Two-step base catalyzed transesterification of chicken fat: optimization of parameters, Energy Sources, Part A Recover. Util. Environ. Eff. 37 (2015) 1861–1866. https://doi.org/10.1080/15567036.2012.654442

[93] C. Ofori-boateng, E.M. Kwofie, M.Y. Mensah, Comparative analysis of the effect of different alkaline catalysts on biodiesel yield, world Appl. Sci. J. 16 (2012) 1445–1449.

[94] T. Singhasiri, N. Tantemsapya, Production of biodiesel from food processing waste using response surface methodology, Energy Sources, Part A Recover. Util. Environ. Eff. 38 (2016) 2799–2808. https://doi.org/10.1080/15567036.2015.1117543

[95] W. Pietrzak, J. Kawa-Rygielska, Simultaneous saccharification and ethanol fermentation of waste wheat-rye bread at very high solids loading: Effect of enzymatic liquefaction conditions, Fuel. 147 (2015) 236–242. https://doi.org/10.1016/j.fuel.2015.01.057

[96] J.N. Nigam, Continuous ethanol production from pineapple cannery waste using immobilized yeast cells, J. Biotechnol. 80 (2000) 189–193. https://doi.org/10.1016/S0168-1656(00)00246-7

[97] J. Ma, T.H. Duong, M. Smits, W. Verstraete, M. Carballa, Enhanced biomethanation of kitchen waste by different pre-treatments, Bioresour. Technol. 102 (2011) 592–599. https://doi.org/10.1016/j.biortech.2010.07.122

[98] A.I. Vavouraki, V. Volioti, M.E. Kornaros, Optimization of thermo-chemical pretreatment and enzymatic hydrolysis of kitchen wastes, Waste Manag. 34 (2014) 167–173. https://doi.org/10.1016/j.wasman.2013.09.027

[99] L. Matsakas, D. Kekos, M. Loizidou, P. Christakopoulos, Utilization of household food waste for the production of ethanol at high dry material content, Biotechnol. Biofuels. 7 (2014) 1–9. https://doi.org/10.1186/1754-6834-7-4

[100] H.S. Oberoi, P. V. Vadlani, A. Nanjundaswamy, S. Bansal, S. Singh, S. Kaur, N. Babbar, Enhanced ethanol production from Kinnow mandarin (*Citrus reticulata*) waste via a statistically optimized simultaneous saccharification and fermentation process, Bioresour. Technol. 102 (2011) 1593–1601. https://doi.org/10.1016/j.biortech.2010.08.111

[101] L.A. Rodríguez, M.E. Toro, F. Vazquez, M.L. Correa-Daneri, S.C. Gouiric, M.D. Vallejo, Bioethanol production from grape and sugar beet pomaces by solid-state fermentation, Int. J. Hydrogen Energy. 35 (2010) 5914–5917. https://doi.org/10.1016/j.ijhydene.2009.12.112

[102] P.M.R. Guimarães, J.A. Teixeira, L. Domingues, Fermentation of lactose to bio-ethanol by yeasts as part of integrated solutions for the valorisation of cheese whey, Biotechnol. Adv. 28 (2010) 375–384. https://doi.org/10.1016/j.biotechadv.2010.02.002

[103] F. Zoppellari, L. Bardi, Production of bioethanol from effluents of the dairy industry by *Kluyveromyces marxianus*, N. Biotechnol. 30 (2013) 607–613. https://doi.org/10.1016/j.nbt.2012.11.017

[104] R.H.S. Diniz, M.Q.R.B. Rodrigues, L.G. Fietto, F.M.L. Passos, W.B. Silveira, Optimizing and validating the production of ethanol from cheese whey permeate by *Kluyveromyces marxianus* UFV-3, Biocatal. Agric. Biotechnol. 3 (2014) 111–117. https://doi.org/10.1016/j.bcab.2013.09.002

[105] D. Arapoglou, T. Varzakas, A. Vlyssides, C. Israilides, Ethanol production from potato peel waste (PPW), Waste Manag. 30 (2010) 1898–1902. https://doi.org/10.1016/j.wasman.2010.04.017

[106] B.J. Khawla, M. Sameh, G. Imen, F. Donyes, G. Dhouha, E.G. Raoudha, N.E. Oumèma, Potato peel as feedstock for bioethanol production: A comparison of acidic and enzymatic hydrolysis, Ind. Crops Prod. 52 (2014) 144–149. https://doi.org/10.1016/j.indcrop.2013.10.025

[107] M. Boluda-Aguilar, L. García-Vidal, F. d P. González-Castañeda, A. López-Gómez, Mandarin peel wastes pretreatment with steam explosion for bioethanol production, Bioresour. Technol. 101 (2010) 3506–3513. https://doi.org/10.1016/j.biortech.2009.12.063

[108] H.S. Oberoi, P. V. Vadlani, L. Saida, S. Bansal, J.D. Hughes, Ethanol production from banana peels using statistically optimized simultaneous saccharification and fermentation process, Waste Manag. 31 (2011) 1576–1584. https://doi.org/10.1016/j.wasman.2011.02.007

[109] J.H. Kim, J.C. Lee, D. Pak, Feasibility of producing ethanol from food waste, Waste Manag. 31 (2011) 2121–2125. https://doi.org/10.1016/j.wasman.2011.04.011

[110] M.II. Nguyen, Alternatives to spray irrigation of starch waste based distillery effluent, J. Food Eng. 60 (2003) 367–374. https://doi.org/10.1016/S0260-8774(03)00059-1

[111] C. Xue, X.Q. Zhao, C.G. Liu, L.J. Chen, F.W. Bai, Prospective and development of butanol as an advanced biofuel, Biotechnol. Adv. 31 (2013) 1575–1584. https://doi.org/10.1016/j.biotechadv.2013.08.004

[112] P. Dürre, Biobutanol: An attractive biofuel, Biotechnol. J. 2 (2007) 1525–1534. https://doi.org/10.1002/biot.200700168

[113] M. Becerra, M.E. Cerdán, M.I. González-Siso, Biobutanol from cheese whey, Microb. Cell Fact. 14 (2015) 1-14. https://doi.org/10.1186/s12934-015-0200-1

[114] T. Lütke-Eversloh, H. Bahl, Metabolic engineering of *Clostridium acetobutylicum*: Recent advances to improve butanol production, Curr. Opin. Biotechnol. 22 (2011) 634–647. https://doi.org/10.1016/j.copbio.2011.01.011

[115] M. Stoeberl, R. Werkmeister, M. Faulstich, W. Russ, Biobutanol from food wastes–fermentative production, use as biofuel an the influence on the emissions, Procedia Food Sci. 1 (2011) 1867–1874. https://doi.org/10.1016/j.profoo.2011.09.274

[116] V. Ujor, A.K. Bharathidasan, K. Cornish, T.C. Ezeji, Feasibility of producing butanol from industrial starchy food wastes, Appl. Energy. 136 (2014) 590–598. https://doi.org/10.1016/j.apenergy.2014.09.040

[117] H. Huang, V. Singh, N. Qureshi, Butanol production from food waste: a novel process for producing sustainable energy and reducing environmental pollution, Biotechnol. Biofuels. 8 (2015) 1–12. https://doi.org/10.1186/s13068-015-0332-x

[118] N. Nishio, Y. Nakashimada, Recent development of anaerobic digestion processes for energy recovery from wastes, J. Biosci. Bioeng. 103 (2007) 105–112. https://doi.org/10.1263/jbb.103.105

[119] C. Mao, Y. Feng, X. Wang, G. Ren, Review on research achievements of biogas from anaerobic digestion, Renew. Sustain. Energy Rev. 45 (2015) 540–555. https://doi.org/10.1016/j.rser.2015.02.032

[120] M. Schlegel, N. Kanswohl, D. Rössel, A. Sakalauskas, Essential technical parameters for effective biogas production, Agron. Res. 6 (2008) 341–348.

[121] A. Singh, A. Kuila, S. Adak, M. Bishai, R. Banerjee, Utilization of Vegetable Wastes for Bioenergy Generation, Agric. Res. 1 (2012) 213–222. https://doi.org/10.1007/s40003-012-0030-x

[122] J. Gomesantunes, R. Mikalsen, A. Roskilly, An investigation of hydrogen-fuelled HCCI engine performance and operation, Int. J. Hydrogen Energy. 33 (2008) 5823–5828. https://doi.org/10.1016/j.ijhydene.2008.07.121

[123] X. Gómez, C. Fernández, J. Fierro, M.E. Sánchez, A. Escapa, A. Morán, Hydrogen production: Two stage processes for waste degradation, Bioresour. Technol. 102 (2011) 8621–8627. https://doi.org/10.1016/j.biortech.2011.03.055

[124] K.Y. Show, D.J. Lee, J.S. Chang, Bioreactor and process design for biohydrogen production, Bioresour. Technol. 102 (2011) 8524–8533. https://doi.org/10.1016/j.biortech.2011.04.055

[125] M.Y. Azwar, M.A. Hussain, A.K. Abdul-Wahab, Development of biohydrogen production by photobiological, fermentation and electrochemical processes: A review, Renew. Sustain. Energy Rev. 31 (2014) 158–173. https://doi.org/10.1016/j.rser.2013.11.022

[126] A. Ghimire, L. Frunzo, F. Pirozzi, E. Trably, R. Escudie, P.N.L. Lens, G. Esposito, A review on dark fermentative biohydrogen production from organic biomass: Process parameters and use of by-products, Appl. Energy. 144 (2015) 73–95. https://doi.org/10.1016/j.apenergy.2015.01.045

[127] G. Cai, B. Jin, P. Monis, C. Saint, Metabolic flux network and analysis of fermentative hydrogen production, Biotechnol. Adv. 29 (2011) 375–387. https://doi.org/10.1016/j.biotechadv.2011.02.001

[128] L. Alibardi, R. Cossu, Effects of carbohydrate, protein and lipid content of organic waste on hydrogen production and fermentation products, Waste Manag. 47 (2016) 69–77. https://doi.org/10.1016/j.wasman.2015.07.049

[129] L. Dong, Y. Zhenhong, S. Yongming, K. Xiaoying, Z. Yu, Hydrogen production characteristics of the organic fraction of municipal solid wastes by anaerobic mixed culture fermentation, Int. J. Hydrogen Energy. 34 (2009) 812–820. https://doi.org/10.1016/j.ijhydene.2008.11.031

[130] B.E. Logan, S.E. Oh, I.S. Kim, S. Van Ginkel, Biological hydrogen production measured in batch anaerobic respirometers, Environ. Sci. Technol. 36 (2002) 2530–2535. https://doi.org/10.1021/es015783i

[131] G. De Gioannis, M. Friargiu, E. Massi, A. Muntoni, A. Polettini, R. Pomi, D. Spiga, Biohydrogen production from dark fermentation of cheese whey: Influence of pH, Int. J. Hydrogen Energy. 39 (2014) 1–12. https://doi.org/10.1016/j.ijhydene.2014.10.046

[132] M. Stoytcheva, G. Montero, Biodiesel- Feedstocks and processing technologies, first ed., InTech, Croatia, 2011.

[133] M. Canakci, The potential of restaurant waste lipids as biodiesel feedstocks, Bioresour. Technol. 98 (2007) 183–190. https://doi.org/10.1016/j.biortech.2005.11.022

[134] A.H. Mouneimne, H. Carrère, N. Bernet, J.P. Delgenès, Effect of saponification on the anaerobic digestion of solid fatty residues, Bioresour. Technol. 90 (2003) 89–94. https://doi.org/10.1016/S0960-8524(03)00091-9

[135] L. Brennan, P. Owende, Biofuels from microalgae-A review of technologies for production, processing, and extractions of biofuels and co-products, Renew. Sustain. Energy Rev. 14 (2010) 557–577. https://doi.org/10.1016/j.rser.2009.10.009

[136] M. Mathiyazhagan, a Ganapathi, Factors Affecting Biodiesel Production, Res. Plant Biol. 1 (2011) 1–5.

[137] A. Demirbas, Progress and recent trends in biodiesel fuels, Energy Convers. Manag. 50 (2009) 14–34. https://doi.org/10.1016/j.enconman.2008.09.001

[138] A. Demirbas, Production of biodiesel fuels from linseed oil using methanol and ethanol in non-catalytic SCF conditions, Biomass and Bioenergy. 33 (2009) 113–118. https://doi.org/10.1016/j.biombioe.2008.04.018

[139] I.A. Musa, The effects of alcohol to oil molar ratios and the type of alcohol on biodiesel production using transesterification process, Egypt. J. Pet. 25 (2016) 21–31. https://doi.org/10.1016/j.ejpe.2015.06.007

[140] D.Y.C. Leung, Y. Guo, Transesterification of neat and used frying oil: Optimization for biodiesel production, Fuel Process. Technol. 87 (2006) 883–890. https://doi.org/10.1016/j.fuproc.2006.06.003

[141] A.B.M.S. Hossain, A.N. Boyce, Biodiesel production from waste sunflower cooking oil as an environmental recycling process and renewable energy, Bulg. J. Agric. Sci. 15 (2009) 312–317. https://doi.org/10.9734/BBJ/2016/22338

[142] U. Rashid, F. Anwar, Production of biodiesel through optimized alkaline-catalyzed transesterification of rapeseed oil, Fuel. 87 (2008) 265–273. https://doi.org/10.1016/j.fuel.2007.05.003

[143] M. Agarwal, K. Singh, S. Upadhyaya, S.P. Chaurasia, Effect of Reaction Parameters on Yield and Characteristics of Biodiesel Obtained from Various Vegetable Oils, Www.Conference.Net.Au/Chemeca. (2011) 11.

[144] A. Gashaw, A. Teshita, Production of biodiesel from waste cooking oil and factors affecting its formation: A review, Int. J. Renew. Sustain. Energy. 3 (2014) 92–98. https://doi.org/10.11648/j.ijrse.20140305.12

[145] T. Eevera, K. Rajendran, S. Saradha, Biodiesel production process optimization and characterization to assess the suitability of the product for varied

environmental conditions, Renew. Energy. 34 (2009) 762–765.
https://doi.org/10.1016/j.renene.2008.04.006

[146] E.C. Abbah, G.I. Nwandikom, C.C. Egwuonwu, N.R. Nwakuba, Effect of reaction temperature on the yield of biodiesel from neem seed oil, Am. J. Energy Sci. 3 (2016) 16–20.

[147] P. Felizardo, M.J. Neiva Correia, I. Raposo, J.F. Mendes, R. Berkemeier, J.M. Bordado, Production of biodiesel from waste frying oils, Waste Manag. 26 (2006) 487–494. https://doi.org/10.1016/j.wasman.2005.02.025

[148] L. Naik, N. Radhika, K. Sravani, A. Hareesha, B. Mohanakumari, K. Bhavanasindhu, Optimized parameters for production of biodiesel from fried oil optimized parameters for production of biodiesel from fried oil, Int. Adv. Res. J. Sci. Eng. Technol. 2 (2015) 62–65. https://doi.org/10.17148/IARJSET.2015.2615

[149] F. Fahed Banihani, Transesterification and production of biodiesel from waste cooking oil: effect of operation variables on fuel properties, Am. J. Chem. Eng. 4 (2016) 154-160. https://doi.org/10.11648/j.ajche.20160406.13

[150] A.B.M.S. Hossain, M.A. Mazen, Effects of catalyst types and concentrations on biodiesel production from waste soybean oil biomass as renewable energy and environmental recycling process, Aust. J. Crop Sci. 4 (2010) 550–555. https://doi.org/10.5897/AJB10.299

[151] S.S. Jagadale, L.M. Jugulkar, Review of various reaction parameters and other factors affecting on production of chicken fat based biodiesel, Int. J. Mod. Eng. Res. 2 (2012) 407–411.

[152] M.P. Dorado, E. Ballesteros, M. Mittelbach, F.J. López, Kinetic parameters affecting the alkali-catalyzed transesterification process of used olive oil, Energy and Fuels. 18 (2004) 1457–1462. https://doi.org/10.1021/ef034088o

[153] N.H.M. Yasin, T. Mumtaz, M.A. Hassan, N. Abd Rahman, Food waste and food processing waste for biohydrogen production: A review, J. Environ. Manage. 130 (2013) 375–385. https://doi.org/10.1016/j.jenvman.2013.09.009

[154] J.J. Lay, K.S. Fan, J. Chang I, C,H, Ku, Influence of chemical nature of organic wastes on their conversion to hydrogen by heat-shock digested sludge, Int. J. Hydrogen Energy. 28 (2003) 1361–1367. https://doi.org/10.1016/S0360-3199(03)00027-2

[155] D.Y. Lee, Y. Ebie, K.Q. Xu, Y.Y. Li, Y. Inamori, Continuous H_2 and CH_4 production from high-solid food waste in the two-stage thermophilic fermentation

process with the recirculation of digester sludge, Bioresour. Technol. 101 (2010) S42–S47. https://doi.org/10.1016/j.biortech.2009.03.037

[156] X. Wang, G. Yang, Y. Feng, G. Ren, X. Han, Optimizing feeding composition and carbon-nitrogen ratios for improved methane yield during anaerobic co-digestion of dairy, chicken manure and wheat straw, Bioresour. Technol. 120 (2012) 78–83. https://doi.org/10.1016/j.biortech.2012.06.058

[157] N.H. Mohd Yasin, N.A. Rahman, H.C. Man, M.Z. Mohd Yusoff, M.A. Hassan, Microbial characterization of hydrogen-producing bacteria in fermented food waste at different pH values, Int. J. Hydrogen Energy. 36 (2011) 9571–9580. https://doi.org/10.1016/j.ijhydene.2011.05.048

[158] S.Y. Adaganti, V.S. Yaliwal, B.M. Kulkarni, G.P. Desai, N.R. Banapurmath, Factors affecting bioethanol production from lignocellulosic biomass (*Calliandra calothyrsus*), Waste and Biomass Valorization. 5 (2014) 963–971. https://doi.org/10.1007/s12649-014-9305-8

[159] Y. Lin, W. Zhang, C. Li, K. Sakakibara, S. Tanaka, H. Kong, Factors affecting ethanol fermentation using *Saccharomyces cerevisiae* BY4742, Biomass and Bioenergy. 47 (2014) 395–401. https://doi.org/10.1016/j.biombioe.2012.09.019

[160] M. Fakruddin, M.A. Quayum, M.M. Ahmad, N. Choudhury, Analysis of key factors affecting ethanol production by *Saccharomyces cerevisiae* IFST-072011, Biotechnology. 11 (2012) 248–252. https://doi.org/10.3923/biotech.2012.248.252

[161] M. Kashid, A. Ghosalkar, Critical factors affecting ethanol production by immobilized Pichia stipitis using corn cob hemicellulosic hydrolysate, Prep. Biochem. Biotechnol. 48 (2018) 288–295. https://doi.org/10.1080/10826068.2018.1425715

[162] V.S. O'Leary, R. Green, B.C. Sullivan, V.H. Holsinger, Alcohol production by selected yeast strains in lactase-hydrolyzed acid whey, Biotechnol. Bioeng. 19 (1977) 1019–1035. https://doi.org/10.1002/bit.260190706

[163] R. Thammasittirong, A. Thirasaktana, T Thammasittirong, M. Srisodsuk, Improvement of ethanol production by ethanol-tolerant *Saccharomyces cerevisiae* UVNR56., Springerplus. 2 (2013) 583-588.

[164] L.P. Yomano, S.W. York, L.O. Ingram, Isolation and characterization of ethanol-tolerant mutants of *Escherichia coli* KO11 for fuel ethanol production, J. Ind. Microbiol. Biotechnol. 20 (1998) 132–138. https://doi.org/10.1038/sj.jim.2900496

[165] J.O. Westman, C.J. Franzén, Current progress in high cell density yeast bioprocesses for bioethanol production, Biotechnol. J. 10 (2015) 1185–1195. https://doi.org/10.1002/biot.201400581

[166] Y.Q. Tang, M.Z. An, Y.L. Zhong, M. Shigeru, X.L. Wu, K. Kida, Continuous ethanol fermentation from non-sulfuric acid-washed molasses using traditional stirred tank reactors and the flocculating yeast strain KF-7, J. Biosci. Bioeng. 109 (2010) 41–46. https://doi.org/10.1016/j.jbiosc.2009.07.002

[167] K. Hegde, N. Chandra, S.J. Sarma, S.K. Brar, V.D. Veeranki, Genetic Engineering strategies for enhanced biodiesel production, Mol. Biotechnol. 57 (2015) 606–624. https://doi.org/10.1007/s12033-015-9869-y

[168] A. Cernac, C. Benning, WRINKLED1 encodes an AP2/EREB domain protein involved in the control of storage compound biosynthesis in Arabidopsis, Plant J. 40 (2004) 575–585. https://doi.org/10.1111/j.1365-313X.2004.02235.x

[169] L. Domingues, P.M.R. Guimarães, C. Oliveira, Metabolic engineering of *Saccharomyces cerevisiae* for lactose / whey fermentation, Bioeng. Bugs. 1 (2010) 1–8. https://doi.org/10.4161/bbug.1.3.10619

[170] C.Y. Chen, M.H. Yang, K.L. Yeh, C.H. Liu, J.S. Chang, Biohydrogen production using sequential two-stage dark and photo fermentation processes, Int. J. Hydrogen Energy. 33 (2008) 4755–4762. https://doi.org/10.1016/j.ijhydene.2008.06.055

[171] W. Han, Y. Yan, Y. Shi, J. Gu, J. Tang, H. Zhao, Biohydrogen production from enzymatic hydrolysis of food waste in batch and continuous systems, Sci. Rep. 6 (2016) 1–9. https://doi.org/10.1038/srep38395

[172] D.E. Algapani, W. Qiao, M. Ricci, D. Bianchi, S. M. Wandera, F. Adani, R. Dong, Bio-hydrogen and bio-methane production from food waste in a two-stage anaerobic digestion process with digestate recirculation, Renew. Energy. 130 (2018) 1108–1115. https://doi.org/10.1016/j.renene.2018.08.079

Microbial Fuel Cells: Materials and Applications   Materials Research Forum LLC
Materials Research Foundations **46** (2019) 289-306   doi: http://dx.doi.org/10.21741/9781644900116-11

Chapter 11

Microbial Fuel Cell for the Treatment of Wastewater

Dibyajyoti Haldar[1], Mriganka Sekhar Manna[1], Dwaipayan Sen[2], Tridib Kumar Bhowmick[3], Kalyan Gayen[1]*

[1]Department of Chemical Engineering, National Institute of Technology Agartala, Tripura-799046, India

[2]Department of Chemical Engineering, Heritage Institute of Technology Kolkata, Kolkata-700107, India

[3]Department of Bioengineering, National Institute of Technology Agartala, Tripura-799046, India

kgayen123@gmail.com, kalyan.chemical@nita.ac.in

Abstract

A microbial fuel cell is a potential alternative for the treatment of wastewater. In this process of water treatment, a substantial amount of energy is produced. Microorganisms are explicitly used in microbial fuel cells to generate electrons and protons that are involved in electrochemical reactions for the treatment of wastewater for the generation of power. Moreover, the efficiency of the treatment process and also the generation of power largely depend on the nature of the substrates used as feed, types of the microorganism used and the configurations of the cells. In view of an effective treatment process, various types of wastewater originated from a number of different sources have extensively been treated in microbial fuel cells using electrogenic microorganisms. Environment-friendly features of microbial fuel cells result in a better technology compared to the existing ones for the purpose. The present chapter of this book comprehensively briefs on the reactions mechanisms involved in different aspects of the technique and discusses extensively the essential changes in the treatment techniques of wastewater from varied sources.

Keywords

Wastewater, Microbial Fuel Cell, Cathode, Anode, COD, Power Density

Contents

Materials Research Forum LLC
doi: http://dx.doi.org/10.21741/9781644900116-11

1.   Introduction

In recent years, a rapid urbanization all across the globe demands continuous exploration of fossil fuel reserves. The obvious issue with such increased use of fossil fuel attributes to global warming because of emission of greenhouse gases, necessitates the search for alternative energy resources with low carbon footprint [1-3]. In the last few decades, venture on renewable sources towards an environment-friendly production of alternative energy has been a top priority [4]. However, several technological improvements and subsequent socio-economic developments lead to cause a severe anthropogenic impact on global sustainability in terms of environmental pollution. As far as environmental pollution is concerned, water has been one of the major issues directly correlated with pollution as it is vulnerable to numerous pollutants originated from different sources [5]. Therefore, treatment of wastewater is of high importance prior to its reuses in industries, agricultural field and for drinking purposes. Especially, drought-prone localities are primarily using treated domestic and industrial wastewaters as potable water for various practices, which mitigate the problem related to water scarcity along with disposal issue of wastewater [6]. However, the presence of organic loads and toxic chemicals such as nitrogen, phosphorus, phenol and heavy metals (for example mercury and cadmium) in the industrial effluent becomes detrimental to soil fertility and crop production. So, sustainable treatment of wastewater necessitates an understanding of the treatment process feasibility prior to its reuse in agricultural lands and others applications.

Microbial Fuel Cells: Materials and Applications   Materials Research Forum LLC
Materials Research Foundations **46** (2019) 289-306   doi: http://dx.doi.org/10.21741/9781644900116-11

Presently, biological treatment process using activated sludge is recognized as the most conventional wastewater treatment scheme as it rapidly produces significant results, though with an investment of high capital and operational cost [7]. However, other methods such as adsorption, coagulation, ion exchange and membrane separation processes are also well recognized for the purpose [8]. Further considering several pros and cons, each of the methods is having limitations in regard to environmental sustainability. The processes like adsorption and coagulation produce a lot of sludge known as a secondary pollutant and global management of this secondary pollutant is one of the typical downstream tasks to achieve. The ion exchange process is expensive while the throughput of a membrane process is usually very low because of a secondary layer formation over the membrane, coined as fouling of the membrane. Therefore, over the years an urge towards a sustainable process development for the treatment of wastewater has been a top research endeavor. One of the most promising and new technologies that demand a critical analysis on the waste to energy concept is the microbial fuel cell, where waste biomaterials are converted into renewable energy. The various issues and mechanisms of the microbial fuel cell are comprehensively presented in this chapter for the treatment of wastewater containing varied contaminations from different sources.

2.   Reaction mechanism

In continuation of the uninterrupted search for a sustainable source of energy, Potter et al. in the early 90s' reported a technique for the conversion of organic waste into electrical energy [9]. Since then, the microbial fuel cell has been demonstrated as a promising biochemical concept towards the production of bioenergy from waste materials with the help of microbial action. Microbes present in the microbial fuel cell are well capable of converting organic substrates into CO_2, water and energy [9]. The electrons produced during the oxidation of biodegradable solid substrates are transferred to a positive electrode (anode) during the electrolysis process. The movement of electrons towards the anode primarily occurs in two mechanistic pathways. Firstly, electrons move through the direct contact of the electrode with proteins like cytochrome C of the cell membrane of microbes and secondly through the influence of any extraneous intermediates, which are known as mediators (Fig. 1). Once the electrons are reached at the anode, these are transferred through an external circuit to the negative electrode (cathode), where available constituents of higher potential (electron acceptor) such as oxygen or metal ions accept the electrons and get themselves reduced. Reduced constituents of wastewater finally bind with the protons (H^+) generated from the oxidation of the same biodegradable substrate at the anode in the anodic chamber and pass through semi-permeable membrane placed in between anode and cathode chambers of a fuel cell. The semi-permeable membrane only allows H^+ ions to pass. A detailed pictorial diagram of the conventional microbial fuel cell is shown in Fig. 2. Microbial fuel cells in comparison to other existing technologies offer few distinct advantages, which make this sustainable scheme more potent towards energy generation directly from organic matter. The advantages are:

(a) The energy present in the substrate is easily converted into electrical energy with high efficiency.

(b) The microbial fuel cell can be operated even at the locations with limited electrical facilities as the process is carried out under ambient conditions.

(c) The biochemical process does not require high energy input if the cathode is already aerated.

(d) Finally, the process can be readily applicable in the sector of bioenergy due to the less production of sludge.

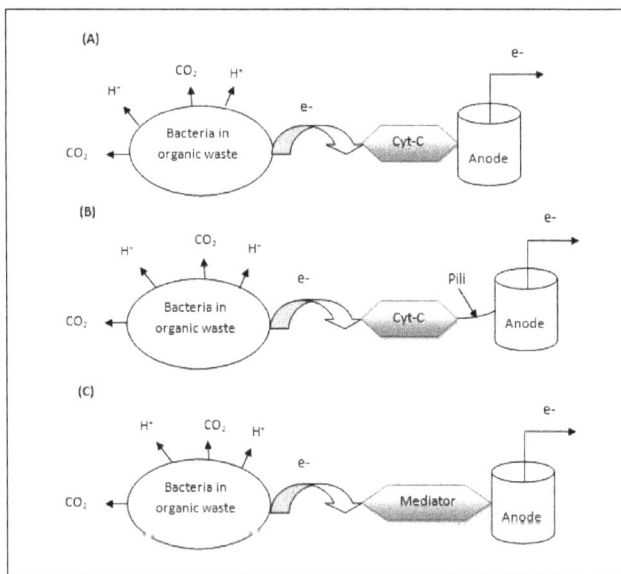

Fig. 1 Different mechanisms of electron transfer from bacteria to anode; (A) Electron (e⁻) is directly transferred to the anode of the fuel cell through cytochrome C, (B) Electron (e⁻) is directly transferred to the anode of the fuel cell through pili or nanowire of the bacteria, (C) Electron (e⁻) is indirectly transferred to the anode of the fuel cell via the external influence of mediator

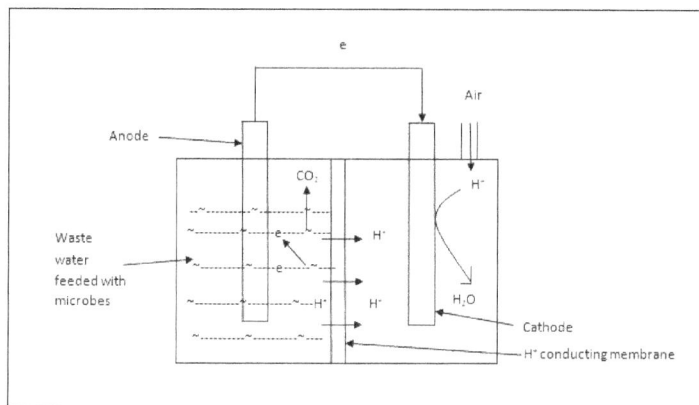

Fig. 2 Pictorial diagram of microbial fuel cell

Now the electrical energy in the microbial fuel cell is primarily generated through the movement of electrons as a result of oxidation and reduction reactions occurred at anode and cathode respectively. Anode acts as an electron donor while cathode acts as an electron acceptor. Substances such as oxygen, metals, nitrite, nitrate accepting the electrons in the cathodic chamber are known as a terminal electron acceptor. The performance of any microbial fuel cell largely depends on its structural configuration. Fig. 3 depicts different configurations of microbial fuel cells based on the essential mechanisms of electrochemical reactions already elaborated above.

2.1   Anodic reaction mechanism

Microorganisms remain in the anode chamber of the microbial fuel cell and they play an important role in generating electrons that are passed through the external circuit to reach the terminal electron acceptors (inorganic and organic contaminants) present in the cathode chamber. The overall performance of microbial fuel cell largely depends on the efficiency of a reaction mechanism that produces electrons in the anode chamber. Consequently, optimum reaction conditions, suitable configuration and constituents (different external agents) of the anode chamber are prerequisite for the performance of microbial fuel cell [10, 11]. Employment of some mediators and modification of anode become essentially important very often in this context. An efficient anode should have the features of high electrical conductivity with low resistance, more surface area, biocompatibility and high resistivity against strong chemical corrosion[11]. Carbonaceous materials being inexpensive, biocompatible and having good conductivity

Microbial Fuel Cells: Materials and Applications   Materials Research Forum LLC
Materials Research Foundations **46** (2019) 289-306   doi: http://dx.doi.org/10.21741/9781644900116-11

are considered as one of the mostly used anode materials in a microbial fuel cell. A lot of research investigations have been carried out with suitably modified configuration of the anode to improve the performance of the system [12-14]. Modified carbon cloth anode with formic acid is an example with increased efficiency of the overall process [15]. Moreover, an electrodeposition of metal oxide like, MnO_2 improves the performance of the microbial fuel cell by around 24% in comparison to the bared carbon anode [16]. Hence, the modification in the texture of the anode is very much necessary to enhance the performance of microbial fuel cells in large-scale industrial applications. Although the metals are much better conductors of electricity the smoother surface of the metals quite often limits the adhesion of bacterial cells leading to the low power density compared to carbon materials [13]. Therefore, the surface modification of the metallic anode is accomplished through several physical and chemical processes [14]. On the contrary, the formation of carboxyl group during the chemical treatment of bacterial cells in the fuel cell generates a strong network of hydrogen bonds with the peptide present in the bacterial cells, thereby improves the electron transfers from microorganisms to surface modified anode surface. The use of different polymeric metal composites as the anode material is another alternative to improve the performance of microbial fuel cells. Polyaniline is widely used as polymeric metal alloy as it is highly able to restrict deactivation of the anode by metabolic products produced by the microbial reactions [17]. Zhang et al. has reported that the power density of 760 mW.m^{-2} is available from a microbial fuel cell employing graphite/PTFE as a composite electrode and *E. coli* as the bacterium [18]. The reaction mechanism for the generation of electrons from acetate rich wastewater at anode chamber is given in Eq. 1

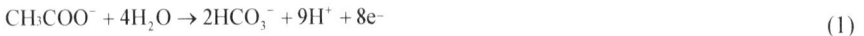

$$CH_3COO^- + 4H_2O \rightarrow 2HCO_3^- + 9H^+ + 8e- \tag{1}$$

2.2 Cathodic reaction mechanism

The negative electrode (cathode) accepts the electrons from the anode through the external circuit and H^+ ions through the proton exchange membrane (PEM). O_2 is purged to the cathode chamber externally. This O_2 and others electron acceptors present in the cathode chamber combined with electrons and H^+ to be reduced to water. The reaction of the formation of water is represented in Eq. 2.

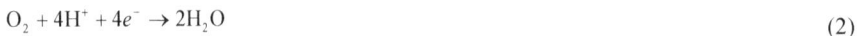

$$O_2 + 4H^+ + 4e^- \rightarrow 2H_2O \tag{2}$$

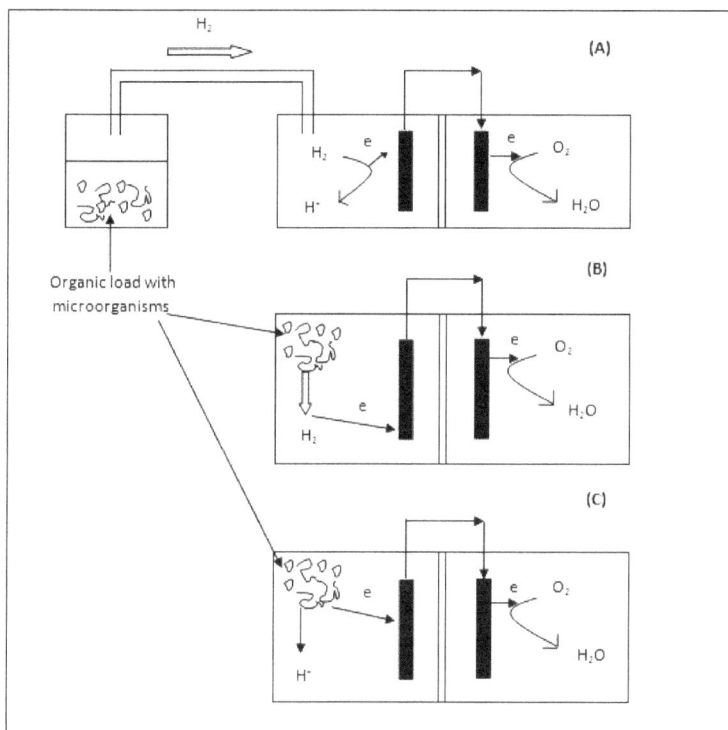

Fig. 3 Different configurations of MFC; (A) H2 is produced in a separate bioreactor and utilized in a fuel cell, (B) Microorganisms produce H2 from organic matter and H2 generates the electrons which leads to produce the electricity, (C) Microbiological activity directly transfers electron to the anode of MFC.

The availability of oxygen and accessibility of protons through PEM largely impact on the performance of the reactions in the cathode. Oxygen is mostly preferred as a terminal electron acceptor in the cathode chamber as it is inexpensive, easily accessible and has high oxidation potential [19]. Two configurations of cathodes namely air and aqueous cathodes are mostly used in microbial fuel cells. Air cathodes are very often adopted in a microbial fuel cell as the configuration involved is most inexpensive. Part of the cathode is directly exposed to air for higher availability of O_2. On the other hand, aqueous cathodes are consisted of carbon cloth coated with a layer of catalyst and very often

Microbial Fuel Cells: Materials and Applications   Materials Research Forum LLC
Materials Research Foundations **46** (2019) 289-306   doi: http://dx.doi.org/10.21741/9781644900116-11

immersed in an aqueous layer with limited oxygen accessibility [20]. Carbon cloth is the supporting material for binding the binder in an air cathode. The binders like Nafion and polytetrafluroethylene (PTFE) are very commonly used to keep the catalyst intact on the cathode plate. More power density is generally achieved when Nafion in comparison with PTFE is used as a binder with other parameters remaining constant in a microbial fuel cell. Nafion is put onto the cathode plate (carbon cloth) as a thick layer of biofilm [21]. Biofouling of cathode is a serious threat to the performance of the microbial fuel cell. With regard to the issue of biofouling, the use of metal composite electrode containing the ingredient having anti-microbes property is the solution to the problem of biofouling. One of the investigations on microbial fuel cell reports that the maximum power density of ~1700 mW.m^{-2} has been achieved using a graphite carbon-based metal composite catalyst (Fe_3O_4/GC) [22] as a cathode in a microbial fuel cell.

3.   Treatment of wastewater in microbial fuel cell

A number of research investigations have been carried out on the application of microbial fuel cell (MFC) in the treatment of different types of wastewater. The configuration of MFC has been the most significant factor in extracting the maximum output in terms of power density and organic removal efficiency. Microbial communities degrade organic components present in the wastewater for an efficient removal of the contaminants. MFCs are well recognized to be capable of removing a number of heavy metals, phenolic compounds and petroleum products etc. The use of synthetic wastewater as the only carbon source in the fuel cells is a common practice in the research on MFC over the years. The results of the investigations predict that the technique can be employed for real wastewater. Various synthetic monomers in the complex mixture of wastewater such as carbohydrates, amino acids and fatty acids have been investigated as substrates in the MFC. Among the carbon sources as a synthetic substrate, glucose is the simplest carbohydrate molecule which has been tested as a carbon source in MFC by a number of investigators [23]. Thereafter, a continuous research activity has been accomplished focusing on other carbohydrates and their derivatives. Catal et al. have achieved a maximum power density of more than 2700 mWm^{-2} using glucuronic acid as a substrate using a single chambered air cathode [24]. Another investigation has been carried out using 30 g/L of glucose as substrate, *S. cerevisiae* as biocatalyst and $KMnO_4$ as an oxidizing agent in a four-layered chamber of MFC to produce electricity with ~2000 mWm^{-2} of power density [25]. Later on, poly-alcohols as carbon source and consortium of microbes as biocatalysts in MFC were investigated for the generation of electricity. It has been usually observed that the power density of the sugar alcohols was in the range of 1500-2600 W.m^{-2}. Nitrogen of proteins has also been proved to be a good substrate for power generation in MFC. Yang et al. have studied that serine among the amino acids, produces a maximum power density of ~770 mW.m^{-2} and alanine produces minimum power density (550 mWm^{-2}) used as a nitrogenous substrate in a single chambered air cathode [26]. Various organic acids like (formic acid, lactic acid and succinic acid) are also capable of producing electricity from fuel cells. Acetate and butyrate are the most common substrates due to their immense potential of generating electricity in microbial fuel cells [27]. When various organic alcohols are tested as a substrate; ethanol has been proved to be more potent compared to methanol. In an experiment using a microbial fuel cell, the maximum power density of ~800 mW.m^{-2} was achieved using ethanol as substrate [28]. After various synthetic substrates were examined for the power generation in fuel cells the research community become interested in working with real wastewater originated from a number of different sources such as; food waste, dairy waste, livestock and even industrial waste characteristically containing a minimum amount of organic components. The relative performance of MFCs for the treatment of different kinds of wastewaters is illustrated in Table 1.

3.1   Treatment of vegetable waste and a waste of food processing

Vegetable wastes contain a substantial amount of carbohydrates and are found as solid waste mostly at different dumping sites of the municipality. An attempt to generate electricity through microbial fuel cell using vegetable waste as substrates results in a power density of 57 mWm^{-2} along with a satisfactory amount of chemical oxygen demand (COD) removal [29]. Waste waters from food processing industries have been investigated in MFCs by a number of researchers [30, 31]. Sangeetha et al. have carried out an experiment in dual-chambered MFC with food processing wastewater and observed high percentage (86%) removal of COD but with a low power density of 124 mWm^{-2} [32]. Substantial amounts of carbohydrates are available in a large volume of wastewater of starch processing industries. Lu et al. were been able to achieve a power density of 240 mWm^{-2} from starch processing wastewater containing an initial COD of 4850 ppm [33]. More than 90% removal of COD of a protein content wastewater containing 600 mg of COD per litre has been achieved in a double chamber fuel cell [34].

Table 1 Application of microbial fuel cells in different types of waste water

Waste water type	Microbial Fuel Cell	Performance	Reference
Beverage distillery	Two chambered MFC made up of carbon fiber paper as electrodes	COD removal of 88% and power density of 124 mWm^{-2} observed	[53]
Manure wash	Three compartment MFC with graphite fiber brush anode and Pt coated carbon cloth as cathode	Maximum power density of 216 mWm^{-2} achieved	[54]
Vegetable wastes waster	Single chambered MFC with graphite plates electrodes	COD removal of 63% and power density of 57 mWm^{-2} obtained	[29]
Dairy waste water	Two chambered MFC with graphite plates electrodes	90% of COD removal and power density of 621 mWm^{-2} of maximum power density obtained	[55]
Domestic waste water	Upflow MFC with carbon fiber brush used as cathode and anode	78% of COD removal and maximum power density of 481 mWm^{-2} achieved	[50]
Food processing waste water	Two chambered MFC separated by PEM and graphic sheets are used as electrodes	86% COD removal and maximum power density of 230 mWm^{-2} realized	[56]
Petroleum refinery waste water	Dual chambered MFC with carbon rod as anode and graphite flake as cathode	COD removal of 64% and maximum power density of 330 mWcm^{-3} obtained	[57]
Pharmaceutical waste water	MFC constructed as single chambered open air cathode with graphite plate as electrodes	85% of COD removal and maximum power density of 206 mWm^{-2} observed	[58]
Sewage waste water	MFC with graphite rod electrodes	Over 82% of COD removal and only around 7 mWm^{-2} of power density obtained.	[59]
Waste water containing pesticides	Single chambered MFC with activated carbon as anode and carbon cloth as cathode	COD removal more than 70% and maximum power density of 77% noted	[60]

3.2   Treatment of wastewater from beverage distilleries

Wastewaters from beverage industries are characterized by high COD and organics mainly consisting of sugars and proteins which are non-toxic in nature. Cost effective biological methods are commonly employed for the treatment of such wastewaters. Microbial fuel cells with various configurations are an alternative technique for the treatment of brewery wastewater as it contains a high load of COD [35, 36]. Single chambered microbial fuel cell results in COD removal of 87% using brewery wastewater containing COD load of ~2250 ppm. Moreover, a maximum power density of 530 mWm^{-2} is achievable using a phosphate buffer [37]. Wen et al. employed air cathode microbial fuel cells using running raw brewery wastewater as substrate and achieved a maximum power density of 670 mWm^{-2} [38]. Few attempts have also been made for the generation of electricity using different wine samples as substrates in MFCs. Air cathode has been employed for the generation of electricity from white and red wine samples. White wines in comparison with red ones showed better efficiency in terms of power density. The power density of 262 mWm^{-2} for white wine samples with higher removal of COD [39] was achieved.

3.3   Dairy wastewater

Dairy wastewater is also a potent feedstock for use in MFC as it contains a huge load of biodegradable organics [40]. Wastewater directly from the dairy field has been treated with single chambered microbial fuel cells with graphite coated anode resulting in a maximum power density of 20 Wm^{-3} [41]. The liquid fraction of whey processing of a real dairy waste contains a substantial amount of carbohydrates, proteins fats, etc. and that can be used as potent feedstock materials in different types of MFCs. Incubated anode for quite a long duration has been observed to sufficiently remove COD and produce a maximum power density of 1800 mWm^{-2} [42]. Tremouli et al. have carried out an investigation using whey permeate as feedstock in two-chambered MFC and found COD removal of around 95% [43].

3.4   Pharmaceutical wastewater

Wastewaters from pharmaceutical industries mostly contain toxic organic matters which have been utilized as a substrate in MFCs in many of the investigations. Synthetic antibiotic penicillin of 50 ppm with 1 g/L glucose was used as a substrate in microbial cells employing air cathode and power density of more than 100 mWm^{-2} was achieved [44]. Paracetamol containing wastewater is also investigated to produce a maximum of ~218 mWm^{-2} power density in another investigation [45]. During acid wash of the drugs, various intermediate toxic chemicals are produced and have been investigated in microbial fuel cells and a maximum power density of 22.3 mWm^{-2} has been achieved [46].

3.5   Petrochemical wastewater

Petrochemical industries produce a large amount of sludge mainly composed of hydrocarbons and toxic effluent and a lesser amount of organics. Treatment of such wastewaters using MFCs has been practised by a number of investigators. Sludge obtained directly from the petroleum industry has been observed to produce more than 50 mWm^{-2} power using microbial fuel cell [47]. Real petrochemical wastewater containing terephthalic acid has been investigated both in acidic and alkaline pH of the substrate using single chambered MFC. The power generation was 50% more in the case of alkaline pH of 8.5 as compared to the acidic substrate [48].

3.6   Treatment of domestic wastewater

As domestic wastewater can be treated in MFCs as it contains a mixture of organic components Cusick et al. employed an air cathode fuel cell with an incorporation of domestic wastewater as a feed material and observed a COD removal of 83% [49]. In a different investigation of combined treatment using a microbial fuel cell with photo bioreactor resulted into phosphate and ammonium removal of 99% [50]. Combined mixture of domestic and oil mill wastewater provided better efficiency in achieving a power density of 125 mWm^{-2} [51]. Moreover, a number of investigations have also accomplished for the treatment of municipal solid waste in MFCs and a substantial amount of electricity generation and COD removal were achieved [52].

Conclusions

Rapid depletion of conventional fuels (viz. petroleum and fossil fuels) continuously urges the search for alternative fuels. Further, emission of greenhouse gases mostly CO and CO_2 and consequently their detrimental effects on the environment, especially global warming have led to the scientific community to find a way out in this regard. On the other hand, more efficient processes of treatment of wastewater are also important for recycling of water to prevent the shortage of water at present and in future. Therefore, environmentally benign techniques for both the purposes are urgently required. MFC technique has been emerged as potent to mitigate the energy crisis without affecting the environment significantly. Moreover, the microbial fuel cell has been an individual kind of sustainable technology which uses microorganism as biocatalysts during the conversion of wastewater into the clean energy. Over the years, a number of attempts have been made to improve the power density from different types of wastewater by altering the basic configuration of microbial fuel cells. However, the practical adaptation of this technology is limited within the laboratory scale and demands for a significant breakthrough to scale up the process into the next level to be feasible for industrial purposes.

References

[1]   D. Haldar, D. Sen, K. Gayen, A review on the production of fermentable sugars from lignocellulosic biomass through conventional and enzymatic route—a comparison, Int. J. Green Energy 13 (2016) 1232-1253. https://doi.org/10.1080/15435075.2016.1181075

[2]   D. Haldar, D. Sen, K. Gayen, Enzymatic hydrolysis of banana stems (*Musa acuminata*): Optimization of process parameters and inhibition characterization, Int. J. Green Energy 15 (2018) 406-413. https://doi.org/10.1080/15435075.2018.1467834

[3]   R. Heede, N. Oreskes, Potential emissions of CO_2 and methane from proved reserves of fossil fuels: An alternative analysis, Global Environmental Change 36 (2016) 12-20. https://doi.org/10.1016/j.gloenvcha.2015.10.005

[4]   S. Chu, A. Majumdar, Opportunities and challenges for a sustainable energy future, Nature 488 (2012) 294. https://doi.org/10.1038/nature11475

[5]   P. Hu, Y. Ouyang, L. Wu, L. Shen, Y. Luo, P. Christie, Effects of water management on arsenic and cadmium speciation and accumulation in an upland rice cultivar, J. Environ. Sci. (China) 27 (2015) 225-231. https://doi.org/10.1016/j.jes.2014.05.048

[6]   J.F. Wang, G.X. Wang, H. Wanyan, Treated wastewater irrigation effect on soil, crop and environment: wastewater recycling in the loess area of China, J. Environ. Sci. (China) 19 (2007) 1093-1099. https://doi.org/10.1016/S1001-0742(07)60178-8

[7]   M. Sustarsic, Wastewater treatment: Understanding the activated sludge process, CEP Magazine, AIChE Publication, November 2009, pp. 26-29.

[8]   P. Rajasulochana, V. Preethy, Comparison on efficiency of various techniques in treatment of waste and sewage water – A comprehensive review, Resour. Effic. Technol. 2 (2016) 175-184. https://doi.org/10.1016/j.reffit.2016.09.004

[9]   M.C. Potter, Electrical effects accompanying the decomposition of organic compounds, Proc. R Soc Lond B Biol Sci. 84 (1911) 260-276. https://doi.org/10.1098/rspb.1911.0073

[10] B.H. Kim, I.S. Chang, G.M. Gadd, Challenges in microbial fuel cell development and operation, Appl. Microbiol. Biotechnol. 76 (2007) 485-494. https://doi.org/10.1007/s00253-007-1027-4

[11] B.E. Logan, B. Hamelers, R. Rozendal, U. Schröder, J. Keller, S. Freguia, P. Aelterman, W. Verstraete, K. Rabaey, Microbial fuel cells: Methodology and technology, Environ. Sci. Technol. 40 (2006) 5181-5192. https://doi.org/10.1021/es0605016

[12] Y. Qiao, C. Li, S.-J. Bao, Q.-L. Bao, Carbon nanotube/polyaniline composit as anode material for microbial fuel cells, J. Power Sources 170 (2007) 79-84. https://doi.org/10.1016/j.jpowsour.2007.03.048

[13] J. Wei, P. Liang, X. Huang, Recent progress in electrodes for microbial fuel cells, Bioresour Technol. 102 (2011) 9335-9344. https://doi.org/10.1016/j.biortech.2011.07.019

[14] M. Zhou, M. Chi, J. Luo, H. He, T. Jin, An overview of electrode materials in microbial fuel cells, J. Power Sources 196 (2011) 4427-4435. https://doi.org/10.1016/j.jpowsour.2011.01.012

[15] W. Liu, S. Cheng, J. Guo, Anode modification with formic acid: A simple and effective method to improve the power generation of microbial fuel cells, Appl. Surf. Sci. 320 (2014) 281–286. https://doi.org/10.1016/j.apsusc.2014.09.088

[16] Z. Changyong, P. Liang, Y. Jiang, X. Huang, Enhanced power generation of microbial fuel cell using manganese dioxide-coated anode in flow-through mode, *J. Power Sourc*. 273 (2015) 580-583. https://doi.org/10.1016/j.jpowsour.2014.09.129

[17] U. Schroder, J. Niessen, F. Scholz, A generation of microbial fuel cells with current outputs boosted by more than one order of magnitude, Angew Chem. Int. Ed. Engl. 42 (2003) 2880-2883. https://doi.org/10.1002/anie.200350918

[18] T. Zhang, Y. Zeng, S. Chen, X. Ai, H. Yang, Improved performances of *E. coli*-catalyzed microbial fuel cells with composite graphite/PTFE anodes, Electrochem. Commun. 9 (2007) 349-353. https://doi.org/10.1016/j.elecom.2006.09.025

[19] K. Watanabe, Recent developments in microbial fuel cell technologies for sustainable bioenergy, J. Biosci. Bioeng. 106 (2008) 528-536. https://doi.org/10.1263/jbb.106.528

[20] K. Scott, I. Cotlarciuc, I. Head, K.P. Katuri, D. Hall, J.B. Lakeman, D. Browning, Fuel cell power generation from marine sediments: Investigation of cathode materials, J. Chem. Technol. Biotechnol. 83 (2008) 1244-1254. https://doi.org/10.1002/jctb.1937

[21] S. Cheng, H. Liu, B.E. Logan, Power densities using different cathode catalysts (Pt and CoTMPP) and polymer binders (Nafion and PTFE) in single chamber microbial fuel cells, Environ. Sci. Technol. 40 (2006) 364-369. https://doi.org/10.1021/es0512071

[22] M. Ma, S. You, G. Xiao-bo, Y. Dai, J. Zou, H. Fu, Silver/iron oxide/graphitic carbon composites as bacteriostatic catalysts for enhancing oxygen reduction in microbial fuel cells, J. Power Sourc. 283 (2015) 74-83. https://doi.org/10.1016/j.jpowsour.2015.02.100

[23] S.K. Chaudhuri, D.R. Lovley, Electricity generation by direct oxidation of glucose in mediatorless microbial fuel cells, Nat. Biotechnol. 21 (2003) 1229-1232. https://doi.org/10.1038/nbt867

[24] T. Catal, K. Li, H. Bermek, H. Liu, Electricity production from twelve monosaccharides using microbial fuel cells, J. Power Sourc. 175 (2008) 196-200. https://doi.org/10.1016/j.jpowsour.2007.09.083

[25] M. Rahimnejad, A.A. Ghoreyshi, G.D. Najafpour, H. Younesi, M. Shakeri, A novel microbial fuel cell stack for continuous production of clean energy, *Int. J. Hydrogen Energ.* 37 (2012) 5992-6000. https://doi.org/10.1016/j.ijhydene.2011.12.154

[26] Q. Yang, X. Wang, Y. Feng, H. Lee, J. Liu, X. Shi, Y. Qu, N. Ren, Electricity generation using eight amino acids by air–cathode microbial fuel cells, Fuel 102 (2012) 478-482. https://doi.org/10.1016/j.fuel.2012.04.020

[27] H. Liu, S. Cheng, B.E. Logan, Production of electricity from acetate or butyrate using a single-chamber microbial fuel cell, Environ. Sci. Technol. 39 (2005) 658-662. https://doi.org/10.1021/es048927c

[28] P.D. Kiely, G. Rader, J.M. Regan, B.E. Logan, Long-term cathode performance and the microbial communities that develop in microbial fuel cells fed different fermentation endproducts, Bioresour. Technol. 102 (2011) 361-366. https://doi.org/10.1016/j.biortech.2010.05.017

[29] S. Venkata Mohan, G. Mohanakrishna, P.N. Sarma, Composite vegetable waste as renewable resource for bioelectricity generation through non-catalyzed open-air cathode microbial fuel cell, Bioresour. Technol. 101 (2010) 970-976. https://doi.org/10.1016/j.biortech.2009.09.005

[30] B. Jin, H.J. van Leeuwen, B. Patel, H.W. Doelle, Q. Yu, Production of fungal protein and glucoamylase by Rhizopus oligosporus from starch processing wastewater, Process Biochem. 34 (1999) 59-65. https://doi.org/10.1016/S0032-9592(98)00069-7

[31] R. Jamuna, S.V. Ramakrishna, SCP production and removal of organic load from cassava starch industry waste by yeasts, J. Ferment. Bioeng. 67 (1989) 126-131. https://doi.org/10.1016/0922-338X(89)90193-1

[32] T. Sangeetha, M. Muthukumar, Catholyte performance as an influencing factor on electricity production in a dual-chambered microbial fuel cell employing food processing wastewater, Energy Sources Part A 33 (2011) 1514-1522. https://doi.org/10.1080/15567030903397966

[33] N. Lu, S.-g. Zhou, L. Zhuang, J.-t. Zhang, J.-r. Ni, Electricity generation from starch processing wastewater using microbial fuel cell technology, Biochem. Eng. J. 43 (2009) 246-251. https://doi.org/10.1016/j.bej.2008.10.005

[34] S. Oh, B.E. Logan, Hydrogen and electricity production from a food processing wastewater using fermentation and microbial fuel cell technologies, Water Res. 39 (2005) 4673-4682. https://doi.org/10.1016/j.watres.2005.09.019

[35] B.K. Ince, O. Ince, P.J. Sallis, G.K. Anderson, Inert COD production in a membrane anaerobic reactor treating brewery wastewater, Water Res. 34 (2000) 3943-3948. https://doi.org/10.1016/S0043-1354(00)00170-6

[36] W. Parawira, I. Kudita, M.G. Nyandoroh, R. Zvauya, A study of industrial anaerobic treatment of opaque beer brewery wastewater in a tropical climate using a full-scale UASB reactor seeded with activated sludge, Process Biochem. 40 (2005) 593-599. https://doi.org/10.1016/j.procbio.2004.01.036

[37] Y. Feng, X. Wang, B.E. Logan, H. Lee, Brewery wastewater treatment using air-cathode microbial fuel cells, Appl. Microbiol. Biotechnol. 78 (2008) 873-880. https://doi.org/10.1007/s00253-008-1360-2

[38] Q. Wen, Y. Wu, L. Zhao, Q. Sun, Production of electricity from the treatment of continuous brewery wastewater using a microbial fuel cell, Fuel 89 (2010) 1381-1385. https://doi.org/10.1016/j.fuel.2009.11.004

[39] T. Pepe Sciarria, G. Merlino, B. Scaglia, A. D'Epifanio, B. Mecheri, S. Borin, S. Licoccia, F. Adani, Electricity generation using white and red wine lees in air cathode microbial fuel cells, J. Power Sourc. 274 (2015) 393-399. https://doi.org/10.1016/j.jpowsour.2014.10.050

[40] E.V. Ramasamy, S.A. Abbasi, Energy recovery from dairy waste-waters: impacts of biofilm support systems on anaerobic CST reactors, Appl. Energy 65 (2000) 91-98. https://doi.org/10.1016/S0306-2619(99)00079-3

[41] M. Mahdi Mardanpour, M. Nasr Esfahany, T. Behzad, R. Sedaqatvand, Single chamber microbial fuel cell with spiral anode for dairy wastewater treatment, Biosens. Bioelectron. 38 (2012) 264-269. https://doi.org/10.1016/j.bios.2012.05.046

[42]   J. Kassongo, C.A. Togo, Performance improvement of whey-driven microbial fuel cells by acclimation of indigenous anodophilic microbes, Afr. J. Biotechnol. 10 (2011) 7846-7852. https://doi.org/10.5897/AJB11.206

[43]   A. Tremouli, G. Antonopoulou, S. Bebelis, G. Lyberatos, Operation and characterization of a microbial fuel cell fed with pretreated cheese whey at different organic loads, Bioresour. Technol. 131 (2013) 380-389. https://doi.org/10.1016/j.biortech.2012.12.173

[44]   Q. Wen, F. Kong, H. Zheng, D. Cao, Y. Ren, J. Yin, Electricity generation from synthetic penicillin wastewater in an air-cathode single chamber microbial fuel cell, Chem. Eng. J. 168 (2011) 572-576. https://doi.org/10.1016/j.cej.2011.01.025

[45]   L. Zhang, X. Yin, S.F.Y. Li, Bio-electrochemical degradation of paracetamol in a microbial fuel cell-fenton system, Chem. Eng. J. 276 (2015) 185-192. https://doi.org/10.1016/j.cej.2015.04.065

[46]   R. Liu, C. Gao, Y.-G. Zhao, A. Wang, S. Lu, M. Wang, F. Maqbool, Q. Huang, Biological treatment of steroidal drug industrial effluent and electricity generation in the microbial fuel cells, Bioresour. Technol. 123 (2012) 86-91. https://doi.org/10.1016/j.biortech.2012.07.094

[47]   K. Chandrasekhar, S. Venkata Mohan, Bio-electrochemical remediation of real field petroleum sludge as an electron donor with simultaneous power generation facilitates biotransformation of PAH: effect of substrate concentration, Bioresour. Technol. 110 (2012) 517-525. https://doi.org/10.1016/j.biortech.2012.01.128

[48]   S.K. Marashi, H.R. Kariminia, I.S. Savizi, Bimodal electricity generation and aromatic compounds removal from purified terephthalic acid plant wastewater in a microbial fuel cell, Biotechnol. Lett. 35 (2013) 197-203. https://doi.org/10.1007/s10529-012-1063-8

[49]   R.D. Cusick, P.D. Kiely, B.E. Logan, A monetary comparison of energy recovered from microbial fuel cells and microbial electrolysis cells fed winery or domestic wastewaters, *Int. J. Hydrogen Energ.* 35 (2010) 8855-8861. https://doi.org/10.1016/j.ijhydene.2010.06.077

[50]   H.-m. Jiang, S.-j. Luo, X.-s. Shi, M. Dai, R.-b. Guo, A system combining microbial fuel cell with photobioreactor for continuous domestic wastewater treatment and bioelectricity generation, J. Cent. South Univ. 20 (2013) 488-494. https://doi.org/10.1007/s11771-013-1510-2

[51]   T.P. Sciarria, A. Tenca, A. D'Epifanio, B. Mecheri, G. Merlino, M. Barbato, S. Borin, S. Licoccia, V. Garavaglia, F. Adani, Using olive mill wastewater to improve performance in producing electricity from domestic wastewater by using

single-chamber microbial fuel cell, Bioresour. Technol. 147 (2013) 246-253. https://doi.org/10.1016/j.biortech.2013.08.033

[52] H.C. Tao, T. Lei, G. Shi, X.N. Sun, X.Y. Wei, L.J. Zhang, W.M. Wu, Removal of heavy metals from fly ash leachate using combined bioelectrochemical systems and electrolysis, J. Hazard. Mater. 264 (2014) 1-7. https://doi.org/10.1016/j.jhazmat.2013.10.057

[53] J. Huang, P. Yang, Y. Guo, K. Zhang, Electricity generation during wastewater treatment: An approach using an AFB-MFC for alcohol distillery wastewater, Desalination 276 (2011) 373-378. https://doi.org/10.1016/j.desal.2011.03.077

[54] X. Zheng, N. Nirmalakhandan, Cattle wastes as substrates for bioelectricity production via microbial fuel cells, *Biotechnol. Lett*. 32 (2010) 1809-1814. https://doi.org/10.1007/s10529-010-0360-3

[55] H.J. Mansoorian, A.H. Mahvi, A.J. Jafari, N. Khanjani, Evaluation of dairy industry wastewater treatment and simultaneous bioelectricity generation in a catalyst-less and mediator-less membrane microbial fuel cell, J. Saudi Chem. Soc. 20 (2016) 88-100. https://doi.org/10.1016/j.jscs.2014.08.002

[56] H.J. Mansoorian, A.H. Mahvi, A.J. Jafari, M.M. Amin, A. Rajabizadeh, N. Khanjani, Bioelectricity generation using two chamber microbial fuel cell treating wastewater from food processing, Enzyme Microb. Technol. 52 (2013) 352-357. https://doi.org/10.1016/j.enzmictec.2013.03.004

[57] X. Guo, Y. Zhan, C. Chen, B. Cai, Y. Wang, S. Guo, Influence of packing material characteristics on the performance of microbial fuel cells using petroleum refinery wastewater as fuel, *Renew. Energ.* 87 (2016) 437-444. https://doi.org/10.1016/j.renene.2015.10.041

[58] G. Velvizhi, S. Venkata Mohan, Electrogenic activity and electron losses under increasing organic load of recalcitrant pharmaceutical wastewater, Int. J. Hydrogen Energ. 37 (2012) 5969-5978. https://doi.org/10.1016/j.ijhydene.2011.12.112

[59] M.M. Ghangrekar, V.B. Shinde, Simultaneous sewage treatment and electricity generation in membrane-less microbial fuel cell, Water Sci. Technol. 58 (2008) 37-43. https://doi.org/10.2166/wst.2008.339

[60] X. Cao, H.L. Song, C.Y. Yu, X.N. Li, Simultaneous degradation of toxic refractory organic pesticide and bioelectricity generation using a soil microbial fuel cell, Bioresour. Technol. 189 (2015) 87-93. https://doi.org/10.1016/j.biortech.2015.03.148

Microbial Fuel Cells: Materials and Applications
Materials Research Foundations 46 (2019) 307-334

Materials Research Forum LLC
doi: http://dx.doi.org/10.21741/9781644900116-12

Chapter 12

Microbial Production of Ethanol

Khush Bakhat Alia[1], Ijaz Rasul[1], Farrukh Azeem[1], Sabir Hussain[2],
Muhammad Hussnain Siddique[1], Saima Muzammil[3], Muhammad Riaz[4], Amna Bari[1],
Sehrish Liaqat[1], Habibullah Nadeem*[1]

[1]Department of Bioinformatics and Biotechnology, Government College University, Faisalabad, Pakistan

[2]Department of Microbiology, Government College University, Faisalabad, Pakistan

[3]Department of Food Sciences, University College of Agriculture, Bahauddin Zakariya University, Multan, Pakistan

[4]Department of Environmental Sciences and Engineering, Government College University, Faisalabad, Pakistan

* habibullah@gcuf.edu.pk

Abstract

With the increase in population, the demand for fuel from renewable resources is increasing as the sources of fossil fuel are exhausting. In this perspective production of biofuels gained importance and now biofuels are obtained by the fermentation of sugarcane and starch-based cereals and lignocellulosic materials including wheat straws, woodchips, agricultural and forest residues, etc. As these sources have certain limitations attention has been shifted towards the use of algae as a feedstock for biofuel. The conversion of these feedstocks into ethanol with the help of microorganisms is in consideration to meet the demand for fuel from renewable resources.

Keywords

Biofuel, Bioethanol, Lignocellulosic Biomass, Renewable Energy

Contents

Microbial Fuel Cells: Materials and Applications   Materials Research Forum LLC
Materials Research Foundations **46** (2019) 307-334   doi: http://dx.doi.org/10.21741/9781644900116-12

1. Introduction

Sustainable energy is a big task in the world of growing population. The demand for energy continuously increases with the increase in population but the sources of fossil fuel are exhausting day by day which causes an increment in petroleum fuel prices [1,2]. The increased use of fossil fuels from petroleum resources have concerns related to environment and energy security issues. Issues, including global climate change due to the emission of greenhouse gases, make researchers find alternative ways for the production of fuels from sustainable bio resources [3]. International political pressure to reduce greenhouse gasses and global energy crisis forced researchers to develop or search alternative renewable energy resources to replace traditional energy sources [4]. In sustainable development, the use of renewable energy sources is in consideration. Liquid fuels like biodiesel, bioethanol and biogases like methane and in solid fuel wood pellets and charcoal are mainly produced from biomass and are known as biofuels [2, 5].

Today the demand for fuel production from renewable resources is high and will be very high in the future [6]. The biofuel production gained importance in this perspective as it is an important type of fuel used in internal combustion engines and powertrains when blended with gasoline [4]. By definition biofuels are fuels, generated by using biological material and ethanol is one of the best-known biofuels [7].

Currently, bioethanol is produced by the fermentation of sugarcane and starch-based cereals but the use of crops in its production leads towards competition between animal feed and food industries and ultimately results in higher food prices and opposition in EU and globally. Lignocellulosic residual materials from agriculture industry and forest have resistance towards fractionation for production of sugars, along with this, the incapability of microbes in the fermentation of lignocellulosic hydrolysates reduces bioethanol production from these abundant materials [4, 6].

2. Classification of biofuels

Biofuels are generally classified into primary and secondary biofuels (Table-1). Biofuels directly obtained by the burning of woody plants and dry waste of animal are known as primary biofuels, while the secondary biofuels are further divided into 1st, 2nd and 3rd generation biofuels. Biofuels produced by the first generation includes fermentation of crop plants like sugar cane or corn. Biofuels production by the first generation has a negative impact on food security when edible parts of plants are in use for biofuel production. This limitation paves way for second-generation biofuel production in which, lignocellulosic material from nonfood biomass like straw or other agricultural waste materials serve as feedstock [8]. These waste materials have a high concentration of complex sugars like cellulose, hemicellulose and other polymers. These complex sugars are usually not accessible for microbes because of the absence of appropriate saccharolytic enzymes [9]. Third generation biofuel is produced from seaweeds, microbes and microalgae. Production of third-generation bioethanol is an effective strategy to meet the global needs [10].

Table 1: Classification of biofuels

Primary Biofuels produced from	Secondary Biofuels		
Wood chips, Firewood, Forest, crops, Animal waste residues	**First generation produced by**	**Second generation**	**Third generation**
	Fermentation of starch (wheat, potato, barley, corn) or sugars (from sugar beet and sugar cane)	Production of bioethanol and biobutanol by lignocellulosic materials (e.g. straw, wood and grass)	Production of bioethanol from microalgae and sea weeds

Plant residues are the most promising low-cost and sustainable biomaterials for the production of biofuels. First generation bioethanol serves as a vehicle fuel and in contrast to fossil fuels, first generation bioethanol reduces CO_2 emission. To meet the demand of first-generation bioethanol higher amount of feedstock is required which can lead to competition between fuel and food. Production of second-generation biofuels may avoid these concerns because second-generation bioethanol production depends upon nonfood bio resources like lignocellulosic material as they are inexpensive and are accessible in large quantities. For second-generation bioethanol production the most common lignocellulosic material is sugarcane bagasse. Both thermal and biological platforms are used for biofuel production from sugarcane bagasse [3].

The production of bioethanol is increasing but depletion of water resources and use of arable land make put it under the scanner. The extensive use of arable land to produce biomass for bioethanol production causes a deficiency of food crops which initiates debate for its sustainability. To eliminate this problem use of algae is gaining importance for the production of biomass for bioethanol. Algae is present in marine, fresh and terrestrial ecosystems. This suggests that algae have the ability to grow in a diverse environment. Therefore, algae as a renewable energy source will ensure energy security and self-sufficiency. As compared to land plants algae produce 5–10 times more biomass because of more photosynthetic efficiencies [11]. Lignin is absent in algae which serve as an obstacle for enzymatic hydrolysis. In ethanol production, this property of algae is helpful in pretreatment and enzymatic hydrolysis [2].

3.   Bioethanol as the fuel of future

Currently, in the world market, the most popular fuel is ethanol. For ethanol, Henry Ford used the term "fuel of future". The use of ethanol as an alternative fuel is because of several reasons, for instance

1. Ethanol can be produced by using agricultural products like molasses, sugar and corn.

2. Unlike other alcoholic fuels, ethanol is less toxic.

3. By-products produced in the result of incomplete oxidation of ethanol are less toxic.

Sugarcane is the mainly used feedstock for ethanol production in tropical regions like Brazil, India and Colombia, while corn is the main feedstock in areas like European Union, the United States and China [12]. From several decades bio-alcohol is used as fuel. Mainly the plant material with high sugar and starch contents like sugarcane and grain crops are a major source of bio-alcohol. As sugarcane and grain crops are food materials and their use results in competition for food and attention for bio-alcohol production has shifted to perennial grasses like miscanthus because they do not serve as food. This cellulosic biomass converted into sugar by fermentation and distillation processes and the sugars are then converted into alcohol [10].

Biomass is a consistent source of energy and the production of biofuel from lignocellulosic feedstock is a well-known concept. It is challenging to convert biomass into liquid fuel because with the complexity in chemical composition the conversion process becomes complicated and expensive. Depletion of oil stocks with increasing demand for energy worldwide increases the interest in the production of biofuels. Another concern in the utilization of biofuel is the environmental impact of liquid fuel consumption. Due to the emission of greenhouse gases use of biofuels is considered eco-friendly option [7].

Bioethanol is a common bioalcohol whereas biopropanol and biobutanol are not common. Microorganisms produce alcohols through the fermentation of feedstock rich in

Microbial Fuel Cells: Materials and Applications   Materials Research Forum LLC
Materials Research Foundations **46** (2019) 307-334   doi: http://dx.doi.org/10.21741/9781644900116-12

carbohydrates. Traditionally ethanol was used in the preparation of alcoholic drinks [10]. A cyanobacterium species, *Synechocystis* sp. PCC 6803 serve as cell factory for the production of fatty acids, biopolymers, ethanol and sugars [4]. Currently, ethanol has been used for other purposes, for instance, bioethanol replaces gasoline as the use of bioethanol reduces the release of CO_2 toxic gases like nitric oxide, carbon mono oxide and other volatile organic compounds. As oxygen is present in molecular structure combustion at low temperature is possible. Sources for bioethanol production are feedstocks like starch from corn, sucrose from sugar beet, sugarcane, wheat and agricultural wastes including lignocellulosic biomass like rice straw, wheat straw, wood and sugarcane bagasse (after juice extraction the remaining dry pulpy material of sugarcane stems). When bagasse is utilized sugarcane produces high quantity bioethanol. In third generation biofuel production, algae is the most important feedstock because it contains almost 50% lipid for biodiesel production and the rest of the components including sugars and proteins are used in the production of bioethanol [10].

4.   Bioethanol production

Production of bioethanol is represented by the following schematic diagram (Fig.1) Biomass when reaching the ethanol plant is stored in a warehouse and prevented from contamination to avoid early fermentation. Pretreatment of the biomass causes extraction of carbohydrates. If sugar from sugarcane juice and molasses is subjected to fermentation then processes such as pretreatment, milling, detoxification and hydrolysis are not necessary but in case of starchy materials, these processes are necessary. Processes like pretreatment, milling and hydrolysis are also required when lignocellulosic material is used. Further, detoxification is not necessary, until a toxic substance is fed to bioreactors [12].

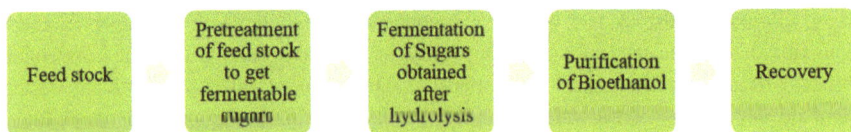

Figure 1: Schematic diagram for the production of biofuels

4.1 First generation biofuels

Production of first-generation biofuels comprises the ethanol production directly from edible sources. Ethanol is generally obtained by the fermentation of six carbon sugars i.e., glucose by the activity of classical or genetically modified yeast strains like *Saccharomyces cerevisiae*. The feedstock, mostly used for the production of first-generation bioethanol is sugarcane or corn. Brazil is one of the leading countries which utilize sugarcane for biofuel production. To produce ethanol, the sugarcane is first crushed in the presence of water in order to remove sucrose, which is then purified to get ethanol. For the production of ethanol, another major feedstock is corn which is a rich source of carbohydrate. In case of corn as feedstock initially, hydrolysis of starch is required in order to release sugars that are then fermented to ethanol. Amylase is the enzyme used in starch hydrolysis [7].

4.2 Second generation biofuels

Biomass used for second-generation biofuels is divided into three categories: homogeneous, like white wood chips, quasi-homogeneous which include agricultural and forest residues and feedstock like municipal solid wastes. The conversion of such biomass to bioethanol is somewhat complicated [7].

Trees are the interesting resource of cellulosic biomass for the production of second-generation biofuels. As trees generally have high carbohydrate contents and therefore, are more suitable for the production of biofuel than food crops. As compared to grain biomass cellulosic biomass is a renewable, cheap resource and does not compete with the food supply. Cellulose is a polysaccharide with high molecular weight and its degradation is a key step in the production of bioethanol. Degradation of cellulose could be achieved by different enzymes and degradation of cellulose is more difficult than starch because of the presence of amorphous and crystalline regions. Cellulase, the generally used cellulose degrading enzymes, includes three different enzymes: exoglucanase or β-glucosidase, cellobiase or cellobiohydrolases and endoglucanases [13, 14]. These enzymes could effectively degrade cellulose chain into soluble cello-oligosaccharides. Sugar molecules obtained from cellulose by enzymes produce lignocellulosic ethanol and reduces the emission of greenhouse gases by 90 % as compared to petroleum fuels. Low-cost lignocellulosic biomasses includes wood chips, grasses, sludge, livestock manure and crop residues that can be hydrolysed by the activity of enzymes to obtain fermentable sugars for biofuel production. The high cost and low yield of the hydrolysis process pose a negative impact for its use in commercial scale production. As compared to sugar-containing materials, feedstocks having cellulose and starch is cheaper but the high cost of cellulase limit the use of processes to convert cellulose or starch to fermentable sugars. Along with this lignin in cellulosic wastes hinders ethanol production as lignin is hydrophobic and covers the cellulose present, therefore, pretreatment is required to remove lignin which increases cost for bioethanol production. Due to these factors, second generation feedstock does not serve the purpose of commercialization, a low-cost fuel for transportation and pave the way towards third generation fuel production [15].

Lignocellulosic biomass obtained from agriculture industry consists of bagasse, coconut shells, wheat straw, rice straw, rice husks, maize cobs, cotton stalks, jute sticks etc. Lignocellulosic biomass also included a fraction of municipal solid waste or could be forestry waste like sawdust, bark and wood chips. Lignocellulosic biomass mainly has hemicellulose (20–40%), cellulose (40– 60%), and lignin (10–25%). Cellulose fibrils are resistant to digestion mediated by enzymes due to their crystalline nature. While in non-crystalline form cellulose is susceptible to enzymatic mediated hydrolysis. Along this lignin adsorb enzymes irreversibly and inhibit the action of the enzyme on cellulose chains. Therefore, certain pretreatments are required to reduce adsorption of enzyme on lignin. For efficient hydrolysis, the accessible area for enzymes should be increased and it is also important to disintegrate the crystalline structure of cellulose [16].

4.2.1 Challenges in second-generation biofuel

Removal of inhibitors and separation of cellulose with high purity without consuming too much energy is a technological challenge. Once macromolecules are isolated, they need hydrolysis to be fermented by yeasts. Hemicelluloses are highly branched carbohydrate having five-carbon sugars and six carbon sugars and make 15 to 25% of dry weight of lignocellulosic biomass. Hemicelluloses contain about 50-75% glucans and xylans. The main advantage of hemicelluloses in ethanol production is that due to its highly branched structure it can be hydrolyzed by water at high temperatures or can be hydrolysed by dilute aqueous acids. The problem with the use of hemicellulose is that classical yeast strains do not ferment five-carbon sugars and require genetic modification in order to produce ethanol. The second most plentiful natural polymer is lignin, comprises 25 to 35% of dry weight in lignocellulosic biomass. This macromolecule has high energy, used as fuel and also used in biorefinery as a source of hydrogen [7].

4.2.2 Pretreatment of lignocellulosic biomass

Pretreatment is an important step in order to enhance the formation of fermentable sugars. The choice of a pretreatment method depends upon the type of feedstock because of difference in chemical composition and physical characteristics of different feedstocks [16]. The main objectives of pre-treatment are

i.   To reduce the degree of polymerization and crystallinity of cellulose besides this pre-treatment increase permeability of lignocellulose to enhance sugars in enzymes mediated hydrolysis.

ii.   Avert sugars degradation especially pentose as well as those derived from hemicellulose.

iii.   To reduce the production of inhibitory products, which hinders the fermentation process.

iv.   To recover lignin.

v.   Minimize the use of energy to make the process cost-effective and easy in operation.

Pretreatment of lignocellulosic biomass is important in order to increase the availability of cellulose. Hemicellulose and cellulose are converted into xylose and glucose by enzymes like hemicellulases and cellulases. These sugar monomers (pentoses and hexoses) formed by saccharification are converted into bioethanol by fermentation. In order to increase the efficiency of bioethanol production, different methods have been proposed to integrate hydrolysis and fermentation processes. Methods used for the treatment of lignocellulosic mass are listed in table-2. Microorganisms used in biological pre-treatment of different lignocellulosic materials are listed in table-3.

Table 2: Pretreatment methods of lignocellulosic biomass with their advantages and disadvantages

Pretreatment	Advantages	Disadvantages	Ref.
Mechanical	Increase in digestibility and surface area of biomass, Reduction in degree of polymerization and crystallinity of cellulose	Economically inefficient due to consumption of high energy.	[16,17]
Extrusion/ Pyrolysis	Fiber shortening	Parameters should be highly efficient in bioreactor.	[16]
Liquid hot water	Better control of pH, Reduction in non-specific polysaccharides degradation, High pentose recovery, Inhibitor production is low, Materials ability to resist corrosion are not required	High demand of energy and water. Commercial scale application is not possible.	[16]

Steam explosion (Auto-hydrolysis)	Better enzymatic hydrolysis, low environmental impact, high yield of sugars, Less harmful chemicals are required.	High cost, less effective method for soft woods, Production of inhibitory products, Partial degradation of lignin and hemicelluloses, For acid addition additional equipment is required	[16,17]
Ammonia fiber expansion (AFEX)	Increases surface area available for enzymes and digestibility Removal of lignin, hemicelluloses and inhibitors	Hemicelluloses content is not significantly reduced High lignin content Expensive because large amount of ammonia is required	[16,17]
Concentrated acid	High yield of glucose, Reduced operational costs because of moderate temperature requirement, Hydrolyse both cellulose and hemicelluloses, Development of degradation products is low, No enzymes requirement	Corrosion of equipment, Formation of inhibitors.	[17]
Dilute acid	Less corrosion problems as compared to concentrated acid, Formation of inhibitors is low Hydrolyses hemicellulose Alternation in lignin structure	High temperature causes degradation products.	[17]
Alkali	Efficient removal of lignin, Improved accessibility of hemicelluloses degrading enzyme.	Downstream processing cost is high, Not appropriate for industrial scale, Long residence times, Irrecoverable salts formation	[16,17]
Organosolv	Improved digestibility of lignin and hemicelluloses by enzymes	High solvent and catalysts cost, Danger of fire and explosions.	[16,17]
Ozonolysis	No production of toxic residues, Room temperature and pressure is required to carry out reaction	Highly expensive procedure because ozone is required in large amount	[16,17]

Biological pretreatment	Low cost because low energy is required and no chemicals are required, Requirement of mild environmental conditions, Control of pH in sugar utilization	Low hydrolysate, Inefficient for industrial purposes because of slow hydrolysis rate, In large scale operation operational costs increases.	[16,17]
Ammonia recycled percolation (ARP)	Delignification is highly selective Lignin removal	Consumption of energy is high.	[17]
Ionic liquids	High digestibility, Green solvents Reduction in crystallinity of cellulose, Removes lignin	Large scale application of ionic liquids is under investigation.	[17]
Supercritical fluid technology	Cost effective, No inhibitors formation, Increase in accessibility of surface area	Have no effect lignin and hemi-cellulose High pressure is required.	[17]
Wet oxidation	Formation of inhibitors is low. Removes lignin	High cost of oxygen and alkaline catalyst	[17]

Table 3: Microorganisms used for pre-treatment of lignocellulosic biomass.

Fungus employed	Material	References
Aspergillus terreus	Sugarcane trash	[17, 18]
Aspergillus awamori		
Trichoderma reesei		
Trichoderma viride		
Ceriporiopsis subvermispora	Corn stover	[17, 19, 20]
Echinodontium taxodii	Softwood, Hardwood	[17, 22]
	Corn stover	[17, 21]
Irpex lacteus	Corn stover	[17, 21, 24]
Phanerochaetechrysosporium	Wheat straw	[17, 25]
	Rice straw	[17, 26]
	Cotton stalks	[17, 27]
Pleurotus ostreatus	Corn stover	[21]
	Rice hull	[17]
Pycnoporus cinnabarinus	Wheat straw	[17, 25]
Pycnoporus sanguineus	Cereal straw	[17, 28]
Fusarium concolor	Wheat straw	[17, 23]

4.2.3 Fermentation process

A number of microorganisms have the ability to ferment pentose and hexose sugars separately as well as simultaneously. The list of microbes and their substrate sugars is given in table-4.

Table 4: Microorganisms having ability to utilize pentoses and hexoses.

Microorganisms		Activity	References
Bacteria	*Aerobacter, Bacillus, Klebsiella, Thermoanerobacter, Aeromonas*	Successfully fermenting xylose to ethanol	[16]
Yeasts	*Pichia stipitis, Candida shehatae, Pachysolen tannophilus*		
Fungi	*Neurospora, Monilia, Fusarium, Mucor, Rhizopus*		

In yeast, glucose is metabolized by the glycolytic pathway into pyruvate. Glucose is phosphorylated to glucose-6-phosphate by hexokinases (HK), then glucose-6-phosphate is isomerized to fructose-6-phosphate by phosphoglucose isomerase (PI), fructose-6-phosphate is then phosphorylated to fructose-1,6-biphosphate by phosphofructokinase (PFK). Yeast fructose-bisphosphate aldolase (FA) reversibly converts fructose-1,6-biphosphate to dihydroxyacetone phosphate and glyceraldehydes-3-phosphate (G3P). These molecules are interconvertible and interconversion is achieved by triosephosphate isomerase (TPI). Later G3P is serially converted to different intermediates and finally into pyruvate. In the conversion of glyceraldehydes-3-phosphate to pyruvate, G3P is converted to 1,3-diphosphoglycerate by the action of 3-phosphate dehydrogenase (GDH). 1,3-diphosphoglycerate is then converted to 3-phosphoglycerate by phosphoglycerate kinase (PGK). Later 3-phosphoglycerate is changed to 2-phosphoglycerate due to the activity of phosphoglycerate mutase (PGM) which causes relocation of a phosphate group from third carbon to second carbon of molecule, then enolase, convert 2-phosphoglycerate into phosphoenolpyruvate. Then this high energy molecule is converted to pyruvate by pyruvate kinase (PK). In anaerobic conditions, a decarboxylation reaction converts pyruvate to acetaldehyde and CO_2 by the activity of pyruvate decarboxylase (PDC). Further reduction of acetaldehyde by alcohol dehydrogenase (ADH) yields ethanol. A gene in *E. coli* codes for pyruvate formate lyase despite pyruvate decarboxylase in order to convert pyruvate into ethanol [16]

4.2.4 Metabolic pathway of D-xylose

In metabolic pathway, all microorganisms, convert D-xylulose to D-xylulose-5-phosphate by the activity of xylulose kinase (XK). Once xylulose-5-phosphate is formed, it is engaged to native pentose phosphate pathway (PPP). In bacteria, D-xylose is directly converted to D-xylulose by the action of xylose isomerase (XI). In yeasts and fungi, the conversion of D-xylulose requires two-step oxidation-reduction pathway. In first step D-xylose is reduced to xylitol by the action of D-xylose reductase (XR) then oxidized to D-xylulose through xylitol dehydrogenase (XDH), convert D-xylulose to D-xylulose-5-phosphate. D-xylose-5-phosphate is further metabolized to G3P and fructose-6-phosphate by the activity of phosphopentose epimerase (PPE), transaldolase (TAL) and transketolase (TK).

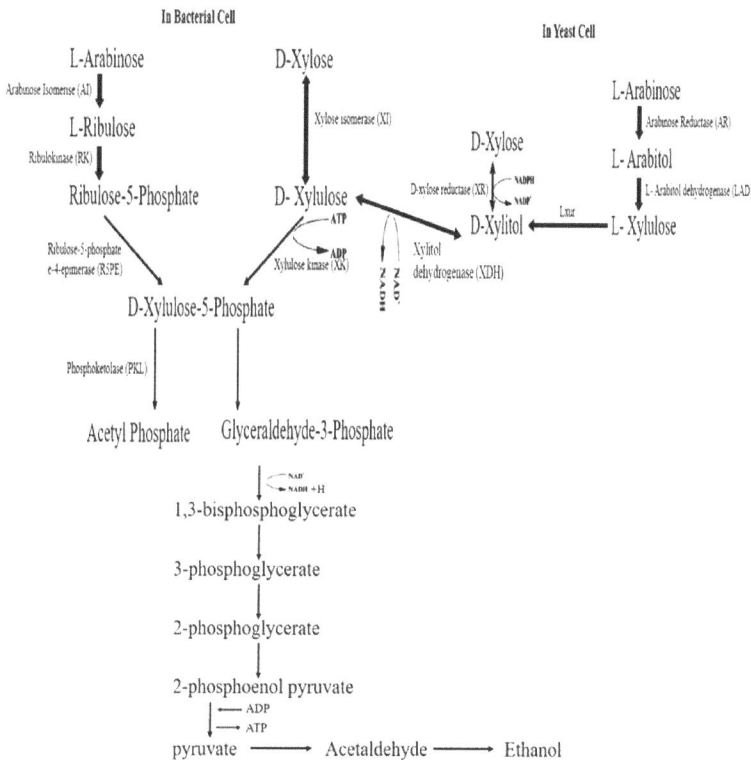

Figure 2: Metabolic pathway of bacteria and yeast to prepare ethanol from xylose and arabinose.

4.2.5 Conversion of L-arabinose in fungi

In most fungi, L-arabinose is converted to D-xylulose-5-phosphate followed by series of oxidation-reduction steps by L-arabitol dehydrogenase (LAD) and arabinose reductase (AR). In bacteria, arabinose is converted to ribulose by arabinose isomerise (AI) after being converted to ribulose-5-phosphate by the activity of ribulokinase (RK). The final epimerization of ribulose-5-phosphate to D-xylulose-5-phosphate is carried out by ribulose-5-phosphate-4-epimerase (R5PE). Xylose and arabinose are converted into ethanol by the pathway as mentioned in figure -2 [16].

4.2.6 Separate hydrolysis and fermentation (SHF)

This process has two stages in which both processes (hydrolysis and fermentation) operate separately. In the first step, the lignocellulosic biomass is pretreated with enzymes to get monomer units like glucose and xylose. In a second step, these sugars (glucose and xylose) are fermented into ethanol by the fermentation process. The advantage of SHF is that enzymatic hydrolysis and fermentation work at their optimum conditions. However, the drawback of this procedure is an accumulation of sugars which inhibits the activity of enzymes [16].

4.2.7 Simultaneous saccharification and fermentation (SSF)

In this method, a single reactor is required for combined saccharification of biomass and sugar fermentation. The advantage of SSF is that extracted sugars from biomass are quickly converted into ethanol by the enzymatic activities of microorganisms and in result prevent accumulation of sugars which have an inhibitory effect in the medium. This process requires fewer equipment as compared to the SHF process. In this process, the presence of ethanol in broth reduces the chances of contamination. The drawback associated with this procedure is the difficulty in optimization process parameters because microorganisms and enzymes are present together in the same fermentor. As cellulolytic enzymes used for enzymatic hydrolysis perform the optimum activity at 50 °C while for microbes, ethanol production has optimal conditions at 28 to 37 °C [16].

4.2.8 Simultaneous saccharification and co-fermentation (SSCF)

This process involves the assimilation of all types of sugars by microorganisms which were released by the pretreatment processes of lignocellulosic biomass. Both pentoses and hexoses can be assimilated by mixed cultures of yeasts. The problem associated with this process is more conversion of hexoses to ethanol because microorganisms which utilize hexoses grow at a faster rate as compared to those microorganisms which utilize pentoses. Another variation is to exploit a single microorganism which has the ability to assimilate hexoses and pentoses to produce ethanol. Shorter operation time, low cost, low risk of contamination and fewer inhibitory effects are the advantages of this process [16].

4.2.9 Consolidated bioprocessing (CBP)

Consolidated bioprocessing (CBP) incorporates all reactions which are required for biomass transformation into ethanol. In this process, a single microbial community carries enzymatic hydrolysis of biomass and fermentation of sugars in a single step. This advantage of the CBP process includes no operational costs for enzyme production [16].

4.3   Third generation biofuels

Third generation biofuels are biofuels obtained from algal biomass [29]. Biofuels refer the production of energy-rich chemicals directly from biological procedures or derived from the chemical conversion of biomass by organisms, mainly photosynthetic organisms (vascular land plants, photosynthetic micro-algae, photosynthetic macro-algae and photosynthetic bacteria). From several decades plant biomass serve as a source of biofuels but currently, the algal biomass is considered as an encouraging source for production of biofuel. Photosynthesis is a distinguishing feature of plants and algae. As photosynthesis is a way for the accumulation of biomass. Biomass obtained from plants and algae serve as raw material for the production of biodiesel, bio-alcohol and bio-hydrogen derived through fermentation [10].

As compared to lignocellulosic biomass algae has distinctive growth yield. Biofuels production from algae usually depends on lipid content, with the use of algal biomass. Algae belong to the group of photosynthetic organisms, can be classified as macroalgae (multicellular) and microalgae (unicellular), in biofuel research microalgae is very important. Usually, microalgae float on the surface of the water and like seaweeds it normally found attached to rocks [29]. Algae have the potential to produce petroleum fuels like triterpenic hydrocarbons, isobutyl alcohol, bioethanol and iso-butyraldehyde. Several bacterial species with ethanol production ability has been identified, along with this genetically modified bacterial species like *Escherichia coli* and *Bacillus subtilis* also have the ability to produce high amounts of derivatives of fatty acids, isoprenoids and bioalcohol. *Thermococcus, Thermotoga, Caldicellulosiruptor* and *Pyrococcus* species have the ability to produce hydrogen in higher amount while these species produce ethanol in less quantity. Ethanol could be efficiently produced by co-culture of *Thermoanaerobacter* species along with cellulolytic organisms. For efficient production of ethanol, a well-known model organism is *Saccharomyces cerevisiae* [10].

4.3.1 Microalgae

Current knowledge suggests that for biofuels, microalgae are an attractive feedstock. Depending on cultivation conditions and species, microalgae have the ability to produce biomethanol, bioethanol, biodiesel, biohydrogen, proteins, carbohydrates and other compounds [10]. Green algae (chlorophyceae), dinoflagellates, diatoms (bacillariophyceae) and golden algae (chryosophyceae) are different types of microalgae. In different species of algae the carbohydrate, lipid and protein contents vary in most of the microalgae the lipid content exceeds 70% by dry weight of biomass [5, 30]. Some species like *Chlorella, Scenedesmus*, and *Chlamydomona* have carbohydrate contents up to 50% of dry weight [31,32]. Factors like O_2, CO_2 nutrient contents, salinity, pH, light, temperature and chemicals have an effect on carbohydrate and lipid contents of microalgae. In the cell wall of microalgae, the common components are pectin, cellulose, hemicelluloses, protein and carbohydrates. By acid or enzymatic hydrolysis, these components can be converted into monomers in order to produce bioethanol.

4.3.2 Macroalgae (seaweed)

Macroalgae are classified into three main groups which are brown (Phaeophyceae), red (Rhodophyceae) and green (Chlorophyceae) [29, 33]. Usually, the cell wall of seaweeds is a matrix of linear sulphated galactan polymers. Seaweeds are the most promising feedstock as their conversion to bioethanol is easy. Brown seaweeds have carbohydrates like fucoidan, laminaran, alginates, mannitol and cellulose. The red seaweed has polysaccharides like a carrageenan cell wall, xylene, agar, mannan and cellulose in the cell wall. Polysaccharides like carrageenan and cellulose in red seaweed and green seaweed have cellulose, xylene and mannose. In comparison, red seaweed has the highest carbohydrate content as compared to other seaweeds. As carbohydrates are present in seaweeds its conversion into bioethanol is important to consider [29].

4.3.3 Algae as a better option for bioethanol production

A variety of algal strains have different carbohydrate contents are used for the production of biofuels (table-5) and there are several characteristics because of them algae is in use for bioethanol production. These characteristics include

i.   Algae are not in competition with food products in land or in water.

ii. The carbohydrate contents in algae are abundant which can be easily fermented to bioethanol.

iii.   Algae have a low level of hemicellulose and have no lignin. Which increases the efficiency of hydrolysis and fermentation yields. Due to these characteristics, the cost for production of bioethanol reduces.

iv.   Algae have the ability to utilize CO_2 from environment and results in the reduction of greenhouse gases.

v.   Algae have rapid growth and it is easier to grow them easily in different aquatic environments like municipal wastewater, fresh water or saline water.

vi.   The microalgae have rapid production and harvesting cycle (1–10 days) as compared to other feedstocks which are harvested once or twice a year due to rapid production and harvesting cycle, algae provide sufficient supplies in order to meet the demand for ethanol [33].

Table 5: Types of algae and carbohydrate contents present in them.

Algal Biomass	Carbohydrate	Reference
Microalgae		
Chlorococcum infusionum	32	[33, 34]
Porphyridium cruentum	40-57	[31,33]
Chlamydomonas reinhardtii	59.7	[33, 35]
Macroalgae		
Brown sea weed		
Undaria pinnatifida	48.5	[33, 36]
Sargassum spp.	41.81	[37]
Red sea weed		
Kappaphycus alvarezii	64	[33, 38]
Gelidium amansii	67.3	[33, 39]
Green sea weed		
Ulva lactuca	54.3	[33, 40]

4.3.4 Challenges with third generation biofuel

Both microalgae and macroalgae are used for the production of bioethanol (Table-6) and a number of challenges including geographical and technical challenges are associated with the production of third-generation biofuels. Typically, a large quantity of water is required for industrial-scale production, the requirement of large volumes of water serve as a problem for countries like Canada where during a significant part of year temperature is below 0 °C. The high water content also serve as a problem when lipid is extracted from algal biomass. The extraction of lipid from algal biomass requires dewatering, which is achieved by either filtration or by centrifugation [7].

Table 6: Algal species used for bioethanol production [29].

Types of algae	Species Used in bioethanol production
Micro algae	*Chlamydomonas reinhardtii* UTEX 90, *Chlorococcum infusionum, Chlorella vulgaris*
Macro algae	**Green:** *Ulvalactuca, Ulvapertusa* **Red:** *Kapphaphycus alvarezii, Gelidiumamansii, Gracilariasalicornia, Gelidiumele- gans,* **Brown:** *Sargassum fulvellum, Laminaria hyperborean, Laminaria japonica, Saccharina latissima, Undari apinnitifida, Alaria crassifolia*

4.3.5 General processes of bioethanol production from algal feedstock

In the production of bioethanol from the algal feedstock, drying is the first step in the handling of algae obtained from water. To reduce the size of feedstock is an important aspect in order to increase the surface area for further analysis. Usually, the powder and slurry of algae are used in hydrolysis followed by fermentation.

4.3.5.1 Hydrolysis

In bioethanol production hydrolysis is a crucial step because, in algae, the cell wall is the structure that needs to be depolymerized to get polysaccharides like carrageenans, alginates, agarans, fucans and ulvans. Polysaccharides are then converted into monomers to be fermented to get bioethanol. For hydrolysis of polysaccharides into monomers, two approaches are in use one approach is chemical hydrolysis and the second approach is enzymatic hydrolysis. Among these two approaches, enzymatic hydrolysis is a recent approach.

4.3.5.1.1 Acid hydrolysis

The release of simple compounds from complex polysaccharides is achieved by chemicals. In this process, a wide range of acids are in use but sulfuric acid (H_2SO_4) is the most important one. It was found that the dilute H_2SO_4 can effectively hydrolyze the polysaccharides from three classes of macroalgae. Initially, in the hydrolysis of polysaccharide, destruction of hydrogen bonds occurs which ruptures the polysaccharide chain and changes into an amorphous state. Then acid causes hydrolysis of glycosidic bonds and in result causes cleavage of the polysaccharide. At the end dilution with water causes complete hydrolysis of hydrolysate into monosaccharide at moderate temperature [41].

4.3.5.1.2 Enzymatic hydrolysis

Algae can be treated both chemically and enzymatically (table-7) to produce sugars which are then converted to bioethanol. In bioethanol production saccharification (conversion of complex carbohydrates into simple monomers) is the most important step. Enzymatic and chemical hydrolysis has been reported to convert algal biomass to sugars. Enzymatic hydrolysis as compared to acid hydrolysis requires mild conditions (pH 4.5–5.0 and 40–50 °C) and less energy. Enzymatic hydrolysis does not have corrosion problem which is a major problem in acid hydrolysis. The main advantage of enzymatic hydrolysis is low toxicity of the hydrolysates. The concentrations of sugars in enzymatic hydrolysis are higher than chemical hydrolysis [33].

Table 7: Comparison between chemical and enzymatic hydrolysis of different types of algae [33]

Chemical Hydrolysis			Enzymatic Hydrolysis	
Algal Source	**Conditions**	**Sugar yield**	**Enzyme mixture**	**Sugar yield**
Ulva pertusa Kjellman	150 °C and 15 MPa for 15 min	Glucose 9.08%	Cellulase and amylo-glucosidase	Glucose 61.1%
Gracilaria salicornia	120 °C and 2% H_2SO_4 for 30 min	4.3 g of glucose/kg	Cellulase	13.8 g of glucose/kg
Gelidium amansii	0.2N H_2SO_4 at 121 °C for 15 min	Sugar content obtained 9.8%	Endo-glucanase and β-glucanase	56.6% Sugar content obtained
Saccharina japonica	40 mM H_2SO_4 at 121 °C for 1h	Production of reducing sugar 13.5 g/L	Termamyl (Industrial enzyme)	20.6 ± 1.9 g/L Reducing sugars
Dunaliella tertiolecta	0.5 M HCl at 121 °C for 15 min	29.5% Production of reducing sugar	Amyloglucosidase	42% reducing sugars obtained

The physical structure and interaction of feedstock with enzymes is an important factor. In enzymatic hydrolysis, the formation of the enzyme-substrate complex is a critical step. The major barrier in enzymatic hydrolysis of cellulosic biomass is the surface accessibility of cellulose. During hydrolysis, the accessibility of enzyme to substrate is by pores present in the cell wall. As this factor is the main contributor to effective hydrolysis. In enzymatic hydrolysis, a class of enzymes called cellulases have a role in the hydrolysis of cellulose. Cellulase is mostly used to degrade polysaccharides and classified into three types like exo-glucanases, endoglucanases and β-glucosidase. In feedstock endo-glucanases hydrolyze the complex sugars by attacking the interior amorphous region of cellulose and exo-glucanases, degrades cellulose by cleaving cellobiose units from the non-reducing end of cellulose fiber. Cellobiose residues finally split into two units of glucose by β-glucosidase [29].

Fungi and bacteria can produce cellulases. For bioethanol production in enzymatic hydrolysis, cellulase causes degradation of cellulose to produce sugars that can be fermented by bacteria or yeasts. In algal biomass starch is present in large quantity therefore, use of amylase can hydrolyze algal biomass. Different factors like substrate concentration, enzyme concentration, pH and temperature are the main factors which influence the yield of sugar monomers. For cellulose, the optimum pH is 4.5–5 [33]. Conversion of complex sugars into simple sugars by enzymatic hydrolysis reduces environmental impact and the ability of enzymes to achieve 80% conversion make this approach attractive for use in bioethanol production [29].

4.3.5.2 Biological pretreatment

Biological treatment is an environmentally friendly method to remove lignin from lingo-cellulosic biomass. As in algae lignin is present in low quantity therefore rarely biological methods are adopted for algal pretreatment. Matsumoto *et al.*, in 2003 saccharify green marine microalga by a bacterium *Pseudoalterimonas undina* NKMB 0074 having ability to produce amylase [42]. In saline suspension the terrestrial glucoamylase and amylase were inactive. While marine amylases in saline conditions are successful for saccharification of microalgae [33].

4.3.5.3 Fermentation

In hydrolysis by the activity of microorganisms the simple sugar molecules released, these molecules can be converted into bioethanol (Fig.3). The main product of fermentation is bioethanol while H_2O and CO_2 are the byproducts. In fermentation yeast, bacteria and fungi are in use. *Saccharomyces cerevisiae* is a frequently used strain of yeast for fermentation of bioethanol because of certain features like its high selectivity, high yield of ethanol, high rate of fermentation and low accumulation of by-products [29].

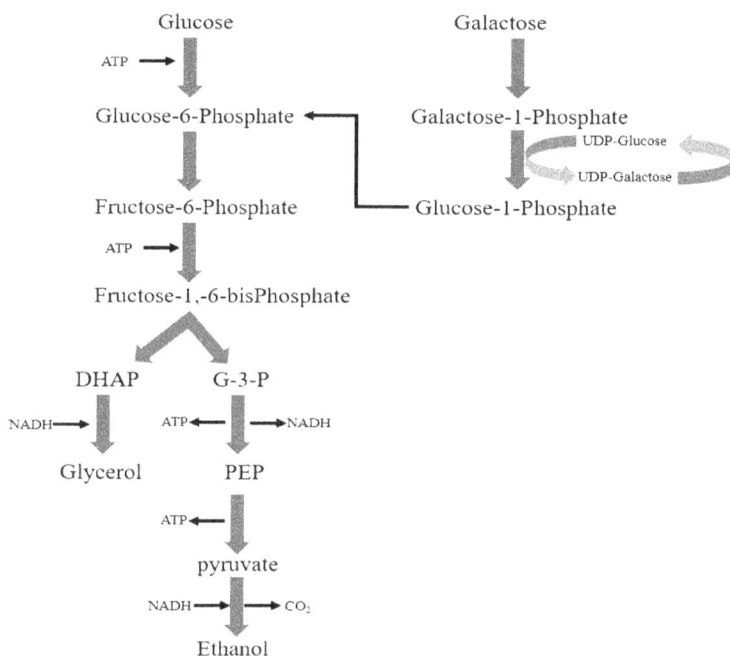

Figure 3: Pathway to prepare bioethanol from glucose and galactose

Fermentation using microorganisms utilize mannose, galactose, glucose, xylose, and arabinose from algal biomass obtained by enzymatic hydrolysis. For ethanol production, the best-known microorganisms are the bacterium *Zymomonas mobilis* and yeast *Saccharomyces cerevisiae* [33, 43]. *Z. mobilis* have the ability to bear high sugar concentrations when compared with *S. cerevisiae* [44]. The lysate of *L. japonica* obtained by acid hydrolysis was subjected to hydrolytic enzymes and fermentation by *Escherichia coli* KO11 and 0.4 g ethanol per gram of sugars was obtained. This yield was obtained due to the ability of *Escherichia coli* KO11 to use mannitol, present in high concentration in hydrolysate [33]. However, when a marine macroalgae named *G. salicornia* was fermented by *Escherichia coli* KO11, the yield of ethanol was low. The low yield of ethanol may be due to mono sugars present in invasive feedstock which were not willingly fermented by *Escherichia coli* KO11.

Surface aeration was developed by researchers in order to improve bioethanol production from hydrolysate having five carbon monosaccharides like xylose. The process of fermentation and enzymatic hydrolysis can be skilled by using three different strategies [33]

    i.   Separate hydrolysis and fermentation (SHF),

    ii.   Simultaneous saccharification and fermentation (SSF),

    iii.   Simultaneous saccharification and co-fermentation (SSCF).

4.3.5.3.1 Separate enzymatic hydrolysis and fermentation

In this process, the hydrolysis of algal biomass to glucose and fermentation of sugars obtained from hydrolysis occurs in separate units. The benefits of this process are short residence time, low cost of chemicals and simple equipment system, which encourage its large-scale application [33].

4.3.5.3.2 Simultaneous saccharification and fermentation

This method combines the process of saccharification and fermentation. In this procedure, microorganisms and enzymes remain in the same processing unit, the fermenting microorganism immediately utilizes sugars like glucose. SSF produces higher ethanol yield as compared to SHF. For efficient production of ethanol, SSF is a good option because it reduces the number of unit operations [33].

4.3.5.3.3 Simultaneous saccharification and co-fermentation

Variety of bacteria and yeast ferment hexose, while most of them cannot easily ferment pentoses, which serve as limiting factor for obtaining ethanol. Genetic engineering is in use to develop microorganisms having the ability to produce ethanol by the fermentation of five and six carbon sugars simultaneously. SSCF involves hydrolysis and co-fermentation of hexoses and pentoses in one vessel [33].

4.3.5.4 Purification

In bioethanol production, the purification step involves techniques like distillation, rectification and dehydration [29].

Table 8: List of genetically modified microorganisms used in ethanol production [17].

Microorganism	Strain	Features
Yeast	*Candida shehatae* NCL-3501	Co-ferment xylose and glucose
	Saccharomyces cerevisiae D5a	Improvement in yield of ethanol
	Saccharomyces cerevisiae 590E1	Ferment glucose and cellobiase
	Saccharomyces cerevisiae RWB217	Ferment glucose and xylose
	Saccharomyces cerevisiae RWB218	
Bacteria	*Zymomonas mobilis* ZM4	Ferment xylose and glucose
	Pichia stipitis BCC15191	
	Zymomonas mobilis AX101	Ferment arabinose, glucose and xylose.
	Thermoanaerobacterium saccharolyticum ALK2	Improved ethanol yield, have ablility to ferment arabinose, glucose, xylose and mannose.
	Thermoanaerobacter mathranii BG1L1	Improved ethanol yield
	Clostridium thermocellum DSM1313 & YD01	
	Escherichia coli KO11	Ferment xylose and glucose
	Escherichia coli FBR5	Ferment xylose and arabinose
	Pichia stipitis A	Adapted at hydrolysate increased concentration
	Pichia stipitis NRRL Y-7124	

5. Recent advances in bioethanol production

There are certain microorganisms which have been modified to enhance the production of ethanol as well as tolerance of ethanol. Instead of this, there is no single commercially available consolidated bioprocessing (CBP) organism reported. For CBP microbial consortium could be used instead of using a single microbe. A consortium consists of an enzyme producing strain having the ability to hydrolyze the available biomass and two other different strains which have potential to ferment five carbon and six carbon sugars into ethanol. Brethauer and Studer (2014) proposed a model by utilizing *Trichoderma reesei, Saccharomyces cerevisiae* and *Scheffersomyces stipitis. Trichoderma reesei*, which secrets enzyme for hydrolysis; *Saccharomyces cerevisiae* has the ability to ferment hexoses while *Scheffersomyces stipitis* have the ability to use pentose sugars. In a biofilm membrane reactor, all these microbes convert ligno-cellulosic biomass to ethanol. This approach seems to be a reasonable one but the major obstacle in this approach is to control the consortium and it is also difficult to find microorganisms having identical fermentation conditions [16].

5.1 Molecular toolings and genetically modified organisms for bioethanol production

In recent years, engineering of organisms has been practised to obtain biofuels from the lingo-cellulosic feedstock. Different bacterial species such as *Saccharolyticum, Escherichia coli, Clostridium thermocellum, Thermoanaero bacterium, Caldicellulosiruptor bescii, C. phytofermentans, Zymomonas mobilis, C. cellulolyticum*; fungal species like *Saccharomyces cerevisiae, Aspergillus oryzae, Trichoderma reesei , Paecilomyces variotii, Fusarium oxysporum*; and yeast like *Clavispora, Kluyveromyces Marxianus and Pichia stipites* have been studied for their application in production of bioethanol [16]. A list of genetically modified organisms to produce ethanol is given in table-8.

Conclusions and Future Perspective

Although the biomass for bioethanol production is renewable, abundant and cost-effective the production of bioethanol from second-generation feedstock is costly due to high recalcitrance of lignocellulosic materials and the development of sustainable society is seriously hindered by environmental pollution and energy crisis. For renewable biofuel production, the algal biomass is unquestionably an eco-friendly source. The commercialization of bioethanol from algae depends upon the economics of the procedure. There are a number of obstacles in bioethanol production from the algal resource, which must be overcome in order to commercialize the bioethanol. In the production of bioethanol, the pretreatment of algae is a costly process. In order to avoid the expenses, the pretreatment must be simple and should avoid high demand for energy.

Furthermore, the algal biomass should be hydrolyzed in a way that fermentation inhibitors produce in less amount. Currently, among different methods, the use of dilute sulfuric acid is most effective. Instead of this, there is a need to further reduce the cost of pretreatment. It is also important to minimize sugar losses in bioethanol production. In pretreatment methodologies, an economically efficient method is biological treatment. One of the fruitful examples is to consume microalgae for bioethanol production after extraction of lipid because after lipid extraction in remaining biomass high concentrations of carbohydrates is present. After ethanol production, the biomass still has leftover residue may contain a good amount of useful minerals and organic matter that could be used as biofertilizer.

References

[1]   L. Brennan, P. Owende, Biofuels from microalgae—a review of technologies for production, processing, and extractions of biofuels and co-products, Renew. Sustain. Energy Rev. 14 (2010) 557–577. https://doi.org/10.1016/j.rser.2009.10.009

[2]   R. Bibi, Z. Ahmad, M. Imran, S. Hussain, A. Ditta, S. Mahmood, A. Khalid, Algal bioethanol production technology: a trend towards sustainable development, Renew. Sustain. Energy Rev. 71 (2017) 976–985. https://doi.org/10.1016/j.rser.2016.12.126

[3]   T.L. Bezerra, A.J. Ragauskas, A review of sugarcane bagasse for second-generation bioethanol and biopower production, Biofuels, Bioprod. Biorefining. 10 (2016) 634–647. https://doi.org/10.1002/bbb.1662

[4]   T.L. da Silva, P.C. Passarinho, R. Galriça, A. Zenóglio, P. Armshaw, J.T. Pembroke, C. Sheahan, A. Reis, F. Gírio, Evaluation of the ethanol tolerance for wild and mutant Synechocystis strains by flow cytometry, Biotechnol. Reports. 17 (2018) 137–147. https://doi.org/10.1016/j.btre.2018.02.005

[5]   T.M. Mata, A.A. Martins, N.S. Caetano, Microalgae for biodiesel production and other applications: a review, Renew. Sustain. Energy Rev. 14 (2010) 217–232. https://doi.org/10.1016/j.rser.2009.07.020

[6]   C. Weber, A. Farwick, F. Benisch, D. Brat, H. Dietz, T. Subtil, E. Boles, Trends and challenges in the microbial production of lignocellulosic bioalcohol fuels, Appl. Microbiol. Biotechnol. 87 (2010) 1303–1315. https://doi.org/10.1007/s00253-010-2707-z

[7]   R.A. Lee, J.-M. Lavoie, From first-to third-generation biofuels: Challenges of producing a commodity from a biomass of increasing complexity, Anim. Front. 3 (2013) 6–11. https://doi.org/10.2527/af.2013-0010

[8]   E.-M. Aro, From first generation biofuels to advanced solar biofuels, Ambio. 45 (2016) 24–31. https://doi.org/10.1007/s13280-015-0730-0

[9]   H.S. Toogood, N.S. Scrutton, Retooling microorganisms for the fermentative production of alcohols, Curr. Opin. Biotechnol. 50 (2018) 1–10. https://doi.org/10.1016/j.copbio.2017.08.010

[10]   M. V Rodionova, R.S. Poudyal, I. Tiwari, R.A. Voloshin, S.K. Zharmukhamedov, H.G. Nam, B.K. Zayadan, B.D. Bruce, H.J.M. Hou, S.I. Allakhverdiev, Biofuel production: challenges and opportunities, Int. J. Hyd. Energy. 42 (2017) 8450–8461. https://doi.org/10.1016/j.ijhydene.2016.11.125

[11]   C.-Y. Chen, X.-Q. Zhao, H.-W. Yen, S.-H. Ho, C.-L. Cheng, D.-J. Lee, F.-W. Bai, J.-S. Chang, Microalgae-based carbohydrates for biofuel production, Biochem. Eng. J. 78 (2013) 1–10. https://doi.org/10.1016/j.bej.2013.03.006

[12]   M. Vohra, J. Manwar, R. Manmode, S. Padgilwar, S. Patil, Bioethanol production: feedstock and current technologies, J. Environ. Chem. Eng. 2 (2014) 573–584. https://doi.org/10.1016/j.jece.2013.10.013

[13]   H. Sakuragi, K. Kuroda, M. Ueda, Molecular breeding of advanced microorganisms for biofuel production, Biomed Res. Int. 2011 (2011).

[14]   C.-C. Fu, T.-C. Hung, J.-Y. Chen, C.-H. Su, W.-T. Wu, Hydrolysis of microalgae cell walls for production of reducing sugar and lipid extraction, Bioresour. Technol. 101 (2010) 8750–8754. https://doi.org/10.1016/j.biortech.2010.06.100

[15]   A.R. Sirajunnisa, D. Surendhiran, Algae–A quintessential and positive resource of bioethanol production: A comprehensive review, Renew. Sustain. Energy Rev. 66 (2016) 248–267. https://doi.org/10.1016/j.rser.2016.07.024

[16]   M. Rastogi, S. Shrivastava, Recent advances in second generation bioethanol production: An insight to pretreatment, saccharification and fermentation processes, Renew. Sustain. Energy Rev. 80 (2017) 330–340. https://doi.org/10.1016/j.rser.2017.05.225

[17]   H.B. Aditiya, T.M.I. Mahlia, W.T. Chong, H. Nur, A.H. Sebayang, Second generation bioethanol production: A critical review, Renew. Sustain. Energy Rev. 66 (2016) 631–653. https://doi.org/10.1016/j.rser.2016.07.015

[18]   P. Singh, A. Suman, P. Tiwari, N. Arya, A. Gaur, A.K. Shrivastava, Biological pretreatment of sugarcane trash for its conversion to fermentable sugars, World J. Microbiol. Biotechnol. 24 (2008) 667–673. https://doi.org/10.1007/s11274-007-9522-4

[19]   C. Wan, Y. Li, Microbial pretreatment of corn stover with Ceriporiopsis subvermispora for enzymatic hydrolysis and ethanol production, Bioresour. Technol. 101 (2010) 6398–6403. https://doi.org/10.1016/j.biortech.2010.03.070

[20]   C. Wan, Y. Li, Microbial delignification of corn stover by Ceriporiopsis subvermispora for improving cellulose digestibility, Enzyme Microb. Technol. 47 (2010) 31–36. https://doi.org/10.1016/j.enzmictec.2010.04.001

[21]   H. Yu, G. Guo, X. Zhang, K. Yan, C. Xu, The effect of biological pretreatment with the selective white-rot fungus Echinodontium taxodii on enzymatic hydrolysis of softwoods and hardwoods, Bioresour. Technol. 100 (2009) 5170–5175. https://doi.org/10.1016/j.biortech.2009.05.049

[22]   X. Yang, Y. Zeng, F. Ma, X. Zhang, H. Yu, Effect of biopretreatment on thermogravimetric and chemical characteristics of corn stover by different white-rot fungi, Bioresour. Technol. 101 (2010) 5475–5479. https://doi.org/10.1016/j.biortech.2010.01.129

[23] X. Yang, F. Ma, Y. Zeng, H. Yu, C. Xu, X. Zhang, Structure alteration of lignin in corn stover degraded by white-rot fungus Irpex lacteus CD2, Int. Biodeterior. Biodegradation. 64 (2010) 119–123. https://doi.org/10.1016/j.ibiod.2009.12.001

[24] S. Kuhar, L.M. Nair, R.C. Kuhad, Pretreatment of lignocellulosic material with fungi capable of higher lignin degradation and lower carbohydrate degradation improves substrate acid hydrolysis and the eventual conversion to ethanol, Can. J. Microbiol. 54 (2008) 305–313. https://doi.org/10.1139/W08-003

[25] J.S. Bak, J.K. Ko, I. Choi, Y. Park, J. Seo, K.H. Kim, Fungal pretreatment of lignocellulose by Phanerochaete chrysosporium to produce ethanol from rice straw, Biotechnol. Bioeng. 104 (2009) 471–482. https://doi.org/10.1002/bit.22423

[26] J. Shi, M.S. Chinn, R.R. Sharma-Shivappa, Microbial pretreatment of cotton stalks by solid state cultivation of Phanerochaete chrysosporium, Bioresour. Technol. 99 (2008) 6556–6564. https://doi.org/10.1016/j.biortech.2007.11.069

[27] C. Lu, H. Wang, Y. Luo, L. Guo, An efficient system for pre-delignification of gramineous biofuel feedstock in vitro: Application of a laccase from Pycnoporus sanguineus H275, Process Biochem. 45 (2010) 1141–1147. https://doi.org/10.1016/j.procbio.2010.04.010

[28] L. Li, X. Li, W. Tang, J. Zhao, Y. Qu, Screening of a fungus capable of powerful and selective delignification on wheat straw, Lett. Appl. Microbiol. 47 (2008) 415–420. https://doi.org/10.1111/j.1472-765X.2008.02447.x

[29] S.A. Jambo, R. Abdulla, S.H.M. Azhar, H. Marbawi, J.A. Gansau, P. Ravindra, A review on third generation bioethanol feedstock, Renew. Sustain. Energy Rev. 65 (2016) 756–769. https://doi.org/10.1016/j.rser.2016.07.064

[30] M. Koller, A. Salerno, P. Tuffner, M. Koinigg, H. Böchzelt, S. Schober, S. Pieber, H. Schnitzer, M. Mittelbach, G. Braunegg, Characteristics and potential of micro algal cultivation strategies: a review, J. Clean. Prod. 37 (2012) 377–388. https://doi.org/10.1016/j.jclepro.2012.07.044

[31] J. Singh, S. Gu, Commercialization potential of microalgae for biofuels production, Renew. Sustain. Energy Rev. 14 (2010) 2596–2610. https://doi.org/10.1016/j.rser.2010.06.014

[32] S.-H. Ho, X. Ye, T. Hasunuma, J.-S. Chang, A. Kondo, Perspectives on engineering strategies for improving biofuel production from microalgae—a critical review, Biotechnol. Adv. 32 (2014) 1448–1459. https://doi.org/10.1016/j.biotechadv.2014.09.002

[33] K. Li, S. Liu, X. Liu, An overview of algae bioethanol production, Int. J. Energy Res. 38 (2014) 965–977. https://doi.org/10.1002/er.3164

[34]   R. Harun, W.S.Y. Jason, T. Cherrington, M.K. Danquah, Exploring alkaline pre-
       treatment of microalgal biomass for bioethanol production, Appl. Energy. 88
       (2011) 3464–3467. https://doi.org/10.1016/j.apenergy.2010.10.048

[35]   S.P. Choi, M.T. Nguyen, S.J. Sim, Enzymatic pretreatment of Chlamydomonas
       reinhardtii biomass for ethanol production, Bioresour. Technol. 101 (2010) 5330–
       5336. https://doi.org/10.1016/j.biortech.2010.02.026

[36]   Y. Cho, H. Kim, S.-K. Kim, Bioethanol production from brown seaweed, Undaria
       pinnatifida, using NaCl acclimated yeast, Bioprocess Biosyst. Eng. 36 (2013) 713–
       719. https://doi.org/10.1007/s00449-013-0895-5

[37]   M.G. Borines, R.L. de Leon, J.L. Cuello, Bioethanol production from the
       macroalgae Sargassum spp., Bioresour. Technol. 138 (2013) 22–29.
       https://doi.org/10.1016/j.biortech.2013.03.108

[38]   M.D.N. Meinita, J.-Y. Kang, G.-T. Jeong, H.M. Koo, S.M. Park, Y.-K. Hong,
       Bioethanol production from the acid hydrolysate of the carrageenophyte
       Kappaphycus alvarezii (cottonii), J. Appl. Phycol. 24 (2012) 857–862.
       https://doi.org/10.1007/s10811-011-9705-0

[39]   J.-H. Park, J.-Y. Hong, H.C. Jang, S.G. Oh, S.-H. Kim, J.-J. Yoon, Y.J. Kim, Use
       of Gelidium amansii as a promising resource for bioethanol: a practical approach
       for continuous dilute-acid hydrolysis and fermentation, Bioresour. Technol. 108
       (2012) 83–88. https://doi.org/10.1016/j.biortech.2011.12.065

[40]   N.-J. Kim, H. Li, K. Jung, H.N. Chang, P.C. Lee, Ethanol production from marine
       algal hydrolysates using Escherichia coli KO11, Bioresour. Technol. 102 (2011)
       7466–7469. https://doi.org/10.1016/j.biortech.2011.04.071

[41]   P. Binod, K.U. Janu, R. Sindhu, A. Pandey, Hydrolysis of lignocellulosic biomass
       for bioethanol production, in: Biofuels, Elsevier, 2011: pp. 229–250.
       https://doi.org/10.1016/B978-0-12-385099-7.00010-3

[42]   M. Matsumoto, H. Yokouchi, N. Suzuki, H. Ohata, T. Matsunaga, Saccharification
       of marine microalgae using marine bacteria for ethanol production, Appl.
       Biochem. Biotechnol. 105 (2003) 247–254. https://doi.org/10.1385/ABAB:105:1-
       3:247

[43]   F. Talebnia, D. Karakashev, I. Angelidaki, Production of bioethanol from wheat
       straw: an overview on pretreatment, hydrolysis and fermentation, Bioresour.
       Technol. 101 (2010) 4744–4753. https://doi.org/10.1016/j.biortech.2009.11.080

[44]   M.S. Sulfahri, E. Sunarto, M.Y. Irvansyah, R.S. Utami, S. Mangkoedihardjo,
       Ethanol production from algae Spirogyra with fermentation by Zymomonas
       mobilis and Saccharomyces cerevisiae, J Basic Appl Sci Res. 1 (2011) 589–593.

Microbial Fuel Cells: Materials and Applications   Materials Research Forum LLC
Materials Research Foundations **46** (2019) 335-353   doi: http://dx.doi.org/10.21741/9781644900116-13

Chapter 13

Microbial Production of Propanol

Mehmet Gülcan[1], Fulya Gülbağça[2], Kubra Sevval Cevik[2], Remziye Kartop[2], Fatih Şen[2]

[1]Department of Chemistry, Faculty of Science, University of Van Yüzüncü Yıl, 65080, Tuşba, Van, Turkey

[2]Sen Research Group, Department of Biochemistry, Faculty of Science, University of Dumlupınar, 43100, Kütahya, Turkey

mehmetgulcan65@gmail.com, fatihsen1980@gmail.com

Abstract

Propanol can be present in two forms, 1-propanol and 2-propanol, and can be used as biofuel. Especially, the reduction of fossil resources required for renewable energy has paid attention for the inorganic synthesis of propanol. In this regard, the microbial production of propanol is very important. This chapter examines the latest developments in microbial production of propanol. Normally, various synthesis ways of propanol have been developed. However, it has been indicated that the optimization of fermentation conditions has enhanced the prevention of toxic propanol accumulation. For this reason, the low cost and high-efficiency propanol production with the help of biosynthetic pathways from microorganisms is one of the major challenges in microbial fuel cells.

Keywords

Biofuel Cells, Enzymes, Graphene Composites, Bioenergy, Propanol

Contents

1. Introduction

Propanol, which is present in two forms, 1-propanol and 2-propanol, is used as chemical fuels for renewable energy systems [1]. The reduction of fossil resources required for renewable energy has been the focus of interest for the inorganic synthesis of propanol. In this regard, this energy problem can be solved with the use of bio-derived propanol as bio-fuel, or with propylene which can be produced chemically by propanol [2–7]. Propanol preference in the development of biofuels in the literature is still under investigation. This molecule, which provides high octane number of electrons, has less energy density compared to ethanol, but its abrasiveness is lower. It has a quite smaller volume compared to alternative biofuels such as butanol or biodiesel [8]. These reasons suggest that pure propanol cannot be used as the ideal biofuel, but the development of microbial production of propanol is an essential development for the economy [8]. Therefore, in recent years, the idea of developing microbial production of propanol has increased steadily. The processing conditions were optimized with the production organisms, and the propanol yield was improved. The engineering of various metabolic pathways has enabled the use of new raw materials.

2. Biosynthesis of Propanol

For his purpose, the preferred reaction in the early 20th century was acetone-butanol-ethanol (ABE) fermentation of Clostridia species. This fermentation provided the production of ammunition during the First World War as a source of acetone. Then, butanol was used as a solvent in the automobile industry [9]. With the development of the industry, butanol was considered to be one of the most efficient biofuels. Butanol, acetone, and ethanol mixtures are produced by Clostridia species at a ratio of 3:6:1, respectively. [10]. The reaction of organic acids re-assimilated at low pH to the alcohol is carried out. Organic acid production can be prevented by a slightly acidic pH adjustment, and the solvents present in the environment can become the major fermentation products. Simultaneous isopropanol deposition and reduced acetone production were also observed in some Clostridia species [10]. The specificity of increased solid NADPH dependent primary/secondary alcohol dehydrogenase for acetone has been shown to occur with isopropanol formation [11 12]. Comparing with the production of butanol and isopropanol, the production of isopropanol remains lower and does not exceed 25% alcohol fraction as shown in Fig. 1 [10].

Microbial Fuel Cells: Materials and Applications   Materials Research Forum LLC

Materials Research Foundations **46** (2019) 335-353   doi: http://dx.doi.org/10.21741/9781644900116-13

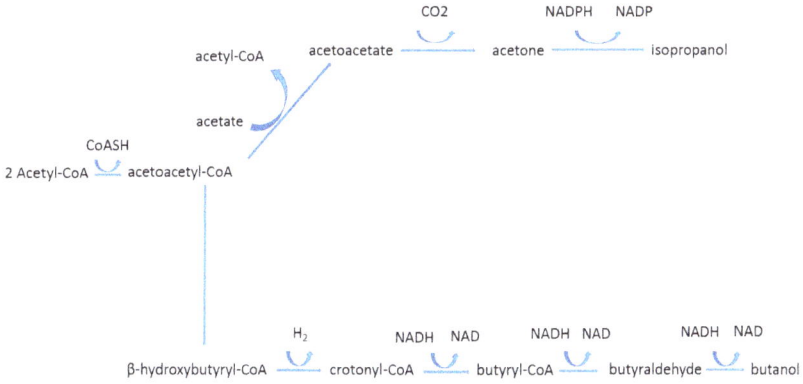

Fig. 1 *Natural routes to isopraponol and butanol in Clostridia species.*

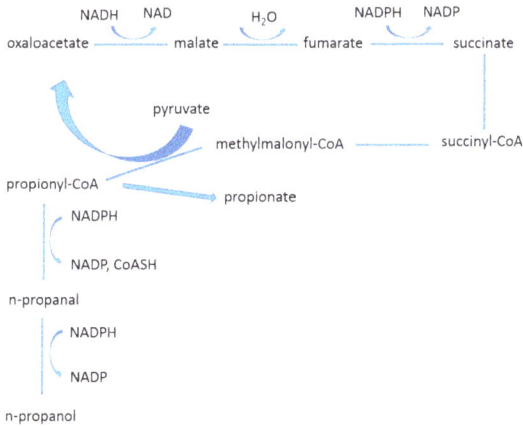

Fig. 2 *Wood-Werkman cycle in propionic acid bacteria.*

Microbial Fuel Cells: Materials and Applications   Materials Research Forum LLC
Materials Research Foundations **46** (2019) 335-353   doi: http://dx.doi.org/10.21741/9781644900116-13

Various anaerobic Propioni Bacteria produce propionic acid as a result of mentioned fermentation and also form isopropanol through the Wood-Werkman cycle according to growing conditions [13–14]. Two production pathways differ from the characteristic labelling models of the C atoms of propionic acid [15–16]. It has been observed that propionic acid bacteria in anaerobic conditions produce much less n-propanol when grown on glucose. On the other hand, when propionic acid was used as a major fermentation product, it was observed that propanol production increased [17]. The expression of the Wood-Werkman reaction in these bacteria can be in two stages as shown in Fig. 2. When propanol is produced, it should be a different reaction than the classic one [18].

In the absence of oxygen, Clostridium propionicum bacteria may use lactate and ethanol instead of growth substrates. The products generally used in the reaction are acetate, propionic acid, and carbon dioxide. Based on the enzymatic path reactions, propanol is formed by a two-step reduction of propionyl-CoA as shown in Fig. 3 [15–16–19].

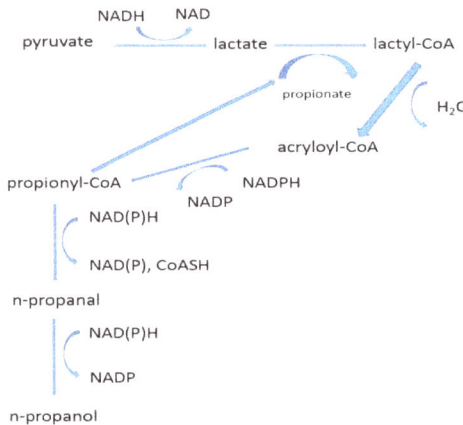

Fig. 3 *Propionic acid and n-propanol path for the biosynthesis of natural acrylate.*

3.   Engineering microorganisms for the path of acetone-dependent isopropanol.

Biofuels are intended to produce alcohol using clostridia species for the potential application of butanol [20]. Acetone cannot be used as a biofuel because it is an unhelpful byproduct that reduces the primary/secondary hydrogen compound of NRRL B593 to

Microbial Fuel Cells: Materials and Applications   Materials Research Forum LLC
Materials Research Foundations **46** (2019) 335-353   doi: http://dx.doi.org/10.21741/9781644900116-13

isopropanol for ABE fermentation and affinity. The most appropriate solution in biofuel applications is the use of fermentation spectrum products. Acetone was produced by E. coli which in turn was created by Bermejo et al. Initially, the acetone pathway produced by C. acetobutylicum is heterogeneous. The attempt to treat and disinfect acetate toxins was interpreted in E. coli cultures, which understand asymmetric proteins, by this study. To measure production capacity, acetone was produced by acetobutylicum C. Coded in different strains of E. coli with different acetates. When using glucose as a carbon source in the production of acetone in the bioreactor, it is best to do it through the following procedure; having 154 mA acetone after 120 hours of culture (undetermined evaporation of acetone) and observing aerobic conditions and pH stabilization at 5.5 [21–22]. Isopropanol was produced in E. coli as a reduction step of the clostridial acetone pathway as shown in Fig. 4 [23–24]. The metabolic pathway was also shown in Fig. 4. The presence of acetyl-CoA for isopropanol synthesis resulted in the lowering the expression of citrate synthesis consuming acetyl-CoA during the production step [25–27].

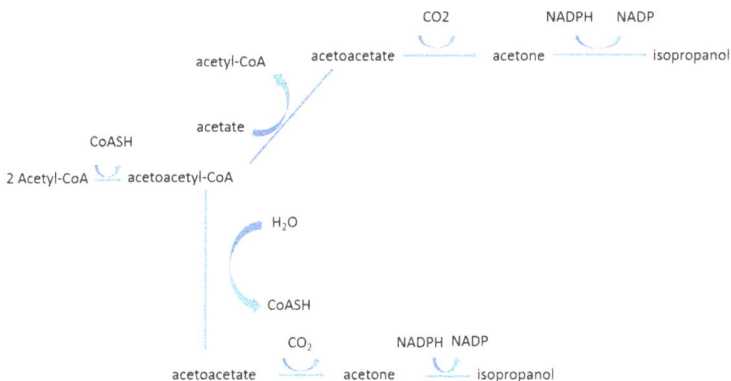

Fig. 4 *The metabolic pathway for isopropanol production.*

To reduce the need for free acetate with bioethanolation of acetone, thioesterase can be used. To prevent the secretion of the secondary product, the acetate must be controlled and normalized due to increased tensile performance.

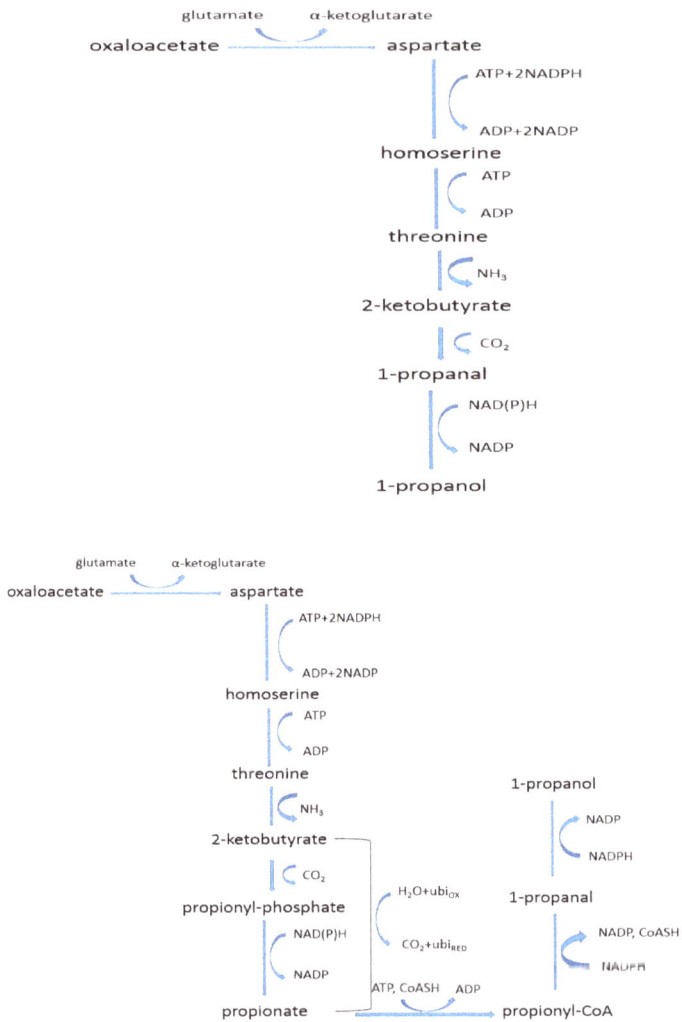

Fig. 5 *Pathway for acetone production in C. acetobutylicum.*

Besides, compared to E. coli, yeasts are characterized by having higher stress tolerance as well as being cultured at lower pH values. Because of these properties of yeasts, it is very important to develop yeast strains for isopropanol production. The feed of yeast for the

Microbial Fuel Cells: Materials and Applications   Materials Research Forum LLC
Materials Research Foundations **46** (2019) 335-353   doi: http://dx.doi.org/10.21741/9781644900116-13

production of isopropanol was carried out using plant hydrolysates. Also, in the case of this commonly used feed, the production occurs even under unfavourable fermentation conditions. In the production of propanol, mild acid treatment was carried out to remove the bacteria in the medium to recover and reuse the yeast cells as shown in Fig. 5 [28–30]. Today, *Candida utilis* was used for the production of isopropanol [31]. E. coli strains using similar expressions were found to be lower than the E. coli strains that produced a difference of 0.73 mol/mol compared to isopropanol yield of 0.41 mol/mol compared to the isopropanol yield. However, the improvement of isopropanol yield obtained from yeast can be achieved by reducing the capacity to regulate carbon flow along the cytosolic isopropanol pathway for microbial activity by advanced metabolic design. There are synthetic routes for the production of alcohol and the possibility of producing n-propanol and other alcohol derivatives from threonine. The intermediate steps during the manufacture of n-propanol proceed in the form of 2-ketobutyrate (2KB) and 1-propanol [32–33] as shown in Fig. 6. By excessive gene expression that symbolizes the threonine enzymes (thRABC) and leucine (leuABCD) and also by deleting the metA and tdh that are encoded to the succinyl transferase homoserine and threonine dehydratase, the production of threonine and 2-ketovalerate were increased. And then, to increase the rate of presence of acetyl-CoA and dehydrogenase in alcohol, the utilization rate was raised to 2-ketobutyrate by deleting acetohydroxy acid synthases [34–35].

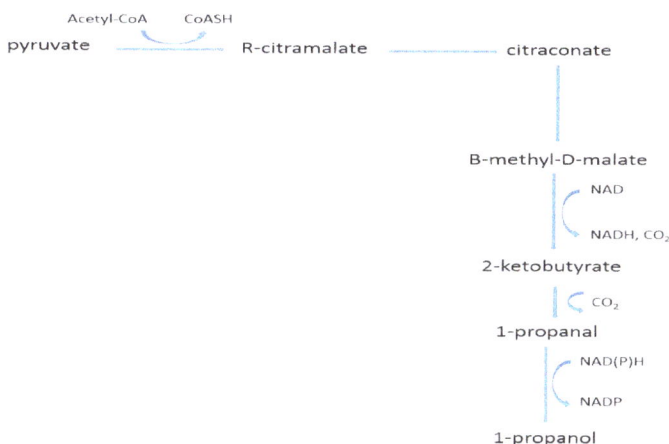

Fig. 6. *A diagram for the production of 1-propanol and 1-butanol by threonine and biosynthetic norvaline in E. coli that has been genetically modified.*

Microbial Fuel Cells: Materials and Applications   Materials Research Forum LLC
Materials Research Foundations **46** (2019) 335-353   doi: http://dx.doi.org/10.21741/9781644900116-13

There is a need for threonine formation on another 2-ketobutyrate expression path described in Archaea [36]. The initiation of the citrate on the 2-ketobutyrate pathway of the Archaea occurs as a result of the intensification of pyruvate and acetyl-CoA. The resulting intermediate has to be treated for citraconate and beta-methyl-malate in 2-ketobutyrate production as shown in Fig. 7. The pathway has been tested for the production of the final product, n-propanol. To produce MJ-Cim1 citramalate synthase of Methanocaldococcus jannaschii, a thermophilic organism, studies have been conducted to obtain air temperatures in E. coli enhanced enzymatic activities [37].

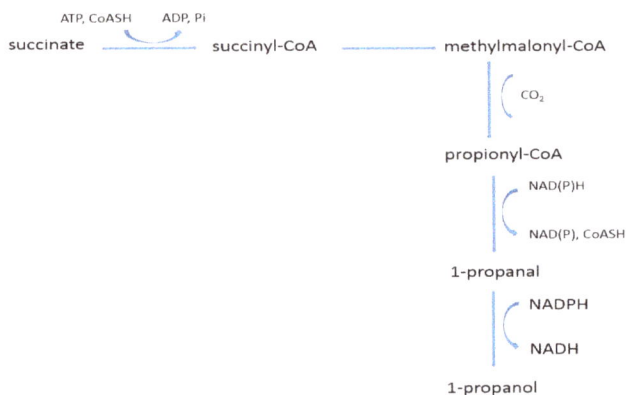

Fig. 7. Schematic representation of the pathway for 1-propanol and 1-butanol production

Threonine and citramalate pathways show some differences in the production of n-propanol according to the common factor requirements. These two syntheses cause a synergistic effect in the case of simultaneous initiation of the propanol production in the organism. Theoretically, n-propanol yield in both paths increases from 1 mol/mol to 1.3 mol/mol for one route. Experimental data support the theoretical data, and n-propanol yield increases. In another study, propionyl phosphate obtained by the conversion of 2-ketobutyrate to n-propanol in the spontaneous (endogenous pyruvate oxidase, PoxB probably catalyzes with spontaneous ox decarboxylation reaction) [38] decarboxylation process which was demonstrated by the conversion of the intermediates to the desired alcohol using propionate, propionyl-CoA, and 1-propanol [39–41].

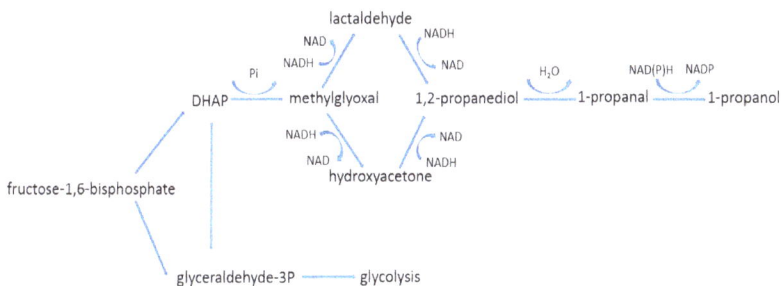

Fig. 8 *Isopropanol production from 1,2-propanediol*

There are alternative reactions to the production of 1-propanol, including the reaction in which propanal-1 reduction occurs which consists of the drought of 1,2-propanediol (1,2 PDO). To make the production of 1,2-propanediol simpler, dihydroxyacetone phosphate (DHAP) is converted to methylglyoxal (MG). Many of the candidate enzymes stimulate these reactions. This interaction between the automatic expression of the activity has been completed [42–44]. By Format oxidation, the NDH-cofactor supply was increased, and the strain was improved by increasing the secretion of hydrogen-Fdh1, from the Candida Candida boidinii as shown in Fig. 8. The applicable maximum thermodynamic efficiency of microbial synthesis product, the product's degree of reduction, (γ) and carbon source [45–47] can be calculated from the ratio. Also, the carbon conserved for the biosynthesis of acetyl-CoA dependent development of the phosphoketolase pathway [28–29], and propanol of synthesis by a great leverage acetone pathways seem to increase the practical and theoretical yield. In previous studies, it was observed that glucose was used to produce carbon. However, it is preferable to use raw materials, which do not cause any side effects for human health, in order to produce propanol. Recently due to the evolution of the biodiesel industry, glycerol has become an alternative to glucose [48–49]. It should be noted that all of these propanol methods are proportional to the use of this carbon source. The use of propanol that produced from glycerol is more advantageous than glucose since the absorption process reduces propanol precursors using additional NADH. As a result, it was observed that propanol production through glycerin had higher carbon yield than the case in which acetone was used. One of the essential components of cygnolucus biomass is cellulose, and it cannot be directly used by the most commonly used propanol producers. Cellulose may be saturated through the cooperative action of cellular enzymes cellobiohydrolase, endoglucanase, and beta-glucosidase (BGL) [50]. Incision step catalyzed by BGL enzyme by infusion cellobiose to glucose. It has been found that these bacteria produce propionic acid from lignocellulosic naturally. To produce propionic acid by propionyl, the threonine or succinate pathway was suggested to be responsible. In addition, it is characteristic for these bacteria that they are grown at elevated temperature under the aerobic conditions and at bear the concentrations of propanol [51]. These unique properties are very useful when producing effective propanol because it requires continuous removal of the product by gas stripping.

There are many types of carbon derivatives and their applications up to now [54–69]. The carbon that contributes to the synthesis of chemicals is syngas [68–69]. There are many advantages to the processes that obscure on hydrolysis and the fermentation of sugars. Therefore it is proposed to generate energy from agricultural waste or direct fermentation of the resulting gas. It is suggested that gasification should bypass the steps of costly biomass treatment and the removal of aquatic toxins. In this case, in terms of the cost, it will be noted that if the genetic material is developed to manipulate microorganisms and improve possible conditions, this will allow the production of active propanol [68]. Interestingly, proteins have been used as a feedstock for high alcohol production, such as butanol, and methyl butanol [2–4]. Alcohol was produced by removing amino acids, removing carboxylic, and reducing the resulting the oxoacids. In this process, a mixture of alcohol is produced from remaining protein residues, or large quantities of carbon dioxide and light are generated from the biomass, and ammonium can also be recycled as fertilizer [2–4]. These features make this process an exciting and sustainable alternative to the production of biofuels by sugar. This process is not entirely appropriate to produce pure alcohol, because the output of these reactions is a complex mixture of alcohol. As it appears from these processes, when using raw materials for protein, the result will be biofuels which require no separation of the product. To produce value-added chemicals, scientists have directed their efforts and research into the use of photosynthetic objects. In this respect, for the production of alcohol from carbon dioxide, Synechococcus elongatus was designed with the isopropanol pathway [70–71]. In another study, by improving the conditions of cultivation, the propanol level was enhanced to 146 mg/L. [71]. There are more significant obstacles to the use of photovoltaic organisms to address and improve chemical production, such as the absence of appropriate genetic engineering tools or the lack of expertise required by effective CO_2 delivery strategies. All these results indicate that more research efforts are still required to increase the production of propanol to industrial levels by photovoltaic organisms.

Materials Research Forum LLC
doi: http://dx.doi.org/10.21741/9781644900116-13

Conclusions

As a conclusion, the use of propanol as biofuel has gained attention in microbial fuel cells recently. In this regard, the microbial production of propanol is very important. This chapter examines the latest developments and synthetic methods which related to the microbial production of propanol. Normally, various synthesis ways of propanol have been developed. However, the low cost and high-efficiency propanol production with the help of biosynthetic pathways from microorganisms is one of the major challenges in microbial fuel cells and these should be solved in near future.

References

[1]   Y. Liu, Z. Yan, X. Lu, D. Xiao, & H. Jiang, Improving The Catalytic Activity of Isopentenyl Phosphate Kinase through Protein Coevolution Analysis. Scientific Reports, 6 (2016) 1–7. https://doi.org/10.1038/srep24117.

[2]   M. Zareian, P. Silcock, & P. Bremer, Effect of Medium Compositions on Microbially Mediated Volatile Organic Compounds Release Profile. Journal of Applied Microbiology, 125 (2018) 813–827. https://doi.org/10.1111/jam.13908.

[3]   W. Zhou, H. Bi, Y. Zhuang, Q. He, H. Yin, T. Liu, & Y. Ma, Production of Cinnamyl Alcohol Glucoside from Glucose in Escherichia coli. Journal of Agricultural and Food Chemistry, 65 (2017) 2129–2135. https://doi.org/10.1021/acs.jafc.7b00076.

[4]   I. M. Mukisa, Y. B. Byaruhanga, C. M. B. K. Muyanja, T. Langsrud, & J. A. Narvhus, Production of Organic Flavor Compounds by Dominant Lactic Acid Bacteria and Yeasts from Obushera, A Traditional Sorghum Malt Fermented Beverage. Food Science and Nutrition, 5 (2017) 702–712. https://doi.org/10.1002/fsn3.450.

[5]   M. Gottardi, J. D. Knudsen, L. Prado, M. Oreb, P. Branduardi, & E. Boles, De novo biosynthesis of Trans-Cinnamic Acid Derivatives in Saccharomyces Cerevisiae. Applied Microbiology and Biotechnology, 101 (2017) 4883–4893. https://doi.org/10.1007/s00253-017-8220-x.

[6]   V. Koppolu & V. K. Vasigala, Role of Escherichia coli in Biofuel Production. Microbiology Insights, 9 (2016) 29–35. https://doi.org/10.4137/MBI.S10878.

[7]   T. Walther & J. M. François, Microbial Production of Propanol. Biotechnology Advances, 34 (2016) 984–996. https://doi.org/10.1016/j.biotechadv.2016.05.011.

[8]   S. Shi, Y. W. Choi, H. Zhao, M. H. Tan, & E. L. Ang, Discovery and Engineering
         of A 1-Butanol Biosensor in Saccharomyces Cerevisiae. Bioresource Technology,
         245 (2017) 1343–1351. https://doi.org/10.1016/j.biortech.2017.06.114.

[9]   A. Kongpol, J. Kato, T. Tajima, T. Pongtharangkul, & A. S. Vangnai, Enhanced 3-
         Methylcatechol Production by Pseudomonas Putida TODE1 in A Two-Phase
         Biotransformation System. The Journal of General and Applied Microbiology, 60
         (2014) 183–190. https://doi.org/10.2323/jgam.60.183.

[10]   V. A. Boumba, N. Kourkoumelis, P. Gousia, V. Economou, C. Papadopoulou, &
         T. Vougiouklakis, Modeling Microbial Ethanol Production By E. Coli under
         Aerobic/Anaerobic Conditions: Applicability to Real Postmortem Cases and to
         Postmortem Blood Derived Microbial Cultures. Forensic Science International,
         232 (2013) 191–198. https://doi.org/10.1016/j.forsciint.2013.07.021.

[11]   L. Fariña, K. Medina, M. Urruty, E. Boido, E. Dellacassa, & F. Carrau, Redox
         Effect on Volatile Compound Formation in Wine During Fermentation by
         Saccharomyces Cerevisiae. Food Chemistry, 134 (2012) 933–939.
         https://doi.org/10.1016/j.foodchem.2012.02.209.

[12]   K. Srirangan, L. Akawi, X. Liu, A. Westbrook, E. J. M. Blondeel, M. G. Aucoin,
         M. Moo-Young, & C. P. Chou, Manipulating The Sleeping Beauty Mutase Operon
         for The Production of 1-Propanol in Engineered Escherichia coli. Biotechnology
         for Biofuels, 6 (2013) 1–14. https://doi.org/10.1186/1754-6834-6-139.

[13]   C. H. Luna-Flores, C. C. Stowers, B. M. Cox, L. K. Nielsen, & E. Marcellin,
         Linking Genotype and Phenotype in An Economically Viable Propionic Acid
         Biosynthesis Process. Biotechnology for Biofuels, 11 (2018) 1–14.
         https://doi.org/10.1186/s13068-018-1222-9.

[14]   L. Navone, T. McCubbin, R. A. Gonzalez-Garcia, L. K. Nielsen, & E. Marcellin,
         Genome-scale Model Guided Design of Propionibacterium for Enhanced
         Propionic Acid Production. Metabolic Engineering Communications, 6 (2018) 1–
         12. https://doi.org/10.1016/j.meteno.2017.11.001.

[15]   F. F. Aburjaile, M, Rohmer, H. Parrinello, M. B. Maillard, E. Beaucher, G. Henry,
         A. Nicolas, M. N. Madec, A. Thierry, S. Parayre, S. M. Deutsch, M. Cocaign-
         Bousquet, A. Miyoshi, V. Azevedo, Y. Le Loir, & H. Falentin, Adaptation of
         Propionibacterium Freudenreichii to Long-Term Survival under Gradual
         Nutritional Shortage. BMC Genomics, 17 (2016) 1–13.
         https://doi.org/10.1186/s12864-016-3367-x.

[16] A. Zhang, J. Sun, Z. Wang, S. T. Yang, & H. Zhou, Effects of Carbon Dioxide on Cell Growth and Propionic Acid Production From Glycerol and Glucose by Propionibacterium Acidipropionici. Bioresource Technology, 175 (2015) 374–381. https://doi.org/10.1016/j.biortech.2014.10.046.

[17] T. Saraoui, S. Parayre, G. Guernec, V. Loux, J. Montfort, A. Cam, G. Boudry, G. Jan, & H. Falentin, A Unique In vivo Experimental Approach Reveals Metabolic Adaptation of The Probiotic Propionibacterium Freudenreichii to The Colon Environment. BMC Genomics, 14 (2013) 911–926. https://doi.org/10.1186/1471-2164-14-911.

[18] V. A. Boumba, V. Economou, N. Kourkoumelis, P. Gousia, C. Papadopoulou, & T. Vougiouklakis, Microbial Ethanol Production: Experimental Study and Multivariate Evaluation. Forensic Science International, 215 (2012) 189–198. https://doi.org/10.1016/j.forsciint.2011.03.003.

[19] A. Thierry, S. M. Deutsch, H. Falentin, M. Dalmasso, F. J. Cousin, & G. Jan, New Insights into Physiology and Metabolism of Propionibacterium Freudenreichii. International Journal of Food Microbiology, 149 (2011) 19–27. https://doi.org/10.1016/j.ijfoodmicro.2011.04.026.

[20] X. Christodoulou & S. B. Velasquez-Orta, Microbial Electrosynthesis and Anaerobic Fermentation: An Economic Evaluation for Acetic Acid Production from CO2 and CO. Environmental Science and Technology, 50 (2016) 11234–11242. https://doi.org/10.1021/acs.est.6b02101.

[21] A. M. El-Nahas, L. A. Heikal, A. H. Mangood, & E. S. E. El-Shereefy, Structures and Energetics of Unimolecular Thermal Degradation of Isopropyl Butanoate as A Model Biofuel: Density Functional Theory and AB Initio Studies. Journal of Physical Chemistry A, 114 (2010) 7996–8002. https://doi.org/10.1021/jp103397f.

[22] B. Andreeen & A. Steinbüchel, Biotechnological Conversion of Glycerol to 2-Amino-1,3-Propanediol (Serinol) in Recombinant Escherichia coli. Applied Microbiology and Biotechnology, 93 (2012) 357–365. https://doi.org/10.1007/s00253-011-3364-6.

[23] Y. Soma, K. Tsuruno, M. Wada, A. Yokota, & T. Hanai, Metabolic Flux Redirection from A Central Metabolic Pathway toward A Synthetic Pathway Using A Metabolic Toggle Switch. Metabolic Engineering, 23 (2014) 175–184. https://doi.org/10.1016/j.ymben.2014.02.008.

[24] T. Horinouchi, A. Sakai, H. Kotani, K. Tanabe, & C. Furusawa, Improvement of Isopropanol Tolerance of Escherichia Coli Using Adaptive Laboratory Evolution

and Omics Technologies. Journal of Biotechnology, 255 (2017) 47–56. https://doi.org/10.1016/j.jbiotec.2017.06.408.

[25] M. Majone & M. Reis, Editorial. New Biotechnology, 31 (2014) 255–256. https://doi.org/10.1016/j.nbt.2014.04.007.

[26] S. Dusséaux, C. Croux, P. Soucaille, & I. Meynial-Salles, Metabolic engineering of Clostridium Acetobutylicum ATCC 824 for The High-Yield Production of A Biofuel Composed of An Isopropanol/Butanol/Ethanol Mixture. Metabolic Engineering, 18 (2013) 1–8. https://doi.org/10.1016/j.ymben.2013.03.003.

[27] B. Ince, G. Koksel, Z. Cetecioglu, N. A. Oz, H. Coban, & O. Ince, Inhibition Effect of Isopropanol on Acetyl-Coa Synthetase Expression Level of Acetoclastic Methanogen, Methanosaeta Concilii. Journal of Biotechnology, 156 (2011) 95–99. https://doi.org/10.1016/j.jbiotec.2011.08.021.

[28] Y. Deng, A. B. Fisher, & S. S. Fong, Systematic Analysis of Intracellular Mechanisms of Propanol Production in The Engineered Thermobifida Fusca B6 Strain. Applied Microbiology and Biotechnology, 99 (2015) 8089–8100. https://doi.org/10.1007/s00253-015-6850-4.

[29] K. Srirangan, X. Liu, A. Westbrook, L. Akawi, M. E. Pyne, M. Moo-Young, & C. P. Chou, Biochemical, Genetic, and Metabolic Engineering Strategies to Enhance Coproduction of 1-Propanol and Ethanol in Engineered Escherichia coli. Applied Microbiology and Biotechnology, 98 (2014) 9499–9515. https://doi.org/10.1007/s00253-014-6093-9.

[30] R. Jain & Y. Yan, Dehydratase Mediated 1-Propanol Production in Metabolically Engineered Escherichia coli. Microbial Cell Factories, 10 (2011) 1–10. https://doi.org/10.1186/1475-2859-10-97.

[31] M. Matsubara, N. Urano, S. Yamada, A. Narutaki, M. Fujii, & M. Kataoka, Fermentative Production of 1-Propanol From D-Glucose, L-Rhamnose and Glycerol Using Recombinant Escherichia coli. Journal of Bioscience and Bioengineering, 122 (2016) 421–426. https://doi.org/10.1016/j.jbiosc.2016.03.011.

[32] V. Shestivska, K. Dryahlina, J. Nunvář, K. Sovová, D. Elhottová, A. Nemec, D. Smith, & P. Španěl, Quantitative Analysis of Volatile Metabolites Released In vitro by Bacteria of The Genus Stenotrophomonas for Identification of Breath Biomarkers of Respiratory Infection in Cystic Fibrosis. Journal of Breath Research, 9 (2015) 027104-24. https://doi.org/10.1088/1752-7155/9/2/027104.

[33]   T. Kusakabe, T. Tatsuke, K. Tsuruno, Y. Hirokawa, S. Atsumi, J. C. Liao, & T. Hanai, Engineering A Synthetic Pathway in Cyanobacteria for Isopropanol Production Directly from Carbon Dioxide and Light. Metabolic Engineering, 20 (2013) 101–108. https://doi.org/10.1016/j.ymben.2013.09.007.

[34]   S. Obruca, I. Marova, O. Snajdar, L. Mravcova, & Z. Svoboda, Production of poly(3-hydroxybutyrate-co-3-hydroxyvalerate) by Cupriavidus Necator from Waste Rapeseed Oil Using Propanol as A Precursor Of 3-Hydroxyvalerate. Biotechnology Letters, 32 (2010) 1925–1932. https://doi.org/10.1007/s10529-010-0376-8.

[35]   S. I. Pavlova, L. Jin, S. R. Gasparovich, & L. Tao, Multiple Alcohol Dehydrogenases but No Functional Acetaldehyde Dehydrogenase Causing Excessive Acetaldehyde Production from Ethanol by Oral Streptococci. Microbiology (United Kingdom), 159 (2013) 1437–1446. https://doi.org/10.1099/mic.0.066258-0.

[36]   R. A. Gonzalez-Garcia, T. McCubbin, A. Wille, M. Plan, L. K. Nielsen, & E. Marcellin, Awakening Sleeping Beauty: Production of Propionic Acid in Escherichia Coli through The SBM Operon Requires The Activity of A Methylmalonyl-Coa Epimerase. Microbial Cell Factories, 16 (2017) 1–14. https://doi.org/10.1186/s12934-017-0735-4.

[37]   S. P. Liu, L. Zhang, J. Mao, Z. Y. Ding, & G. Y. Shi, Metabolic Engineering of Escherichia Coli for The Production of Phenylpyruvate Derivatives. Metabolic Engineering, 32 (2015) 55–65. https://doi.org/10.1016/j.ymben.2015.09.007.

[38]   R. Fasan, N. C. Crook, M. W. Peters, P. Meinhold, T. Buelter, M. Landwehr, P. C. Cirino, & F. H. Arnold, Improved Product-Per-Glucose Yields in P450-Dependent Propane Biotransformations Using Engineered Escherichia coli. Biotechnology and Bioengineering, 108 (2011) 500–510. https://doi.org/10.1002/bit.22984.

[39]   Q. Guo, J. Chu, Y. Zhuang, & Y. Gao, Controlling The Feed Rate of Propanol to Optimize Erythromycin Fermentation by On-Line Capacitance and Oxygen uptake Rate Measurement. Bioprocess and Biosystems Engineering, 39 (2016) 255–265. https://doi.org/10.1007/s00449-015-1509-1.

[40]   Y. Hirokawa, I. Suzuki, & T. Hanai, Optimization of Isopropanol Production by Engineered Cyanobacteria with A Synthetic Metabolic Pathway. Journal of Bioscience and Bioengineering, 119 (2015) 585–590. https://doi.org/10.1016/j.jbiosc.2014.10.005.

[41]   Y. Hirokawa, Y. Dempo, E. Fukusaki, & T. Hanai, Metabolic engineering for
       Isopropanol Production by An Engineered Cyanobacterium, Synechococcus
       Elongatus PCC 7942, under Photosynthetic Conditions. Journal of Bioscience and
       Bioengineering, 123 (2017) 39–45. https://doi.org/10.1016/j.jbiosc.2016.07.005.

[42]   N. Urano, M. Fujii, H. Kaino, M. Matsubara, & M. Kataoka, Fermentative
       Production of 1-Propanol from Sugars Using Wild-Type and Recombinant
       Shimwellia Blattae. Applied Microbiology and Biotechnology, 99 (2014) 2001–
       2008. https://doi.org/10.1007/s00253-014-6330-2.

[43]   S. Arai, K. Hayashihara, Y. Kanamoto, K. Shimizu, Y. Hirokawa, T. Hanai, A.
       Murakami, & H. Honda, Alcohol-Tolerant Mutants of Cyanobacterium
       Synechococcus Elongatus PCC 7942 Obtained by Single-Cell Mutant Screening
       System. Biotechnology and Bioengineering, 114 (2017) 1771–1778.
       https://doi.org/10.1002/bit.26307.

[44]   E. M. Ammar, Z. Wang, & S. T. Yang, Metabolic Engineering of
       Propionibacterium Freudenreichii for N-Propanol Production. Applied
       Microbiology and Biotechnology, 97 (2013) 4677–4690.
       https://doi.org/10.1007/s00253-013-4861-6.

[45]   K. Liu, H. K. Atiyeh, B. S. Stevenson, R. S. Tanner, M. R. Wilkins, & R. L.
       Huhnke, Continuous Syngas Fermentation for The Production of Ethanol, N-
       Propanol and N-Butanol. Bioresource Technology, 151 (2014) 69–77.
       https://doi.org/10.1016/j.biortech.2013.10.059.

[46]   C. R. Shen & J. C. Liao, Synergy As Design Principle for Metabolic Engineering
       of 1-Propanol Production in Escherichia coli. Metabolic Engineering, 17 (2013)
       12–22. https://doi.org/10.1016/j.ymben.2013.01.008.

[47]   Y. Chen, M. Huang, Z. Wang, J. Chu, Y. Zhuang, & S. Zhang, Controlling The
       Feed Rate of Glucose and Propanol for The Enhancement of Erythromycin
       Production and Exploration of Propanol Metabolism Fate by Quantitative
       Metabolic Flux Analysis. Bioprocess and Biosystems Engineering, 36 (2013)
       1445–1453. https://doi.org/10.1007/s00449-013-0883-9

[48]   T. Hanai, S. Atsumi, & J. C. Liao, Engineered Synthetic Pathway for Isopropanol
       Production in Escherichia coli. Applied and Environmental Microbiology, 73
       (2007) 7814–7818. https://doi.org/10.1128/AEM.01140-07.

[49]   H. Tamakawa, T. Mita, A. Yokoyama, S. Ikushima, & S. Yoshida, Metabolic
       Engineering of Candida Utilis for Isopropanol Production. Applied Microbiology

and Biotechnology, 97 (2013) 6231–6239. https://doi.org/10.1007/s00253-013-4964-0.

[50] Y. Soma, K. Inokuma, T. Tanaka, C. Ogino, A. Kondo, M. Okamoto, & T. Hanai, Direct Isopropanol Production from Cellobiose by Engineered Escherichia coli Using A Synthetic Pathway and A Cell Surface Display System. Journal of Bioscience and Bioengineering, 114 (2012) 80–85. https://doi.org/10.1016/j.jbiosc.2012.02.019.

[51] H. Graber & H. J. La Roche, Mutations Responsible for Alcohol Tolerance in The Mutant of Synechococcus elongatus PCC 7942 (SY1043) Obtained by Single-Cell Screening System. Bell Labs Technical Journal, 8 (2003) 111–127. https://doi.org/10.1016/j.jbiosc.2017.11.012.

[52] B. Sen, B. Demirkan, A. Şavk, S. Karahan Gülbay, & F. Sen, Trimetallic PdRuNi Nanocomposites Decorated on Graphene Oxide: A Superior Catalyst for The Hydrogen Evolution Reaction. International Journal of Hydrogen Energy, 43 (2018) 17984–17992. https://doi.org/10.1016/j.ijhydene.2018.07.122.

[53] S. Eris, Z. Daşdelen, Y. Yıldız, & F. Sen, Nanostructured Polyaniline-rGO Decorated Platinum Catalyst with Enhanced Activity and Durability for Methanol Oxidation. International Journal of Hydrogen Energy, 43 (2018) 1337–1343. https://doi.org/10.1016/j.ijhydene.2017.11.051.

[54] S. Eris, Z. Daşdelen, & F. Sen, Enhanced Electrocatalytic Activity and Stability of Monodisperse Pt Nanocomposites for Direct Methanol Fuel Cells. Journal of Colloid and Interface Science, 513 (2018) 767–773. https://doi.org/10.1016/j.jcis.2017.11.085.

[55] B. Şen, E. H. Akdere, A. Şavk, E. Gültekin, Ö. Paralı, H. Göksu, & F. Şen, A Novel Thiocarbamide Functionalized Graphene Oxide Supported Bimetallic Monodisperse Rh-Pt Nanoparticles (RhPt/TC@GO NPs) for Knoevenagel Condensation of Aryl Aldehydes together with Malononitrile. Applied Catalysis B: Environmental, 225 (2018) 148–153. https://doi.org/10.1016/j.apcatb.2017.11.067.

[56] S. Eris, Z. Daşdelen, & F. Sen, Investigation of Electrocatalytic Activity and Stability of Pt@f-VC Catalyst Prepared by In-Situ Synthesis for Methanol Electrooxidation. International Journal of Hydrogen Energy, 43 (2018) 385–390. https://doi.org/10.1016/j.ijhydene.2017.11.063.

[57] B. Şen, B. Demirkan, M. Levent, A. Şavk, & F. Şen, Silica-based Monodisperse PdCo Nanohybrids as Highly Efficient and Stable Nanocatalyst for Hydrogen

Evolution Reaction. International Journal of Hydrogen Energy, 43 (2018) 20234–20242. https://doi.org/10.1016/j.ijhydene.2018.07.080.

[58] Y. Koskun, A. Şavk, B. Şen, & F. Şen, Highly Sensitive Glucose Sensor Based on Monodisperse Palladium Nickel/Activated Carbon Nanocomposites. Analytica Chimica Acta, 1010 (2018) 37–43. https://doi.org/10.1016/j.aca.2018.01.035.

[59] B. Şen, A. Aygün, A. Şavk, S. Akocak, & F. Şen, Bimetallic Palladium–Iridium Alloy Nanoparticles as Highly Efficient and Stable Catalyst for The Hydrogen Evolution Reaction. International Journal of Hydrogen Energy, 43 (2018) 20183–20191. https://doi.org/10.1016/j.ijhydene.2018.07.081.

[60] S. Günbatar, A. Aygun, Y. Karataş, M. Gülcan, & F. Şen, Carbon-nanotube-based Rhodium Nanoparticles as Highly-Active Catalyst For Hydrolytic Dehydrogenation of Dimethylamineborane at Room Temperature. Journal of Colloid and Interface Science, 530 (2018) 321–327. https://doi.org/10.1016/j.jcis.2018.06.100.

[61] B. Sen, A. Şavk, & F. Sen, Highly Efficient Monodisperse Pt Nanoparticles Confined in The Carbon Black Hybrid Material for Hydrogen Liberation. Journal of Colloid and Interface Science, 520 (2018) 112–118. https://doi.org/10.1016/j.jcis.2018.03.004.

[62] B. Sen, E. Kuyuldar, B. Demirkan, T. Onal Okyay, A. Şavk, & F. Sen, Highly Efficient Polymer Supported Monodisperse Ruthenium-Nickel Nanocomposites for Dehydrocoupling of Dimethylamine Borane. Journal of Colloid and Interface Science, 526 (2018) 480–486. https://doi.org/10.1016/j.jcis.2018.05.021.

[63] B. Sen, B. Demirkan, B. Şimşek, A. Savk, & F. Sen, Monodisperse Palladium Nanocatalysts for Dehydrocoupling of Dimethylamineborane. Nano-Structures and Nano-Objects, 16 (2018) 209–214. https://doi.org/10.1016/j.nanoso.2018.07.008.

[64] B. Şen, B. Demirkan, A. Savk, R. Kartop, M. S. Nas, M. H. Alma, S. Sürdem, & F. Şen, High-Performance Graphite-Supported Ruthenium Nanocatalyst for Hydrogen Evolution Reaction. Journal of Molecular Liquids, 268 (2018) 807–812. https://doi.org/10.1016/j.molliq.2018.07.117.

[65] B. Şen, A. Aygün, T. O. Okyay, A. Şavk, R. Kartop, & F. Şen, Monodisperse Palladium Nanoparticles Assembled on Graphene Oxide with The High Catalytic Activity and Reusability in The Dehydrogenation of Dimethylamine-borane. International Journal of Hydrogen Energy, 3 (2018) 2–8. https://doi.org/10.1016/j.ijhydene.2018.03.175.

[66]   R. Ayranci, G. Başkaya, M. Güzel, S. Bozkurt, F. Şen, & M. Ak, Carbon Based
       Nanomaterials for High Performance Optoelectrochemical Systems.
       ChemistrySelect, 2 (2017) 1548–1555. https://doi.org/10.1002/slct.201601632.

[67]   G. Başkaya, Y. Yıldız, A. Savk, T. O. Okyay, S. Eriş, H. Sert, & F. Şen, Rapid,
       Sensitive, and Reusable Detection of Glucose by Highly Monodisperse Nickel
       Nanoparticles Decorated Functionalized Multi-Walled Carbon Nanotubes.
       Biosensors and Bioelectronics, 91 (2017) 728–733.
       https://doi.org/10.1016/j.bios.2017.01.045.

[68]   F. R. Bengelsdorf, A. Poehlein, S. Linder, C. Erz, T. Hummel, S. Hoffmeister, R.
       Daniel, & P. Dürre, Industrial Acetogenic Biocatalysts: A Comparative Metabolic
       and Genomic Analysis. Frontiers in Microbiology, 7 (2016) 1–15.
       https://doi.org/10.3389/fmicb.2016.01036.

[69]   B. M. L. Raun & N. B. Kristensen, Metabolic Effects of Feeding Ethanol or
       Propanol to Postpartum Transition Holstein Cows. Journal of Dairy Science, 94
       (2011) 2566–2580. https://doi.org/10.3168/jds.2010-3999.

[70]   C. Jun, Y. Xue, R. Liu, & M. Wang, Study on The Toxic Interaction of Methanol,
       Ethanol and Propanol against The Bovine Hemoglobin (BHb) on Molecular Level.
       Spectrochimica Acta - Part A: Molecular and Biomolecular Spectroscopy, 79
       (2011) 1406–1410. https://doi.org/10.1016/j.saa.2011.04.076.

[71]   J. Li, F. Che, Y. Pang, C. Zou, J. Y. Howe, T. Burdyny, J. P. Edwards, Y. Wang,
       F. Li, Z. Wang, P. De Luna, C.-T. Dinh, T.-T. Zhuang, M. I. Saidaminov, S.
       Cheng, T. Wu, Y. Z. Finfrock, L. Ma, S.-H. Hsieh, Y.-S. Liu, G. A. Botton, W.-F.
       Pong, X. Du, J. Guo, T.-K. Sham, E. H. Sargent, & D. Sinton, Copper Adparticle
       Enabled Selective Electrosynthesis of n-Propanol. Nature Communications, 9
       (2018) 4614–4623. https://doi.org/10.1038/s41467-018-07032-0.

Keyword Index

About the Editors

Dr. Inamuddin is currently working as Assistant Professor in the Chemistry Department, Faculty of Science, King Abdulaziz University, Jeddah, Saudi Arabia. He is a permanent faculty member (Assistant Professor) at the Department of Applied Chemistry, Aligarh Muslim University, Aligarh, India. He obtained the Master of Science degree in Organic Chemistry from Chaudhary Charan Singh (CCS) University, Meerut, India, in 2002. He received his Master of Philosophy and Doctor of Philosophy degrees in Applied Chemistry from Aligarh Muslim University (AMU), India, in 2004 and 2007, respectively. He has extensive research experience in multidisciplinary fields of Analytical Chemistry, Materials Chemistry, and Electrochemistry and, more specifically, Renewable Energy and Environment. He has worked on different research projects as project fellow and senior research fellow funded by the University Grants Commission (UGC), Government of India, and the Council of Scientific and Industrial Research (CSIR), Government of India. He has received the Fast Track Young Scientist Award from the Department of Science and Technology, India, to work in the area of bending actuators and artificial muscles. He has completed four major research projects sanctioned by the University Grant Commission, Department of Science and Technology, Council of Scientific and Industrial Research, and Council of Science and Technology, India. He has published 133 research articles in international journals of repute and eighteen book chapters in knowledge-based book editions published by renowned international publishers. He has published forty two edited books with Springer, United Kingdom, Elsevier, Nova Science Publishers, Inc. U.S.A., CRC Press Taylor & Francis Asia Pacific, Trans Tech Publications Ltd., Switzerland and Materials Research Forum LLC, U.S.A. He is the member of various editorial boards of journals and is serving as associate editor for journals such as Environmental Chemistry Letter, Applied Water Science, Euro-Mediterranean Journal for Environmental Integration, Springer-Nature, Frontiers Section Editor of Current Analytical Chemistry, published by Bentham Science Publishers, editorial board member for Scientific Reports-Nature and editor for Eurasian Journal of Analytical Chemistry. He has attended as well as chaired sessions in various international and national conferences. He has worked as a Postdoctoral Fellow, leading a research team at the Creative Research Initiative Center for Bio-Artificial Muscle, Hanyang University, South Korea, in the field of renewable energy, especially biofuel cells. He has also worked as a Postdoctoral Fellow at the Center of Research Excellence in Renewable Energy, King Fahd University of Petroleum and Minerals, Saudi Arabia, in the field of polymer electrolyte membrane fuel cells and computational fluid dynamics of polymer electrolyte membrane fuel cells. He is a life member of the Journal of the Indian

Chemical Society. His research interest includes ion exchange materials, a sensor for heavy metal ions, biofuel cells, supercapacitors and bending actuators.

Dr. Mohammad Faraz Ahmer is presently working as Assistant Professor in the Department of Electrical Engineering, Mewat Engineering College, Nuh Haryana, India, since 2012 after working as Guest Faculty in University Polytechnic, Aligarh Muslim University Aligarh, India, during 2009-2011. He completed M.Tech. (2009) and Bachelor of Engineering (2007) degrees in Electrical Engineering from the Aligarh Muslim University, Aligarh in the first division. He obtained a Ph.D. degree in 2016 on his thesis entitled "Studies on Electrochemical Capacitor Electrodes". He has published six research papers in reputed scientific journals. He has edited three books with Materials Research Forum LLC, U.S.A. His scientific interests include electrospun nano-composites and supercapacitors. He has presented his work at several conferences. He is actively engaged in searching of new methodologies involving the development of organic composite materials for energy storage systems.

Prof. Abdullah M. Asiri is the Head of the Chemistry Department at the King Abdulaziz University since October 2009 and he is the founder and the Director of the Center of Excellence for Advanced Materials Research (CEAMR) since 2010 till date. He is the Professor of Organic Photochemistry. He graduated from King Abdulaziz University (KAU) with B.Sc. in Chemistry in 1990 and received his Ph.D. from the University of Wales, College of Cardiff, U.K. in 1995. His research interest covers color chemistry, synthesis of novel photochromic and thermochromic systems, synthesis of novel coloring matters and dyeing of textiles, materials chemistry, nanochemistry and nanotechnology, polymers and plastics. Prof. Asiri is the principal supervisors of more than 20 M.Sc. and six Ph.D. theses. He is the main author of ten books of different chemistry disciplines. Prof. Asiri is the Editor-in-Chief of King Abdulaziz University Journal of Science. A major achievement of Prof. Asiri is the discovery of tribochromic compounds, a new class of compounds which change from slightly or colorless to deep colored when subjected to small pressure or when grind. This discovery was introduced to the scientific community as a new terminology published by IUPAC in 2000. This discovery was awarded a patent from the European Patent office and from the UK patent. Prof. Asiri is involved in many committees at the KAU level and on the national level. He took a major role in the advanced materials committee working for KACST to identify the national plan for science and technology in 2007. Prof. Asiri played a major role in advancing the chemistry education and research in KAU. He has been awarded the best researchers from KAU for the past five years. He also awarded the Young Scientist Award from the Saudi Chemical Society in 2009 and also the first prize for the distinction in science from the Saudi Chemical Society in 2012. He futher received a recognition certificate from the

American Chemical Society (Gulf region Chapter) for the advancement of chemical science in the Kingdome. He received a Scopus certificate for the most publishing scientist in Saudi Arabia in chemistry in 2008. He is a member of the editorial board of various journals of international repute. He is the Vice- President of the Saudi Chemical Society (Western Province Branch). He holds four USA patents, more than one thousand publications in international journals, several book chapters and edited books.

www.ingramcontent.com/pod-product-compliance
Lightning Source LLC
Chambersburg PA
CBHW060801220326
41598CB00022B/2505